Molten Salt Techniques

Volume 3

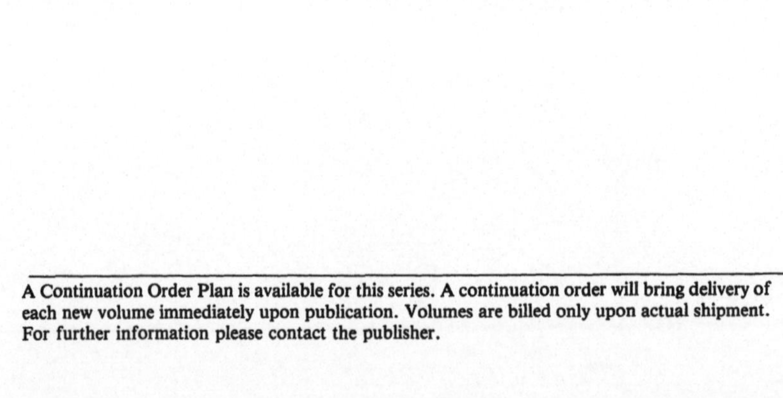

Molten Salt Techniques

Volume 3

Edited by

David G. Lovering

The David Graham Consultancy
Swindon, Wiltshire, England

and

Robert J. Gale

Louisiana State University
Baton Rouge, Louisiana

Plenum Press • New York and London

Library of Congress Cataloging in Publication Data

(Revised for vol. 3)

Molten salt techniques.

Includes bibliographies and indexes.
1. Fused salts. I. Lovering, D. G. (David G.) II. Gale, Robert J., 1942–
QD189.M59 1983 546′.343 83-9582
ISBN-13: 978-1-4612-9031-5 e-ISBN-13: 978-1-4613-1847-7
DOI: 10.1007/ 978-1-4613-1847-7

Softcover reprint of the hardcover 1st edition 1987
A Division of Plenum Publishing Corporation
233 Spring Street, New York, N.Y. 10013

TO THE MEMORY OF
DEPARTED FAMILY AND FRIENDS

REMEMBERING OUR CONTRIBUTORS
PAOLO FRANZOSINI
AND
VINCE NORVELL

Felix qui potuit
rerum cognoscere causas
—Lucretius

Contributors

Duane E. Bartak ● Department of Chemistry, University of North Dakota, Grand Forks, North Dakota

Susan Biggin ● H. H. Wills Physics Laboratory, University of Bristol, Bristol, United Kingdom

Dagfinn Bratland ● Institute of Inorganic Chemistry, Norwegian Institute of Technology, University of Trondheim, Trondheim, Norway

Robert J. Gale ● Department of Chemistry, Louisiana State University, Baton Rouge, Louisiana

David G. Lovering ● The David Graham Consultancy, Swindon, United Kingdom

Nguyen Quang Minh ● Argonne National Laboratory, Chemical Technology Division, Argonne, Illinois

James G. Reavis ● Los Alamos National Laboratory, Materials Science and Technology Division, Los Alamos, New Mexico

Laszlo Redey ● Argonne National Laboratory, Chemical Technology Division, Argonne, Illinois

Foreword

The physicist Kamerlingh Onnes, who was the first to liquify helium (1908), had written on the walls of his laboratory in Leiden: "From measuring to knowing." As true as this is at very low temperatures, it is just as applicable at the high temperatures of molten salts. Only on the basis of exact measurements by a plethora of experimental methods can any real understanding be reached of both classes of liquids. In both temperature ranges experimental difficulties are much greater than those encountered around ambient temperature.

Molten salts often present a formidable challenge to the experimentalist, for example, because of corrosion and other materials problems. Applications of molten salts were for a long time based on empirical knowledge alone. This was true for the first application of molten salts in 1807, when Davy obtained sodium and potassium by electrolysis of the molten hydroxides. For 100 years the winning of aluminum has been based on the very nearly simultaneous invention by Hall and Héroult (1886) of the electrolysis of molten cryolite. The process, though essentially unchanged, has since been perfected owing to an improvement in our understanding of what actually happens, based on difficult measurements of the many variables. However, even now there are gaps in our knowledge.

Chemical purity of the salts and sufficient accuracy and reliability of the apparatus are, although obvious, not the only conditions for meaningful measurements; in most cases the atmosphere is just as important. The absence in the melt of impurities resulting from the ubiquitous water vapor is of utmost importance; this is true for all systems, not only those containing hygroscopic lithium and aluminum compounds. Every melt has in principle a second boiling point just below the triple point where water vapor will be absorbed, often leading to hydrolysis. The use of dry boxes and of specific drying methods is a prerequisite for reliable measurements. In addition, stable, reproducible reference electrodes are necessary with electrochemical techniques.

Understanding and theory can be extrapolated whereas empirical data can at best only be interpolated. Molten salts, as ionic liquids, are theoretically much simpler than, e.g., aqueous solutions. Chemical and most physical properties depend far more on the electronic configuration than on nuclear properties. As an example, liquid KCl is isoelectronic with liquid argon. Theoretical experiments by computer simulation have succeeded in calculating the properties of both liquids within the limits of accuracy of the experimental values, yet it is now fruitful to predict these values at still higher temperatures and pressures where experimental determinations become difficult or even impossible.

Authoritative reviews of measuring methods and of data collections, such as those presented in this volume and in previous volumes, will be of great service to every scientist interested in molten salts and also to engineers working on the technologies of molten salts in e.g., electrolysis and fuel cells, and as heat transfer and reaction media.

J. A. A. Ketelaar

Preface

Within the last decade, operations in molten salts and probably other nonaqueous systems appear to have become more readily accepted as the serious option they always were for our advancing technology. Although we have been advocates of their many virtues for rather longer, it is always rewarding to have the message understood. Having reached the present state of sophistication in civilization, it transpires that the easy routes to materials preparation and processing have mostly been exploited. It now remains for molten salts—as well as organic liquids and liquid gases, among other fluids—to assume their proper role in our research, development, and advancement. It is hoped that this series will continue to follow and record progress in how molten salts may be employed to achieve our goals.

In the present volume, a similar mix of physical techniques and melts systems is offered as in the last two volumes. Some overlap may be apparent between the contributions of Reavis and Bartak, for example, as well as between current chapters and those appearing in earlier volumes. It was considered that this could be beneficial in providing more than a single viewpoint or experimental approach, for no two researchers ever arrive at exactly the same solution to a given problem. In the present instance, Bartak's chapter will be of wide and general interest to many molten salt workers, whereas Reavis's objectives were clearly different, even though the techniques he has developed are generally applicable for the handling of radioactive materials.

While the technique of neutron diffraction is only available to a select few, it does represent one of the most definitive structural probes for all materials. The inclusion of Biggin's chapter now seems especially timely with the commissioning of the new spallation source in the U.K. and other developments in France and the U.S.A.

Although the fortunes of the aluminum industry have fluctuated somewhat wildly in recent years, the metal is strategically important in both civil and military areas. Since it is a major consumer of secondary power there will be considerable scope for future research on its smelting, purification, and recovery. Bratland's contribution on cryolite melt handling could prove most valuable in years to come in a field where the benefit of saving a millivolt of overpotential is measured in thousands if not millions of dollars. The *tour de force* on reference electrodes by Minh and Redey will serve us all for many years to come and is invaluable as a contribution to the future.

Although no schedule of appearance was announced for volumes in the series, this volume is later than we would have liked. We apologize. Many events have conspired to cause this delay, but it is hoped that the series will continue. Feedback from readers is always welcome. Please send any comments or suggestions to

David G. Lovering
The David Graham Consultancy
76 Manor Crescent
Swindon SN2 2LF, U.K.

Robert J. Gale
Department of Chemistry
Louisiana State University
Baton Rouge, Louisiana 70803-1804
U.S.A.

Contents

Chapter 6. Dry Boxes and Inert Atmosphere Techniques

Duane E. Bartak

1

Introduction

David G. Lovering and Robert J. Gale

Operations utilizing molten salts impinge on a wide range of contemporary technologies and offer considerable scope and promise for future advancement. No shaded or color chart based on the Periodic Table can usefully be constructed to illustrate the extent to which molten systems might be employed for materials processing or melt choice—a uniform gray would obscure virtually all elements save for the noble gases! Furthermore, Table II of our Introduction to the first volume in this series adequately highlighted the applications in metals and materials processing, power generation and storage as well as engineering, chemical, and nuclear aspects which are currently used and are envisaged as ripe for development.

Although this text aims to provide practical assistance for those contemplating working with molten milieu, it may be helpful to review briefly some areas of current and future technologies likely to benefit from their application.

In Group I, even hydrogen may be electrolytically generated from melts[1] and it can certainly be used as fuel in the molten carbonate (or any other) fuel cell. The alkali metals are best recovered and purified in molten salts,[2] as is well recognized. The same is true of the lighter members of the alkaline earths as well as aluminum.

Lithium is of special interest in this group. Its method of extraction might benefit from a new approach.[1] This is particularly desirable at the present time since new applications in aerospace alloys, batteries, and nuclear fusion seem to suggest growth in demand for the metal. There could always be scope for advances in the recovery, recycling, and application of

David G. Lovering • The David Graham Consultancy, Swindon SN2 2LF, United Kingdom. *Robert J. Gale* • Chemistry Department, Louisiana State University, Baton Rouge, Louisiana 70803.

sodium, magnesium, and aluminum. It is not clear yet whether the sodium-sulfur battery or the aluminum-air cell will be sufficiently successful to encourage new demand for these metals; the Alcoa chloroaluminate route for aluminum extraction has not proved commercially viable, though.

Extensive scope certainly exists for improving and devising new methods for the extraction and (electro)plating of many of the transition and lanthanide elements. Of these, perhaps most benefit would be derived if better routes for titanium winning and coating could be developed. The remaining refractory elements are also somewhat intractable to manipulate, so improved (molten salt) methods of forming and plating these materials might offer more resistant metals for vital aerospace applications, for example. The lighter group VIII ferromagnetic metals should not be ignored either—thin electroplates of these prepared from melts might prove interesting.

The precious metals are also recovered and recycled from cyanide melts. Here, less toxic and lower melting systems could bring clear advantage.

Chalcogenide, interchalcogenide, halogens, interhalogens, and intergroup combinations may either be extracted from melts (the only route for fluorine?) or form interesting and underresearched molten mixtures, i.e., Ovshinsky switches. New methods for chlorine production from melts have been proposed.[1]

In group IV, the extent to which hydrocarbons may be catenated, cracked, recovered, and reformed using molten media is only just being appreciated. Developments in this area are certain.

Only a handful of papers have so far described the properties of molten salts in contact with semiconductors from group IV or III–V composites. New routes for the preparation of thin films of these materials from melts have barely been conceived at this point.

Traditionally, the lanthanides had been extracted from their ores via molten salt routes, although these methods have presently been displaced. Apart from the timeliness of a new search for even better extractive procedures, exciting developments in laser, magnet, and optical fields suggest that the electroplating of thin films of these elements (from melts, presumably) could be of interest.

The actinides cannot go without a mention. Reprocessing of these hazardous materials may often be accomplished through a molten salt method, as is amply illustrated by the chapter by Reavis in this volume. Additionally, the storage of nuclear waste in glasses must properly be viewed as a molten salt discipline, for a glass is no more than a supercooled melt and its preparation owes much to molten salt technology.[2]

Interdisciplinary technologies such as metals treatment, fabrication, and joining may all draw on molten salts in some aspects of their operations—new developments are appearing in these areas continuously.

Finally, when our machines are afflicted by corrosion at high temperatures, in some cases, condensed salt vapors forming liquid melts will surely aggravate undesirable decay; studies in this important area of our technology are continuing apace.

References

1. Y. Ito and S. Yoshizawa, in: *Advances in Molten Salt Chemistry*, Vol. 4 (G. Mamantov and J. Braunstein, eds.), Plenum Press, New York (1981).
2. D. G. Lovering, ed., *Molten Salt Technology*, Plenum Press, New York (1982).

2

Actinides

James G. Reavis

1. Introduction

1.1. Perspective

The importance of the actinides to the evolution of society in the near future (the next 100 years) can hardly be overemphasized. The potentially destructive aspects of energy release from actinides have been widely discussed by the popular communications media. It is less widely known that the useful energy that may be derived from economically recoverable actinides (uranium and thorium) is 5 to 20 times as great as the energy available from all economically recoverable fossil fuels.[1] This fact establishes the economic importance of development of efficient processes for preparation of thorium, uranium, plutonium, and their compounds.

The study of the actinides not only has a great economic incentive but also is fascinating from the technical point of view. The actinide series is, in many ways, similar to the lanthanide series of the Periodic chart. There are similarities and differences between homologs of the two series just as there are interesting and unexplained changes in properties during progression through the respective series. The actinides are chemically active metals, only slightly less active than the alkaline-earth and rare-earth elements.

Molten salt techniques used in studies of actinide compounds are in general similar to those used in studies of rare-earth and alkaline-earth compounds, but there are important differences. All of the actinides are radioactive, and work with them requires special health protection measures. Another hazard of working with plutonium and enriched uranium is that

James G. Reavis • Los Alamos National Laboratory, Materials Science and Technology Division, Los Alamos, New Mexico 87545.

a critical mass may be inadvertently assembled. Many of the techniques and examples discussed in this chapter apply equally well to work with nonactinides, and there may be significant duplication of discussions found in previous chapters of this series, but it is hoped that there will be sufficient new material in this discussion to reward the reader.

Although there are 15 elements (including actinium) in the actinide series, only about six of them have been used in molten salt studies. The list of anions observed in molten salt studies is similarly limited. As an example of the cation/anion combinations studied at high temperatures, the number of combinations found in an extensive compilation[(2)] of phase diagrams is listed in Table I. In addition to the oxides, fluorides, and chlorides listed in Table I, 14 other phase diagrams were given for bromides, sulfates, oxychlorides, phosphates, silicates, molybdates, and tungstates of the actinides. These 14 other diagrams are included in the last column of Table I. This distribution is typical of the distribution of actinide–anion combinations appearing in other molten salt studies. Because oxides are not generally classified as salts, this chapter deals primarily with halides of the most abundant actinides simply because other compounds have not been studied at high temperatures. One reason for this lack is that the oxyanion compounds of actinides are often unstable.

1.2. Historical Availability of Actinides

Of the 15 members of the actinide series, only four (actinium, protactinium, thorium, and uranium) exist in nature in concentrations detectable by any other than the most sophisticated techniques. Actinium was discovered in 1899 by Devierne and, apparently, independently in 1902 by Geisel, but was not isolated in pure form until 1947, when milligram quantities were separated from neutron-irradiated radium.[(3)] Thorium was discovered by Berzelius in 1828 and was used commercially before 1900. Protactinium was discovered in 1918 and was first isolated in milligram amounts in 1927. Uranium was discovered in 1789 and was well known to scientists before 1900. The remaining actinides are all synthetic elements

TABLE I. Number of Actinide/Anion Combinations Listed in a Compilation of Phase Diagrams

Actinide	Oxide	Fluoride	Chloride	Total
Cm	1	0	0	1
Np	1	0	0	3
Pu	4	3	9	16
Th	36	11	0	53
U	34	30	10	80

and were unknown before about 1940. Plutonium was first isolated in visible quantities as an oxide in September 1942. Neptunium, americium, curium, berkelium, and californium were isolated by 1950. The other five actinides were discovered after 1950.

1.3. Current Availability of Actinides

Studies of the chemistry of the actinides are limited to some extent by the restricted availability of the elements. These restrictions are due to a variety of factors. Aside from government regulations, there are certain physical problems such as limited rate of creation and isolation of the elements, coupled with short half-lives, which limit availability. Only microgram quantities of transcalifornium elements will exist in the foreseeable future. Another problem is high radiation intensities of emissions of most of the actinides and their daughter elements, requiring shielding and containment. These are available in only a few commercial and university laboratories outside the small number of government laboratories built for those specific studies. Order-of-magnitude values of isolated and purified supplies of the most abundant actinides and half-lives of their most useful isotopes are listed in Table II. The actinides not listed in Table II are available only in microgram or submicrogram quantities and probably will never be studied extensively by molten salt techniques. Studies of actinium, berkelium, and californium will be limited because of lack of availability and their intense radioactivity. The high levels of radioactivity of americium and curium discourage their study. Protactinium is unique in that it is naturally occurring and has a long half-life but is very expensive to recover because the richest mineral source contains only a few parts per million of the element. About 125 g of the element were isolated from about 60 tons

TABLE II. Availability and Half-Lives of the Most Abundant Actinides

Element	Quantities available	Prevalent isotope	Half-life (yr)
Ac	Milligrams	227	22
Th	Megagrams	232	1.4×10^{10}
Pa	Grams	231	3.2×10^{4}
U	Megagrams	238	4.5×10^{9}
Np	Kilograms	237	2.1×10^{6}
Pu	Megagrams	239	2.4×10^{4}
Am	Grams[a]	241	458
Cm	Grams[a]	244	18
Bk	Miligrams	249	0.86
Cf	Milligrams	249	360

[a] It is planned that kilograms of compounds of americium and curium will be isolated, but batch sizes of only 10 g or less are amenable to studies of molten salts without extensive radiation shielding.

of ore by workers[4] in the U.K. A sufficient quantity of element is therefore available for experimentation using ordinary techniques, but its high cost dictates conservation measures.

Anyone wishing to conduct experiments with actinides or actinide compounds must face not only problems of availability due to limited amounts in existence, but also regulation of their possession and use by government agencies. The experimenter may be forced to deal with the International Atomic Energy Agency, the United States Nuclear Regulatory Commission and Department of Energy, state government agencies, and local government agencies.[5–7] Some of the regulations limit mere possession of these elements, much less their use.

2. Special Enclosures Required for Actinides

2.1. Health and Safety Standards

Selection of the type of enclosure to be used for a specific project involving actinides is in large measure left up to the experimenter and his employer. This is not to say that the experimenter has a large measure of freedom in this matter. The various regulatory agencies listed in Section 1.3 will insist that certain guidelines be obeyed before operating licences are issued, and these guidelines will dictate certain tradeoffs between amounts of actinides to be used and the complexity of the facility that must be constructed to handle the material. These guidelines dictate levels of irradiation allowable for various body parts of the operators; degree of protection of the environment from radiological and toxic chemical release during normal operation and in case of fire, explosion, tornado, and natural disasters, and other items too numerous to mention here. Some of the well-known sources of these guidelines are the International Atomic Energy Agency Safety Standards, the Scientific Committee on the Effects of Atomic Radiation of the United Nations, the (U.S.) National Council on Radiation Protection, the International Commission on Radiation Protection, the Advisory Committee on the Biological Effects of Ionizing Radiation of the National Academy of Sciences National Research Council, and others. The various regulatory agencies may specify compliance with the regulations they choose. In the United States, the Nuclear Regulatory Commission regulations appear in the Code of Federal Regulations, Title 10, Chapter 1—Energy. The regulations of other U.S. agencies such as the Environmental Protection Agency and the Department of Transportation also govern handling of the actinides and are listed in other "Titles" of the Code. Additional information concerning health and safety standards and practices can be found in Refs. 5–7.

2.2. Benchtop and Chemical Hood Enclosures

Before about 1940 little thought was given to respiratory protection of workers handling massive amounts of the then commonly available actinides, thorium and uranium. Work with these compounds was commonly performed on open benchtops without limitation. During the 1940s, more intensely radioactive actinides became available, and there was an increasing general awareness of the nature of radioactivity. Consequently, government regulations were set up in an attempt to legislate safety for workers.

There appear to be wide differences in interpretation of these regulations by managers of various laboratories. The nature of the enclosure required is determined in part by regulations, in part by the nature of the operation, and in part by the amount of actinide involved. Aqueous chemistry experiments involving micrograms of even very highly active members of the series are performed in chemical hoods, while larger quantities of the same element in powder form must be handled in an enclosure of much higher integrity. Within certain guidelines the safety of each series of experiments must be evaluated to determine the enclosure or hood requirements.

2.3. Glovebox Enclosures

Glovebox enclosures have been used for a large fraction of the research, development, and production work with actinides. The basic enclosure may be considered as a cube measuring about 75 cm on each edge. It has a window and a pair of elastomer gloves about 75 cm long with 20-cm-diam cuffs. Basic building blocks of this sort may be made taller and/or combined side by side and back to back (omitting side or back walls from the combination as appropriate) to accommodate the size of equipment required for the operations to be performed. An airlock or antechamber in which radioactive contamination is kept at a low level should be provided for introduction of equipment from the laboratory in such a manner as to avoid escape of radioactive material to the environment. The enclosure must be "leak free" and must have a system to control the internal pressure at some value slightly below ambient laboratory pressure. The materials of construction must be chosen to be compatible with the operation to be performed and to conform with standards of resistance to fire, eathquake, pressure differential, radiation shielding, etc. Early gloveboxes were constructed of plywood, ordinary window glass, and obstetrical gloves. There has been a long period of evolution of the designs in current use at various laboratories. The basic glovebox has evolved with many variations in each laboratory where particular processes are used, but certain design features have been generally used repetitively in a given facility. These unique design features have evolved because of stronger emphasis on different criteria at the different facilities. For instance, designers at one facility may place emphasis on fire safety, another on resistance to failure at high pressure differential, another on operator comfort, another on extreme leak tightness, and so on.

Design features of gloveboxes used at the Los Alamos National Laboratory will be discussed in some depth as an example, not because these gloveboxes are the best for all processes or experiments, but because they meet or exceed current U.S. safety criteria and have proved useful for molten salt experiments as well as for pyrochemical processing operations. Design features of these enclosures can be seen in Figs. 1 and 2. The boxes are fabricated from stainless steel type 304 or 316 chosen for the appropriate corrosion resistance. They are of welded construction, with all welds polished to the smoothness of the adjacent metal. Floors have a No. 4 finish and the walls and tops have a 2B finish. Sharp corners are avoided wherever possible. The inside corners of the enclosure have a radius of 25 mm. These features aid in cleanliness of the enclosures, which helps in two ways to reduce the radiation background: First, because rough surfaces are difficult to clean, they accumulate thin layers of the actinides that are handled inside the enclosure. Thin films of this sort do not have the benefit of self-absorption that dense, massive samples have so that a small amount of actinide in the film contributes an inordinate amount of radiation to the general background. Second, if the actinide film remains on the wall for a long period (longer than the normal glovebox residence time for the bulk of the material) it may accumulate decay products which are also radioactive, often with a greater energy of radiation than was emitted by the parent.

The standard window is 0.25 in. thick and is held in place by a neoprene gasket of one of two designs, one of which requires a stainless steel frame and the other of which requires no frame. The stainless steel walls and floors are approximately 5 mm thick except in enclosures containing very heavy equipment requiring a stronger floor. The gloves are neoprene, or similar elastomer, and are at least 0.38 mm thick.

Gloveboxes constructed as described above give adequate protection for handling large quantities of thorium, natural uranium, and highly enriched uranium. They are also adequate for handling up to a kilogram of plutonium containing less than approximately 1% ^{241}Pu and having had americium separated from it within the previous few months. If multikilogram amounts of plutonium are being processed, or if the ^{241}Am decay product has been allowed to accumulate several months from decay of ^{241}Pu present at concentrations on the order of 1%, the additional radiation emission of ^{241}Am requires extra shielding to protect personnel adequately. Such usage should be anticipated and 0.25 in. of lead shielding should be included in the walls of the enclosure at the time of fabrication. The detail

FIGURE 1. A modern plutonium laboratory.

of Fig. 2 shows how the lead is covered by a thin layer of stainless steel that is welded to the shell of the box to make a more easily cleanable outside surface. This construction eliminates exposed cracks at the edge of the lead and makes decontamination much easier when inadvertent contamination of the laboratory occurs. Additional shielding of windows is easily accomplished by adding lead glass outside the safety glass. Additional shielding for the hands may be provided by lead-impregnated gloves that are available in thicknesses up to 1.65 mm. When these modifications are made, the enclosure is adequate for handling multigram quantities of americium, even in dilute solution where self-shielding is minimized. Multigram quantities

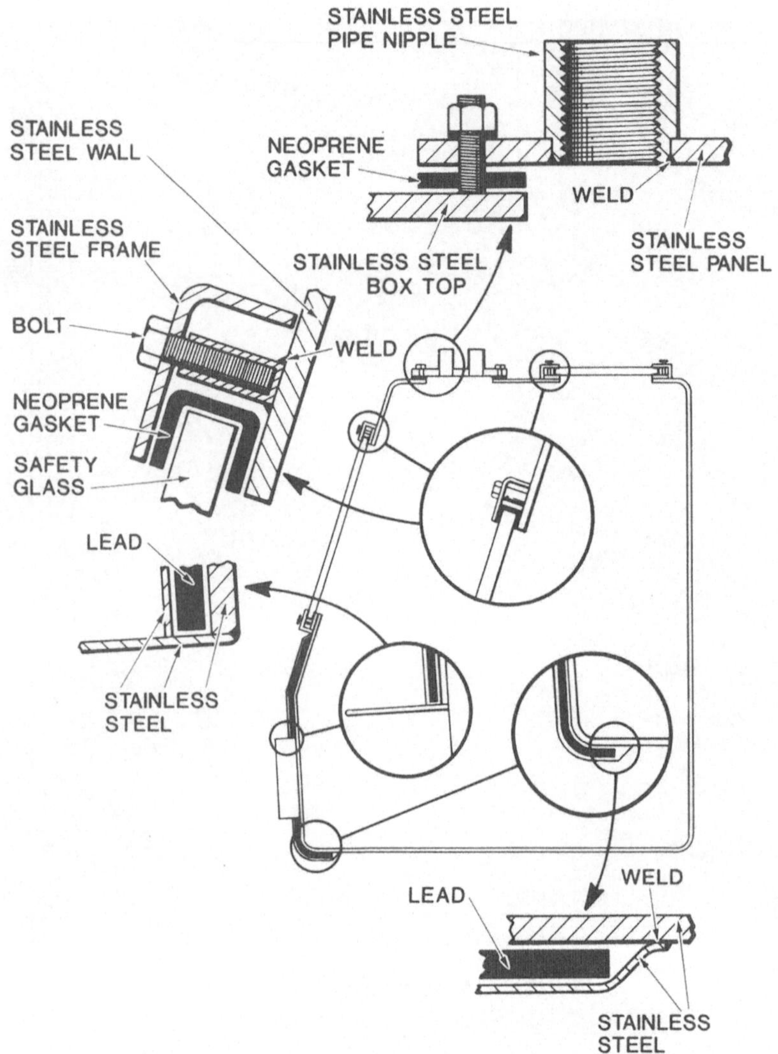

FIGURE 2. **Cross section of a glovebox enclosure showing some of the details of construction.**

of ^{237}Np and ^{233}U may also be handled in such an enclosure if exposure times are kept short. For routine operations with these isotopes, more lead shielding must be added. Plutonium-238 processing is routinely performed in the same type of enclosure with the addition of about 15-cm-thick hydrogenous shielding (either tanks of water or slabs of plastic). The hydrogenous shielding is for protection against neutrons produced by alpha-neutron reactions from the intense alpha particle emission of ^{238}Pu.

Glovebox enclosures have been used for a large variety of pyrochemical research and processing operations involving molten salts. This type of enclosure has many advantages over benchtop and open-hood enclosures when corrosive and hygroscopic materials are being used (even if they are not radioactive) because the atmosphere can be controlled to eliminate undesirable side reactions. Commercial, regenerable air dryers using dual beds of molecular sieve material can be used to maintain air atmospheres in enclosures with water concentrations on the order of 1 ppm. Similarly, "boil off" gas from liquid argon or nitrogen can be used in a once-through flow system to maintain atmospheres having oxygen and water concentrations less than 10 ppm. Recirculating purifiers use dual, regenerable sorbent beds containing molecular sieves and molecular sieves loaded with activated copper or nickel. To maintain atmospheres of this quality, the enclosure must be virtually leak free. Very sensitive leak detection instruments such as a commercial helium mass spectrometer leak detector must be used to ensure freedom from leaks. The allowable leak rate of enclosures operating at oxygen and water concentrations of 10 ppm are orders-of-magnitude smaller than the allowable leak rates for air atmosphere enclosures for intense alpha emitters such as plutonium and americium. If one develops even a pinhole in a glove of an inert atmosphere enclosure, instrumentation for detection of leakage of oxygen and water responds much more quickly than instrumentation for detecting escape of radioactive particles.

2.4. Remote Handling

The actinide researcher can almost always avoid the use of remotely operated hot cells if the proper isotopes are available and if the amount of actinide can be kept small and repurified often to remove highly active decay products. The small amounts of berkelium and heavier actinides that are available almost force the use of microchemical techniques for their studies so that the question of use of remote operation and intermediate-shielding facilities for research is moot.

Separation of transuranium elements from irradiated sources is quite a different matter. The transuranium actinides are commonly produced by irradiation of actinides in high-flux reactors. During irradiation, a wide variety of intensely radioactive elements are created. Operations for isolating the desired actinides must be performed in highly shielded hot cells. Currently most or all of these separations are done through aqueous procedures, but there are arguments[8] that molten salt processes should be developed. The expected benefits of pyrochemical processing include freedom from radiation damage of solvents, extractants, or ion exchange resins, and production of a smaller volume of radioactive waste products whose disposal is quite expensive. Pyrochemical processing should not only be considered for preparation of research quantities of actinides, but should also assume great commercial importance in separating fertile and fissionable fuel from spent reactor fuels.

Figure 3 is an illustration of the cross-section and plan views of a typical research hot cell facility[9] with four cells containing individual alpha containment enclosures. These cells are separated by thick steel doors from

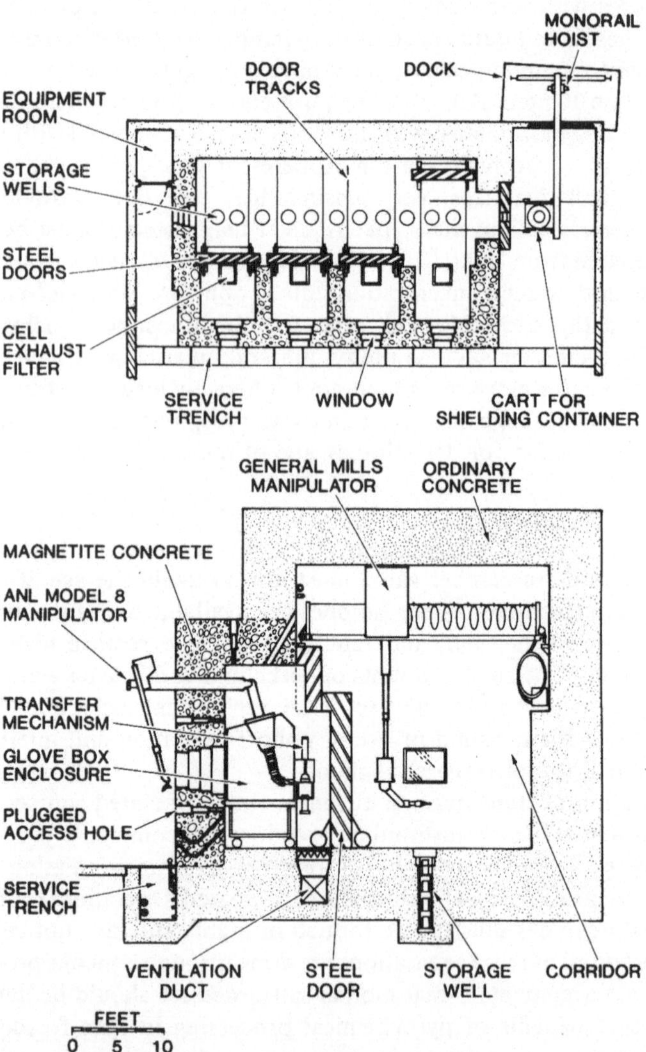

FIGURE 3. Plan and cross-section views of four hot cells used for research and small-scale production operations with actinides containing intense gamma radiation emitters.

a corridor used for transfer of highly active samples from shielded shipping casks to an alpha enclosure or a storage well. These cells are typical for research operations with highly active gamma emitters. Cells of this size can also be used for processing kilogram quantities of irradiated fuel or isotope source material by pyrochemical methods where the reactants are kept in a very dense, compact form. Work with manipulators is extremely time consuming and maintenance of hot cells is very expensive. Hot cells are also expensive to construct and they require large amounts of space in expensive buildings that meet the criteria for actinide containment. It is generally less expensive to set up equipment for microchemical measurements of chemical and physical properties. These are possible when the pure actinides are available and do not require separation from large amounts of other radioactive materials. Less time is often required for "hands-on" microchemical experiments than for cumbersome operations performed with manipulators. There are measurements in the field of molten salt studies, however, that are extremely difficult to make by microchemical techniques and that may be done more efficiently in a hot cell. Additional information concerning design and operation of enclosures for handling the actinides and their compounds is given in Refs. 5, 7, and 10.

3. Materials Problems

3.1. Irradiation Degradation of Materials

Irradiation degradation of chemicals, elastomers, and glass is severe in hot cells where irradiated materials are handled, but because reprocessing does not extensively employ molten salts or pure actinides (except as an end product), materials degradation in high-intensity gamma fields will not be discussed here. This subject is thoroughly covered in many publications, including Refs. 5, 11, and 12. Less well-known is the severity of radiation damage by intense alpha emissions of some of the actinides. The intensity of alpha particle emission by protactinium, thorium, natural uranium, and the most commonly used isotope of plutonium (mass 239) is sufficiently low to cause almost negligible effects. One exception is that care must be used in long-term storage of plutonium in plastic containers, which may suffer alpha-radiation degradation over a period of several months. The isotopes having higher intensities of alpha emission produce significant deterioration of most organic materials. While it is quite practical to use 0.4-mm-thick neoprene gloves on enclosures for ^{239}Pu, ^{238}Pu causes rapid deterioration of these gloves and requires Hypalon-coated gloves at least 0.6 mm thick. Hypalon is a modified polyethylene manufactured by Du Pont. Silicone grease may be used for periods of days to weeks on ground glass joints and stopcocks of an experimental apparatus for handling ^{239}Pu, but ^{238}Pu in the same apparatus produces unacceptably rapid increases in viscosity of the grease. Exhaust filters on enclosures used for handling multigram quantities of powdered compounds of the more intense alpha emitters must be fabricated with radiation-resistant glues. If silicone or other types of oils are used in pressure-relief devices on such enclosures, they must be checked periodically to ascertain that the oil has not become so viscous that the device no longer functions properly. These effects are seen even in enclosures provided with inert atmospheres. If the enclosure atmosphere is air containing water vapor at ambient relative humidities, these effects are accelerated and corrosion of metals and other corrosion-resistant materials becomes serious. There is speculation that ozone, formed in air subjected to alpha radiation, may play an important part in the acceleration of corrosion. Some workers have attempted to remove ozone from enclosure atmospheres by decomposition on MnO_2 to reduce the rate of corrosion. The lighter, halogenated hydrocarbons (such as freons) are similarly degraded and produce corrosive products.

Storage of the less active actinides, even in kilogram amounts as solutions or dry compounds in polyethylene or glass containers, presents no particular problems (subject to critical-mass limitations). Plutonium-238 and more active actinide isotopes do cause significant container degradation. It is advisable that even milligram amounts of these actinides be stored in nonreactive metal containers. Plastics are degraded quickly, not only by the intense alpha irradiation, but also because of the heat generated locally. Glass vials containing fractional gram amounts of ^{238}Pu develop cracks, probably because of large thermal gradients rather than because of radiation damage.

As was mentioned previously, one way to keep radiation damage (either to personnel or to equipment and enclosures) to a minimum is to keep actinides and their compounds confined to the minimum volume (within thermal and criticality constraints) to utilize self-shielding. Avoidance of thin deposits of alpha emitters on elastomers such as glovebox gloves is very important. Good housekeeping can hardly be overemphasized.

3.2. Container Compatibility at High Temperatures

Problems of containment of molten actinide salts are very similar to problems of containment of salts of other active metals such as the alkali, alkaline-earth, and rare-earth elements. The radioactivity of the actinides adds little to the experimental problems, other than the general problems of handling radioactive substances discussed in preceding sections. A large body of information has been generated, however, in research on actinide salts, in particular metal/salt systems and fluoride systems. The latter have been studied extensively in the Molten Salt Reactor Experiment program.[13,14]

Two types of containment problems are encountered in molten salt work. One is reaction of the container with the contents, which contaminates the contents to make the reaction product undesirable and/or changes the property being measured in an experiment. The second type of problem is seepage of the liquid into pores or through cracks in the container without significant contamination of the contents. The latter type of problem often has much greater relative significance in work with actinides because of their value and because the actinide must be recovered from the container at the end of the operation in accordance with Environmental Protection Agency regulations. The effort expended to recover the actinide often is much greater than the effort expended in making a measurement or preparing a compound. Because there is no such thing as absolute recovery of the actinides, the scrap from the original container, as well as wastes generated during recovery operations, will all have low-level contamination and must be disposed of by elaborate and very expensive methods.

Misadventures involving container breakage and release of actinides may have the gravest of consequences. If a salt or salt/metal mixture at high temperatures is released, not only from its primary container but also from the glovebox enclosure, reaction with the atmosphere may occur, leading to dispersal of finely divided radioactive material. This may be inhaled by personnel or may result in severe contamination of the building. If hundreds of grams of anhydrous fissile material is being handled, care must be taken to avoid the possibility of mixing with water or other hydrogenous material, which event could lead to a criticality incident.

These constraints lead to choices of materials and designs often more expensive and conservative than would be used for containment of non-radioactive materials. Double containment is often specified for actinides where single containment would be used for other elements. Limited-volume, circulating cooling water systems are used to prevent flooding of enclosures and possible criticality incidents. High-density vitrified ceramic crucibles are used to minimize loss of accountable material to crucible scrap. Additional design time and safety analyses are required for work with these elements.

Chemical reactions with container materials must be carefully considered before starting work with molten salt systems containing actinides. Experiments with pure molten chloride systems may be performed in borosilicate (Pyrex) glass at temperatures up to about 500°C, but an additional measure of resistance to breakage due to thermal shock or to melting in the case of temperature control malfunction is afforded by fused quartz or Vycor (96% SiO_2). Quartz may show significant attack if air and water are not rigorously excluded from the systems, but these should be excluded anyway to prevent direct reaction with the actinide halides to form oxides and oxyhalides. All salts put into the system must be rigorously treated by well-known methods to eliminate traces of water and oxyhalides. Much less work has been done with actinide bromides and iodides and they may be expected to be thermally less stable than the chlorides, but quartz is not expected to contribute significantly to their decomposition. Fluorides, on the other hand, are expected to react with siliceous materials at elevated temperatures, although a report[15] of studies of spectra of actinides in mixed fluoride/chloride molten salts did not mention reaction with quartz cells. Platinum and gold have been used as containers for experiments with fluorides, and extensive corrosion studies of fluoride systems containing thorium and uranium fluorides were carried out in the Molten Salt Reactor Experiment program.[13,14] The best containment for flowing molten fluoride salts in large systems was provided by Hastelloy N (7.4% Cr, 4.5% Fe, 17.2% Mo, and 70% Ni) and by titanium-modified Hastelloy N (7.3% Cr, 13.6% Mo, 77% Ni, balance Ti).[16]

Molten salt studies of actinides in oxyanion systems have been much less extensive than in halide systems. Pyrex optical cells showed no evidence of reaction with $LiNO_3$–KNO_3 solutions of several of the actinide nitrates when observed at temperatures up to 250°C. Molybdates, tungstates, phosphates, and silicates have been studied in platinum and gold containers.

The choices of containers for work with metal/molten salt systems are much more difficult than the choices for pure salts. The actinide metals are very reactive and react with many ordinary ceramic materials. Table III contains a list of the approximate free energies of formation of selected oxides used as container materials for actinide metal/salt mixtures, at a common temperature of observation. The free energies of formation of oxides that may form from the metals being studied are also included in

TABLE III. Free Energies of Formation of Selected Oxides[a] at 725°C (Stated in kcal/g-atom of Oxygen)

Oxide	$-\Delta F$	Oxide	$-\Delta F$	Oxide	$-\Delta F$
Ac_2O_3	129	SrO	118	ZrO_2	108
CaO	127	MgO	118	CeO_2	108
ThO_2	123	BaO	111	NpO_2	102
La_2O_3	122	HfO_2	110	PaO_2	101
Ce_2O_3	122	Li_2O	110	TiO_2	91
BeO	120	Al_2O_3	109	SiO_2	83
Am_2O_3	120	UO_2	109	Ta_2O_5	78
Y_2O_3	119	Pu_2O_3	109	Na_2O	66

[a] Reference 17.

this tabulation. Examination of the values listed in Table III leads one to choose as containers the oxides of calcium, thorium, lanthanum, cerium, beryllium, yttrium, magnesium, and aluminum, in that order. One must also consider other chemical, historical, and economic factors that govern the availability of containers made from these materials. Calcium oxide reacts too readily with water, thoria is radioactive, the rare earths are expensive, and beryllium presents health hazards. Thorium and the rare earths are very difficult to separate from the actinides in waste recovery operations. Beryllium produces neutron radiation problems by the alpha-neutron reaction when mixed with the intense alpha emitters. On the other hand, magnesia and alumina are commonly available, present no appreciable health hazards, and have been studied for many years, with the result that fabrication techniques for these ceramics are well known. Also, magnesium and aluminum are relatively easily separated from the actinides by traditional waste recovery techniques. Because of these and other reasons, something other than the thermodynamically "best" container material is often chosen to contain molten actinide metal/salt mixtures.

In the selection of fabricated containers, the physical properties as well as chemical properties must be considered. If the time of contact with liquid phases is short and the container will be subject to thermal shock, the sintered density should be low to avoid breakage during rapid thermal cycling. If the heating cycle is slow and the liquid contact time is long, the container should be vitrified (sintered to a density approaching 100% of the theoretical density) to minimize the surface area available for reaction and penetration of the wall by the liquid phase, which contains accountable actinides. Regardless of the density requirements of the container, undesirable reactions can be minimized by reducing roughness (and therefore reactant area) of the interior of the container.

Specifications for typical magnesia containers used for routine plutonium metal preparation by molten salt reductions (see Sections 6.2 and 6.3 for process descriptions) are listed in Table IV. Only recognized significant properties are specified. Obviously, not all properties of these containers can be specified. If one manufacturer's product meets these specifications and is found to be satisfactory, there is no guarantee that another manufacturer's product made to the same specifications will be satisfactory, because an unknown, unspecified property may be critical. It is suspected that this factor contributes to contradictory reports from different laboratories concerning the suitability or otherwise of containers. For instance, it is the experience of the Los Alamos National Laboratory that magnesia crucibles perform better than alumina crucibles for preparation of plutonium,[18] while experiments at the Atomic Weapons Research Establishment found the opposite to be true.[19] With the present state of the art, one may select a container material for molten salt/metal systems

TABLE IV. Partial List of Specifications for Magnesia Crucibles Used for Plutonium Metal Preparation at Los Alamos by Bomb Reduction of the Fluoride and Direct PuO_2 Reduction in $CaCl_2$

Item specified	Bomb reduction	Direct oxide reduction
MgO[a]	97% (min)	98.5% (min)
CaO	—	1.0% (max)
SiO_2	2.0% (max)	0.5% (max)
SiO_2/CaO wt. ratio	1.0 (min)	—
Fe_2O_3	0.15% (max)	0.15% (max)
Al_2O_3	0.35% (max)	0.35% (max)
Be	15 ppm (max)	15 ppm (max)
Pb, Cu	25 ppm (max)	25 ppm (max)
Ti	30 ppm (max)	30 ppm (max)
Ga, C	50 ppm (max)	50 ppm (max)
B, Mn, Ni, Zr, Cr	100 ppm (max)	100 ppm (max)
Porosity:	—	No open porosity; no penetration by ethanol when crucible is filled
Density:	2.8 g/cm^3 (78% TD)	Vitrified
Cracks:	—	None detectable by dye test

[a] Direct substitution of 3 wt% Y_2O_3 for MgO is permissible.

based on thermodynamic and special chemical considerations, write specifications of physical properties that are expected to best fit the use, try samples supplied by several fabricators, and finally select the one that gives the best performance.

Refractory metal containers should not be overlooked when one is considering containers for molten actinide salt/metal mixtures. The metals that have been most often used for such containment are tantalum, tungsten, molybdenum, and alloys of these elements. The solubility of these elements in molten actinide metals is low, and freedom from anion contamination, often contributed by ceramics, is a significant advantage gained by use of metal containers. The most serious problem encountered with use of these container materials is removal of the metal phase from the container after the experimental measurement or reaction. In some cases, surface treatments such as oxidation of tantalum crucible surfaces are effective in preventing wetting of the container by the pure actinide metals, but molten salts often break down such interfaces so that the actinide metal wets the crucible and cannot be removed after solidification except by aqueous dissolution, an undesirable procedure. A long-term goal for actinide pyroprocessing is the transfer of molten salt and metal products from metal reaction containers as liquids,[20] but success of efforts to achieve this goal has been limited.

4. Microtechniques

For purposes of this discussion, the term "microtechnique" will be expanded to include techniques employed in handling microgram to milligram quantities of the element being studied because an apparatus built to study microgram samples can often be easily modified for study of milligram quantities. For a number of reasons, microtechniques are well suited to studies of the actinide elements. With the exception of thorium, uranium, neptunium, plutonium, americium, and curium, world supplies are very small, and the individual laboratory may not be able to procure even milligram amounts. With the exception of natural thorium and uranium, the actinides are all very expensive. Health, safety, and security regulations may allow work with milligram quantities of some of the elements in a certain facility, while possession of gram or multigram quantities is not permitted. Even if available quantities of actinides were not limited by these external factors, there are other reasons for keeping their quantities to a minimum. The intense radioactivity of many of the actinides gives rise to problems of personnel irradiation, equipment degradation, and generation of heat which can cause difficulties in control of the sample temperature.

Even though there are many reasons for using microtechniques in molten salt studies of the actinides, their application has been limited. This is probably because work with actinides (particularly those that are the most highly radioactive and in shortest supply) is very expensive in terms of money and time, and the rewards are limited to extension of knowledge with few perceived industrial applications. Microtechniques using molten salts have been applied almost exclusively in the area of preparative chemistry and almost not at all in study of properties of the molten salts themselves.

Study of actinide compounds by microtechniques began in about 1942 because only microgram quantities of the synthetic elements were available. One of the pioneers in the field was B. B. Cunningham. His review article[21] published in 1961 is a valuable reference. Later reviews[22,23] are equally valuable. The earliest work was a study of both aqueous and dry chemistry of plutonium and its compounds. Since that time transplutonium metals up to and including californium have been prepared by microtechniques. Two generalized reactions have been employed to prepare these metals:

$$AnF + M_{(g)} \rightarrow An + MF \qquad (1)$$

$$AnO + M' \rightarrow An_{(g)} + M'O \qquad (2)$$

where An is an actinide, M is a volatile active metal (generally Li or Ba), and M′ is a nonvolatile active metal (generally La or Th). A sketch of a furnace for the reduction step of metal preparation is shown by Fig. 4.

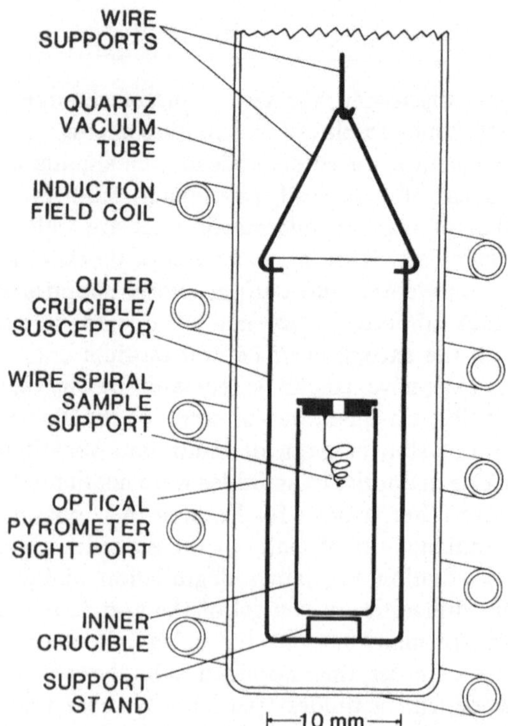

FIGURE 4. Cross section of a typical induction-heated double-wall microfurnace.

When the reductant is lithium, it vaporizes at the temperature of reduction and reacts as a gas with the (molten) fluoride supported by the wire spiral (or cup), producing volatile lithium fluoride and a bead of actinide metal that sticks to the wire spiral or cup. The lithium fluoride emerges as a gas from the hole in the top of the crucible lid. The operator removes the cooled metal product from its support by peeling away the refractory wire or foil. When barium is used as the gas-phase reductant, the "slag" (barium fluoride) is chipped away from the metal product.

When a nonvolatile reductant (lanthanum or thorium metal) is used to reduce oxides, the actinide oxide and the reductant are mixed in the bottom of the crucible, the susceptor shown by Fig. 4 is omitted, and the volatile actinide metal product vaporizes and is collected on a target suspended above the effusion hole in the crucible lid.

Figure 4 may be considered to represent a generalized apparatus for preparation and reduction of actinide salts. The actinide often emerges from aqueous purification absorbed on ion exchange beads. These beads can be calcined in the same type of furnace, except that the material of construction must be platinum. The resulting oxide is converted to fluoride in a similar

Monel furnace by treatment with $HF-H_2$. Other halides may be prepared similarly by the use of other hydrogen halides. Such experiments with small amounts of samples must be carefully planned to minimize sample transfers to avoid losing or contaminating the sample. Unfortunately, transfers may be required because few container materials are suitable for both reducing and oxidizing conditions at high temperatures.

Physical properties of molten salts that may be measured in very small samples include vapor pressures, melting points, and absorption spectra. Measurements of these properties by conventional techniques will be discussed in a later section of this chapter, but the application of microtechniques to melting-point and absorption spectra measurements will be discussed here. Melting points of high-melting (above 750°C), high-purity, irregularly shaped salt crystals as small as 1 mm^3 can be easily estimated if the operator simultaneously views slumping during heating and measures its temperature by use of a hot-wire telemicroscope optical pyrometer. This technique has two major faults. The first is that the crystal and its background must be at different temperatures to provide contrast to make the crystal visible, thereby violating a primary rule of optical pyrometry that exact temperatures can be measured only under blackbody conditions. The other problem is encountered in the case of relatively volatile compounds, which appear to slump when the sharp corners are sublimed. Nevertheless, reasonably accurate (often within a few degrees) determinations of melting point may be made simply with a properly designed oven and a good telemicroscope pyrometer. Another method of melting point determination that has been used on microsamples of metals is observation of the temperature at the instant of collapse of a bead held in compression between the two arms of a tungsten wire clip.[(24)] This technique should work equally well for determination of melting points of pure salts.

Absorption spectra of molten salts may also be measured by utilization of microtechniques.[(25,26)] Light from a source is focused by a microscope objective lens on the sample in a quartz capillary or other suitable microsample holder. The transmitted light then is passed through another microscope objective or condensing lens and is focused on the aperture of a monochromator, which transmits the spectrum to a suitable detector. Systems of this type have been built from components on optical benches or have been used in the sample compartments of commercial spectrophotometers. In the only report[(26)] of molten salt observation, the sample was contained in a quartz capillary that was heated by a coil of platinum wire. Conversion of other types of sample holders, such as thin metal sample holders on which the sample is mounted in a pinhole of an electrically heated sample holder, should be a relatively easy task. Recently introduced spectrophotometers[(27)] that utilize fiber optics to direct the analyzing light beam might further simplify measurement of absorption spectra because the analyzing beam can be conducted through a furnace in any position in an enclosure while the light supply, dispersion, and detecting systems remain outside the enclosure.

Although microtechniques have some advantages over larger-scale work, they also have some drawbacks. Microsamples must be kept scrupulously clean. Unseen impurities added to microsamples may produce mixtures having significant mole fractions of unwanted elements in the sample. Although corrosion of a sample support may not be visually detectable, even with optical aids, microscopic corrosion may have introduced significant impurities. Microsamples are so small as to defy accurate analyses after the preparation or experimental measurement so that one can seldom prove that the material used was really pure.

Another problem with microchemical work at high temperatures is accurate measurement of temperatures. When the double container technique is used (Fig. 4), it is relatively easy to measure temperatures accurately by use of the optical pyrometer, but if the sample is suspended in an unshielded position, unknown emissivity corrections give rise to significant uncertainties of sample temperatures. When thermocouples are used to measure temperatures of unshielded samples, it may be impossible to attach the thermocouple junction to the apparatus in sufficiently close proximity to the sample to give an accurate temperature indication. This is because temperature gradients are often high in small furnaces with minimal radiation reflection. Heat conduction from the junction by the thermocouple leads may cause significant errors.

In spite of these limitations, many of the measurements are surprisingly accurate. For instance, it was reported[(24)] in 1951 that the melting point of neptunium metal determined by observing two samples weighing 200 μg each was 640 ± 1°C. Later measurements on multigram samples of this element indicated a melting point of 637°C. On the other hand, the melting point of two samples of americium weighing on the order of 1 mg each was observed[(28)] to be 994 ± 7°C. Later determinations of the melting point using multigram samples indicate that the correct melting point is 1175 ± 3°C.[(29)] These examples illustrate that microtechniques can be very useful and can produce accurate data but that extra caution may be required in their use.

5. *Techniques for Purification of Multigram Quantities of Actinides and Their Salts*

The actinides that are used on the scale familiar to most chemists in universities and industry are thorium, uranium, and plutonium. Intermediate quantities of protactinium and neptunium are available, and smaller amounts of americium and curium are used, not necessarily because the world supply is small, but because of their intense radioactivity. These elements are discussed in this section; the remaining actinides are best studied by the microtechniques of section 4.

5.1. Aqueous Methods of Purification

The traditional aqueous separation and purification methods of ion exchange, solvent extraction, and precipitation are used in separation of actinides from anionic and cationic impurities. As was pointed out in the discussion of microtechniques, the end product of microchemical-scale purification of actinides was often one or a few ion exchange beads loaded with the desired element. Aqueous actinide preparation and purification on the multigram scale have as their predominating end product an oxide or a compound that is readily converted to an oxide by calcination. Compounds for molten salt work are then prepared by pyrochemical techniques. Among the limited number of exceptions are nitrates and some fluorides. Very brief discussions of aqueous purification techniques are presented here.

5.1.1. Thorium

It is estimated that the thorium abundance in the earth's crust is on the order of 10 ppm. The most important commercial source is a rare-earth phosphate mineral called monazite. Two processes[30,31] have been used to recover and purify thorium on the scale of tons per year. One process consists of digestion in H_2SO_4, dissolution with H_2O, selective precipitation by NH_4OH, dissolution in acid, purification by solvent extraction, and precipitation as a hydroxide. The second process involves digestion in NaOH, dissolution by HCl, and selective hydroxide precipitation. For complete separation of thorium from uranium and rare earths, additional solvent extraction stages may be required. The final product from aqueous processing is usually thorium nitrate tetrahydrate, thorium hydroxide, or thorium oxalate, depending on the intended use. Thorium tetrafluoride suitable for molten salt operations has been prepared by precipitation from aqueous solution, drying, and finally heating to 400°C in an HF atmosphere. Care is required to prevent double salt formation during precipitation of the fluoride if potassium or ammonium ions are present.

5.1.2. Protactinium

Protactinium-231 occurs naturally as a member of the actinium series that has ^{235}U as the parent isotope. Since the half-life of ^{231}Pa is shorter than that of ^{235}U by a factor of 2×10^4, the equilibrium concentration of this element in uranium ores is extremely small. Nevertheless, workers in the UK mounted a tremendous effort to recover protactinium from about 60 tons of uranium-processing sludge.[4] This effort culminated in 1961 in recovery of over 100 g of protactinium in the form of pure solutions and compounds. After much research and development, the recovery process finally adopted was leaching of the sludge with HNO_3–HF, separation of uranium by solvent extraction, coprecipitation by adding $AlCl_3$, removing aluminum by dissolution with NaOH, dissolving the remaining solid in HCl, and recovering the protactinium by solvent extraction.

The aqueous chemistry of protactinium appears to be quite complex,[32–34] with hydrolysis playing a major role. The chemical behavior appears to be much more similar to that of niobium and tantalum than to its actinide neighbors thorium and uranium. Protactinium assumes valences of 4 and 5 in a variety of complex compounds with other metals and nonmetals. Precipitation as the iodate, which can be thermally decomposed to Pa_2O_5, appears to be the best route for beginning of preparation of compounds for molten salt studies.

5.1.3. Uranium

It is estimated that the abundance of uranium in the earth's crust is about 3 ppm. There are many minerals that contain economically recoverable amounts of uranium; therefore, there are many variations of extraction processes.[35–37] The crushed mineral may be leached with sulfuric acid and purified by ion exchange or solvent extraction, followed by precipitation as a peroxide or diuranate. Other processes involve leaching the ore with sodium carbonate solution. Products of these processes are usually uranyl nitrate hexahydrate, ammonium diuranate, or oxide. Uranium dioxide, which may be formed by heating the higher oxides under flowing hydrogen, is most often chosen as the feed material for preparation of uranium compounds for molten salt work.

5.1.4. Neptunium

Neptunium is a synthetic element prepared in nuclear reactors by the $n, 2n$ reaction with ^{238}U, forming ^{237}U, which decays by beta emission to ^{237}Np. This process is shown by Fig. 5. Because almost all reactors contain ^{238}U, it is obvious that many kilograms of neptunium have been created in reactors all around the world. The major use of neptunium is irradiation in reactors to make ^{238}Pu for heat sources. The effort spent to recover neptunium during processing of irradiated reactor fuels is dependent on the perceived demand for ^{238}Pu. Parameters of the solvent-extraction and ion-exchange processes normally used in processing irradiated fuels can be changed to divert neptunium to the waste stream or to the uranium stream,

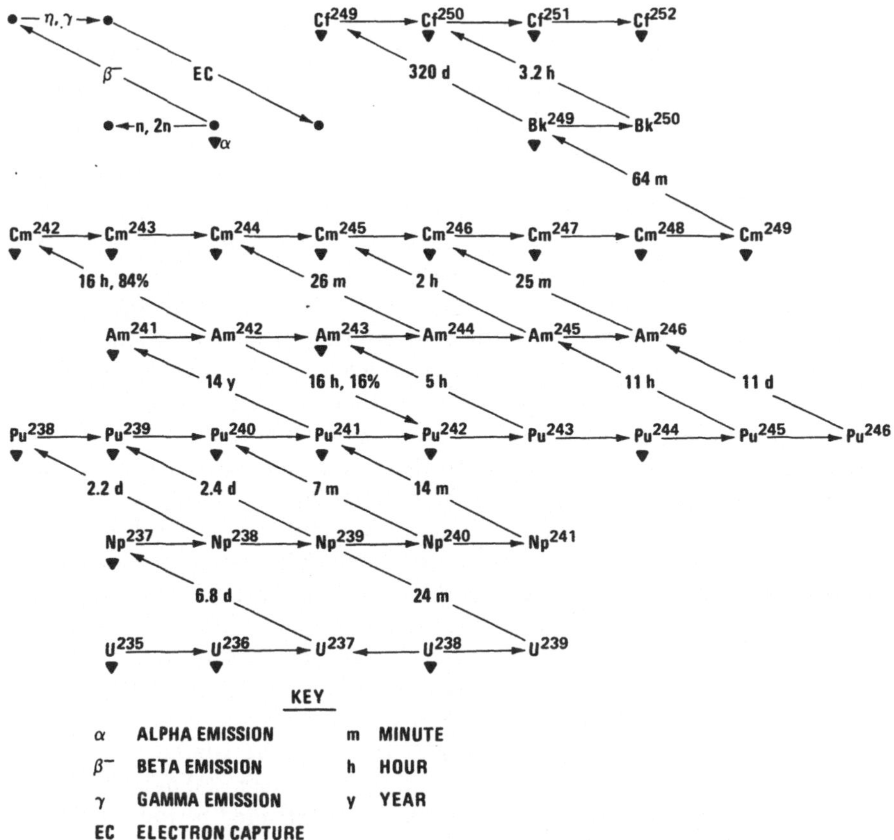

FIGURE 5. Modes of formation and decay of some of the transuranium elements.

where the neptunium is separated and recovered. The final product of separation and purification is a solution from which neptunium(IV) oxalate is usually precipitated.[38] This oxalate is calcined at about 500°C for purposes of target fabrication for generation of ^{238}Pu, or for neptunium metal or compound preparation.

5.1.5. *Plutonium*

Plutonium-239 (which is the most commonly encountered isotope of plutonium) is a synthetic isotope formed in reactors by neutron capture by ^{238}U to form ^{239}U. As can be seen from Fig. 5, plutonium formed by neutron absorption will always consist of a mixture of isotopes whose composition depends on the time and flux of irradiation. Plutonium-239 is the most desirable isotope (with the exception of ^{244}Pu, whose formation in macro amounts is impractical) for chemical studies because it has a long half-life and decays by alpha emission to ^{235}U, which is a relatively innocuous daughter element. Since ^{239}Pu is the first long-lived isotope formed during neutron irradiation of ^{238}U, plutonium formed by short-term irradiations of uranium is the most desirable.

Many pyrochemical schemes have been proposed for processing irradiated reactor fuels, but almost all reprocessing that is done worldwide is based on solvent extraction and ion exchange. Uranium and plutonium are first separated from fission products (which go to a waste stream) and are finally separated from each other. The plutonium usually emerges from separation and purification in nitric acid solution in the tetravalent state, although some processes deliver the product in the trivalent state. The tetravalent plutonium may be precipitated, dried, and calcined to PuO_2 for further processing. Another common treatment of the product solution is reduction of the plutonium to the trivalent state and precipitation of the oxalate, which may be calcined to form a reactive PuO_2. The only commonly used treatment for direct conversion to a salt useful in molten salt work is reduction of the plutonium to the trivalent state, followed by precipitation of the trifluoride, which may be dried by carefully heating under flowing argon or helium to 600°C. Direct precipitation of the tetravalent oxalate or fluoride is almost never done because of problems of filtration of these compounds if conditions of precipitation are not closely controlled. The oxide formed on calcination of the peroxide and oxalate is reactive and can be used in most preparative work. This is to be contrasted to the nonreactive refractory oxide formed by oxidation of the metal or by heating reactive oxide to temperatures above 1000°C for extended periods of time.

5.1.6. Americium

Americium is being created at the rate of hundreds of kilograms per year, but its rate of separation and purification is considerably lower because of lack of demand. It can be seen from Fig. 5 that all isotopes of americium will be created in power reactors, but careful consideration of neutron cross sections, half-lives, and decay modes reveals that the most abundant americium isotopes in spent fuel for reprocessing are ^{241}Am and ^{243}Am. Only a small fraction of the americium created in this way is separated and purified, since very little power reactor fuel is being processed, and of that which is processed much of the americium is diverted into waste streams. Plans[39] have been made to separate kilogram amounts of americium from a waste by combined solvent extraction, oxalate precipitation, and ion exchange. The viability of this process will depend on the demand for americium and on perceived advantages of separation of long-lived ^{243}Am from high-level wastes before disposal.

An even richer source of americium is the decay of ^{241}Pu to ^{241}Am in existing plutonium. Plutonium from spent reactor fuel contains on the order of 1% ^{241}Pu, which decays by beta emission (Fig. 5) to ^{241}Am. During recycle of plutonium, this americium is separated and either purified or discarded to waste. Several laboratories[40–42] are separating and purifying ^{241}Am from this source in multigram to kilogram quantities annually. The aqueous methods used vary somewhat from laboratory to laboratory, depending on their previous experience, the availability of equipment, and the composition of the americium source. The separation and purification schemes use combinations of solvent extraction, anion exchange, cation exchange, peroxide precipitation, and oxalate precipitation. In all aqueous schemes the final step is precipitation of the oxalate, which is calcined to AmO_2. This oxide is the feed material for preparation of compounds used in molten salt studies. In addition to the aqueous methods of separation and purification mentioned here, one pyrochemical process[43,44] that has been used for americium separation will be discussed in a later section.

5.1.7. Curium

The aqueous preparation of curium generally involves separation from americium because the two common uses of curium utilize the combination of the two elements. A popular method[45] of preparation of "medical grade" ^{238}Pu is neutron irradiation of ^{241}Am to form ^{242}Am, which undergoes beta decay to ^{242}Cm, which in turn forms pure ^{238}Pu by alpha decay. Another use[46] of curium is in mixed targets of $^{243}Am/^{244}Cm$, which are irradiated for the production of ^{252}Cf neutron sources, which are widely used in industry. Curium is also found in high-level wastes from reactor fuel reprocessing. Curium, in all of these systems, is closely associated with americium. The chemistry of the two is very similar, and curium follows americium through the processing outlined in Section 5.1.6. Mixtures of the two are separated[47] by oxidation of americium to the pentavalent state with ozone or similar oxidant so that it can be precipitated as $K_5AmO_2(CO_3)_3$, while the curium remains in solution in the trivalent state. Curium is finally precipitated as the oxalate and calcined to the oxide. The oxide is used as the starting point for preparation of curium salts.

5.1.8. Other Actinides

The transcurium elements are processed by using highly specialized techniques involving solvent extraction and ion exchange in a hot-cell facility.[48] The products are isolated in milligram or smaller amounts in a variety of special forms.

5.2. Pyrochemical Preparative Chemistry

For purposes of this discussion, oxides are not considered salts and will be referred to only in an incidental fashion. As mentioned previously, most of the actinide compounds of interest in molten salt chemistry are halides, and as discussed in the preceding section, "reactive" oxides are the prevalent product of aqueous actinide separation and purification procedures. It will be seen in this section that preparation of most of the anhydrous actinide salts used in molten salt work starts with a reactive oxide, or an oxalate that decomposes to oxide during early stages of heating.

5.2.1. Halide Preparation

It is estimated that over 90% of the physical properties studies and commercial handling of actinide salts involve halides. The predominant method of preparation of these halides is treatment of oxides (or oxalates or hydroxides that decompose to oxides during the early stages of heating) with halogens or hydrohalogens at elevated temperatures. This can be seen from the methods of preparation listed in Table V. Other useful methods of preparation of halides shown in Table V include conversion of one halide composition to another (either different valence states of a given halide or from one halide to another). Other preparations start with the actinide in the form of metal or hydride.

With the exception of the compounds of actinium and curium, most of the preparatory reactions listed in Table V have been performed on the multigram scale. An effort was made to select those methods applicable to multigram preparations that are most useful to the experimenter, but the citations deal with the entire range of micrograms to multikilograms.

A simple, but also effective, apparatus used for preparation of fluorides is shown by Fig. 6. Nickel is the preferred furnace tube material; however, Monel (an alloy of nickel, copper, and small concentrations of other elements) fittings, valves, and connecting tubing are satisfactory and are more readily available. Platinum "boats" or trays have given hundreds of hours of service in hydrofluorination reactions, but nickel has been preferred for use with hot fluorine. Excess hydrogen fluoride may be sparged from the furnace exhaust with aqueous potassium hydroxide. Fluorine is often diluted with an inert gas to prevent excessive temperature rise from the heat of reaction. This gas mixture may be pumped in a closed loop to fluorinate some oxides because several percent of oxygen in the gas does not appear to affect significantly reaction rates. This recycle procedure alleviates problems of fluorine supply and disposal. Large quantities of fluorine may be disposed of by charcoal trapping[83,84] (there may be an explosion hazard!), and smaller quantities may be disposed of by trapping in beds of molecular

TABLE V. Pyrochemical Methods of Preparation of Actinide Compounds Used in Molten Salt Work

Compound	Method of preparation	Reference
AcF_3	$Ac(OH)_3 + HF$, 700°C	49
AcF_3	$Ac^3 + HF_{(aq)}$, 25°C	49
$AcCl_3$	$Ac(OH)_3 + CCl_{4(g)}$	49
$AcCl_3$	$Ac_2(C_2O_4)_3 + NH_4Cl$	50
$AcBr_3$	$Ac_2O_3 + AlBr_3$	49
$AcBr_3$	$Ac_2(C_2O_4)_3 + AlBr_3$	49
AcI_3	Ac_2O_3 or $Ac_2(C_2O_4)_3 + AlI_3$ or NH_4I	49
ThF_4	$ThO_2 + HF$, 550°C	51. 52
$ThCl_4$	$ThO_2 + CCl_4 + Cl_2$, 600°C	53, 54
$ThBr_4$	$ThO_2 + C + Br_2$	52, 55
ThI_4	$Th + I_2 + H_2$	52, 55
PaF_4	$Pa_2O_5 + H_2 + HF$, 400°C	56, 57
PaF_5	$PaF_4 + F_2$, 700°C	58
$PaCl_4$	$PaO_2 + CCl_4$, 300°C	58
$PaCl_4$	$PaCl_5 + Al + H_2$, 450°C	59, 60
$PaCl_5$	$Pa_2O_3 + SOCl_2$, 400°C	60, 61
$PaBr_4$	$PaBr_5 + Al$, 450°C	59
$PaBr_5$	$PaCl_5 + BBr_3$, 25°C	62
$PaBr_5$	$Pa_2O_5 + C + Br_2$, 700°C	63
PaI_4	$PaI_5 + Al$, 450°C	59
PaI_5	$Pa + I_2$, 450°C	64
PaI_5	$PaCl_5 + SiI_4$, 600°C	64
UF_3	$UF_4 + H_2$, 1000°C	65
UF_4	$UO_2 + HF$, 550°C	66, 67
UF_5	$UF_6 + UF_4$, 100–200°C	65
UF_6	$UO_2 + F_2$, 500°C	65
UF_6	$UO_2 + BrF_3, ClF_3$, 25°C	65
UCl_3	$UH_3 + HCl$, 250–300°C	65
UCl_4	$UO_2 + CCl_4, COCl_2, SOCl_2$	65
UCl_5	$UCl_4 + Cl_2$, 550°C	65
UCl_6	$UCl_5 + Cl_2$, 350°C	65
UBr_3	$UH_3 + HBr$, 300°C	65
UBr_4	$UH_3 + Br_2$, 300°C	65
UI_3	UI_4 thermal decomp.	65
UI_4	$U + I_2$, 535°C	65
NpF_3	$NpO_2 + H_2 + HF$, 500°C	68
NpF_4	$NpF_3 + O_2 + HF$, 500°C	68
NpF_4	$NpO_2 + O_2 + HF$, 500°C	68
NpF_6	$NpF_4 + F_2(BrF_3, BrF_5)$, 240–400°C	69
$NpCl_3$	$NpCl_4 + H_2$, 450°C	68
$NpCl_4$	$NpO_2 + CCl_4$, 500°C	68
$NpBr_3$	$NpO_2 + Al + Br_2$, 600–800°C	68
$NpBr_4$	$NpO_2 + AlBr_3$, 350°C	68
NpI_3	$NpO_2 + AlI_3$	68

(*cont.*)

TABLE V. (cont.)

Compound	Method of preparation	Reference
AcF_3	$Ac(OH)_3 + HF$, 700°C	49
PuF_3	$Pu_2(C_2O_4)_3 + HF + H_2$, 600°C	70
PuF_4	$PuO_2 + HF + O_2$, 600°C	70
PuF_6	$PuO_2 + F_2$, 100–600°C	71
PuF_6	$PuF_4 + F_2$, 100–600°C	71
$PuCl_3$	$PuH_{2.7} + HCl$, 450°C	70
$PuBr_3$	$PuH_{2.7} + HBr$, 600°C	70
PuI_3	$Pu + HI$, 450°C	72
AmF_3	$AmO_2 + HF$, 650°C	73
AmF_4	$AmO_2 + F_2$, 100–400°C	74
$AmCl_2$	$Am + HgCl_2$, 300°C	75
$AmCl_3$	$AmO_2 + HCl$, 850°C	73
$AmCl_3$	$AmO_2 + CCl_4$, 550°C	76
$AmBr_2$	$Am + HgBr_2$, 300°C	75
$AmBr_3$	$AmO_2 + AlBr_3$, 500°C	73, 77
$AmBr_3$	$AmCl_3 + NH_4Br$, 450°C	78
AmI_2	$Am + HgI_2$, 300°C	79
AmI_3	$AmO_2 + AlI_3$, 500°C	77
AmI_3	$AmO_2 + Al + I_2$	73
AmI_3	$AmCl_3 + NH_4I$, 400–900°C	78, 80
$Am_2(MoO_4)_3$	$MoO_3 + AmO_2$, 900°C	81
$Am_2W_3O_{12}$	$WO_3 + AmO_2$, 900°C	81
$Am(VO_4)$	$AmO_2 + V_2O_5$, 1000°C	81
CmF_3	$CmF_{3(aq)} + HF$, 400°C	78
CmF_4	$CmF_3 + F_2$, 400°C	82
$CmCl_3$	$CmCl_{3(aq)} + NH_4Cl$, 400°C	78
$CmBr_3$	$CmCl_3 + NH_4Br$, 400°C	78
CmI_3	$CmCl_3 + NH_4I$, 400°C	78

sieve or soda line (plugging problems to be expected). All apparatus components and trapping material must be closely monitored for alpha activity before disposal because fine powders or volatile actinide fluorides may be transported in the gas stream.

Halides other than the fluorides may be prepared in quartz or borosilicate glass apparatus with less risk of contamination by metallic impurities from trays or furnace tubes of metal systems. Another advantage of glass apparatus is its transparency, which allows observation of the progress of a reaction. An apparatus used on the 100-g scale for preparation of actinide halides is shown by Fig. 7. This design of apparatus has been used (with halogenating agents other than fluorides)[70] for preparation of halides by reaction of oxides, oxalates, metal, and hydrides. The system should be set up to allow either upflow or downflow through the furnace tube, although downflow is preferred to reduce channeling and to give more rapid and

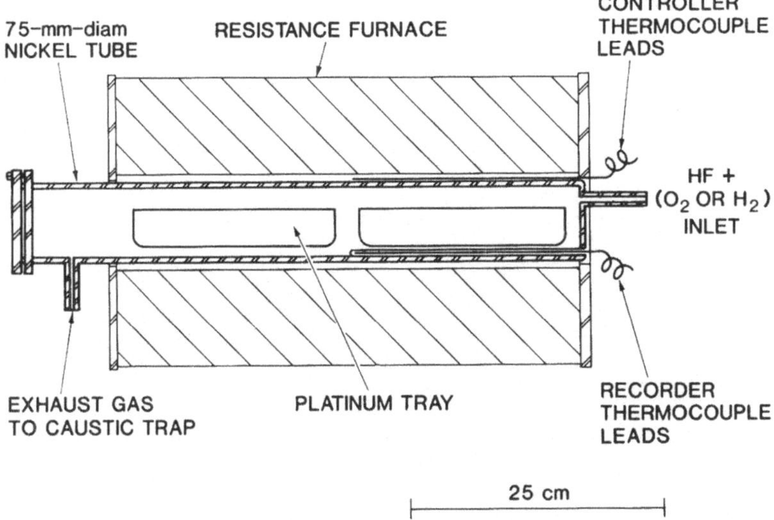

FIGURE 6. Apparatus used for preparation of up to 100-g batches of actinide fluorides.

complete removal of reaction products (such as water) from the reaction zone. A major advantage that this design has over the horizontal tube and tray design is much higher efficiency of use of the halogenating agent, thereby reducing the excess reagent disposal problem. In some instances, the actinide was introduced into the furnace tube as metal, converted to hydride, which was then thermally decomposed and converted to the halide. Hydride procedures are undesirable because of the hazards of handling hydrogen and the irreproducibility of hydride formation, but the halide produced may have lower levels of carbon and oxygen impurities, often present in halides prepared from oxalates or oxides.

5.2.2. Oxyhalide Preparation

Oxyhalides of actinides may have been present as impurities in many molten salt studies or applications, but intentional studies of oxyhalides in molten systems are very limited.[85] These compounds can be prepared by several methods[86-88] and have been characterized by techniques such as x-ray diffraction, although without complete chemical analysis to determine the purity rigorously. Pure oxyhalides probably are best prepared by treating pure powdered halides at controlled elevated temperatures with a reactant such as a hydrohalogen carrying a known concentration of water vapor. Protactinium seems to have a greater tendency toward oxyhalide formation[88] than the other actinides do, and many of its oxyhalides have been identified.

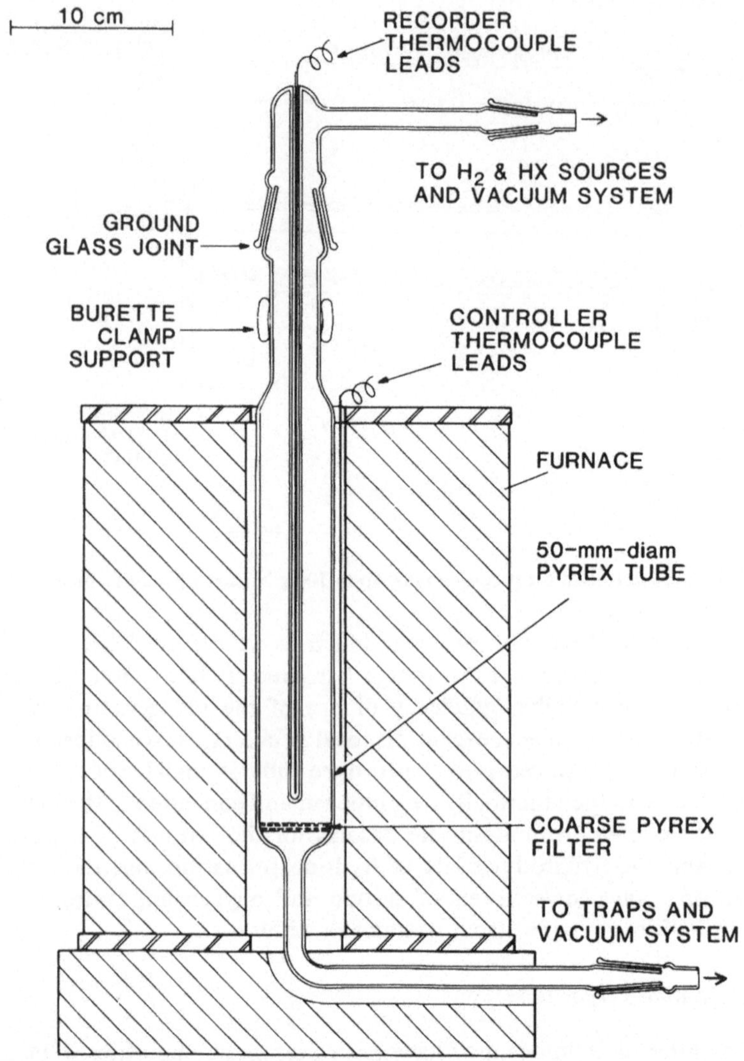

FIGURE 7. Apparatus used for preparation of actinide halides other than fluorides in batches of up to 100 g.

5.2.3. Preparation of Other Oxycompounds

Other actinide oxycompounds such as sulfates and nitrates have been prepared by precipitation from or evaporation of aqueous solution. It is generally difficult to remove water of hydration from these compounds without decomposing them. One alternative to preparation of these pure compounds for molten salt studies is dissolution of small amounts of other compounds (such as halides) in molten salts such as nitrate eutectics. The effect of the foreign anion may be negligible if the required actinide concentration is low. This may be true for studies such as observation of absorption spectra.

Another small group of oxygen-containing actinides has been prepared in the course of phase diagram studies.[(2)] Apparently, little has been done with these compounds, other than preparation and characterization by application of x-ray diffraction techniques and microscopy. This group includes tungstates, molybdates, and silicates of thorium, neptunium, and uranium. Not all members of this cation/anion matrix have been prepared, and they are often considered part of pseudobinary systems with the corresponding alkali metal compounds. For instance, the NpO_2–Li_2O–MoO_3 system is treated as the Li_2MoO_4–$Np(MoO_4)_2$ system.[(89)] These compounds have been prepared by sealing such combinations as Li_2O, MoO_3, and NpO_2 in platinum tubes and holding them at elevated temperatures for extended periods of time.

5.3. *Final Steps of Purification*

Even though care is exercised to use the purest reagents available in the actinide salt preparation, and salts of the highest purity available are used as diluents, small amounts of impurities that may affect experimental measurements often remain with the molten salt. Some of these can be removed by chemical or physical treatment, even after they have been introduced into the apparatus for the observation.

5.3.1. *Distillation Methods*

Purification by distillation is easily applied to many of the halides prepared on the microgram-to-gram scale. An apparatus of the size used for these small batches is easily evacuated to very low pressures and is easily heated to the temperatures required for distillation. These distillations (or sublimations) are often accomplished by heating the salt in the furnace where it was formed and condensing it as a solid or liquid on the cooler walls of the furnace tube just outside the furnace. This method was used in the preparation of the protactinium halides and oxyhalides,[(59)] the more volatile halides being distilled away from the oxyhalides.

Distillation is not limited to small batches, however. The simple apparatus shown by Fig. 8 was used to distill plutonium halide batches as large as 100 g. The distillation tube used was an ordinary quartz tube, closed at one end, with an indentation halfway up the side of the tube to form a dam to contain the liquid product of distillation. The pool at the bottom of the furnace was maintained at about 900°C and the collected pool was at about 800°C. Volatile impurities collected on the walls of the tube outside

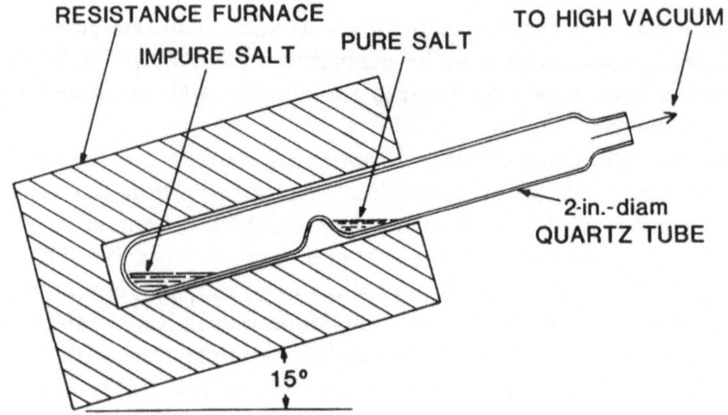

FIGURE 8. Apparatus used for distillation of actinide halides other than fluorides.

the furnace. Even larger quantities of more volatile halides can be purified by distillation methods. Kilogram quantities of uranium hexafluoride are purified by vacuum distillation between traps that are alternately cooled by liquid nitrogen (or dry ice) and warmed to temperatures near ambient. Nickel, Monel, graphite, and precious metals are used in construction of apparatus for handling the fluorides.

5.3.2. Sparging with Halogens or Hydrohalogens

The favored methods of preparation of actinide halides often lead to the appearance of traces of oxide and oxyhalide in the product. Even if these compounds are not present at the time of preparation, they may be introduced by exposure of the salt to air and water vapor. Diluent salts, including those that are not hygroscopic and do not have water of hydration, almost always adsorb some water, which will form oxyhalides when heated with actinide halides. The halides of calcium, lithium, magnesium, and zinc are often used as diluent or reactant in actinide molten salt work, and all of these are very difficult to prepare in truly anhydrous form.

One example of a rigorous purification procedure is the method used for preparation of salts for phase diagram experiments with plutonium trichloride, lithium chloride, and sodium chloride.[90] The apparatus used for removing water, oxychloride, hydroxide, oxide, and any material insoluble in the molten salt is shown by Fig. 9. The salt to be purified was contained in a quartz crucible in a quartz furnace tube. This was closed at the top by a neoprene stopper through which passed a quartz thermocouple well, a quartz filtration assembly, and a small tube for evacuation of the furnace tube. The most hygroscopic salt (lithium chloride) was vacuum-dried during slow heating to 400°C, then sparged with anhydrous hydrogen

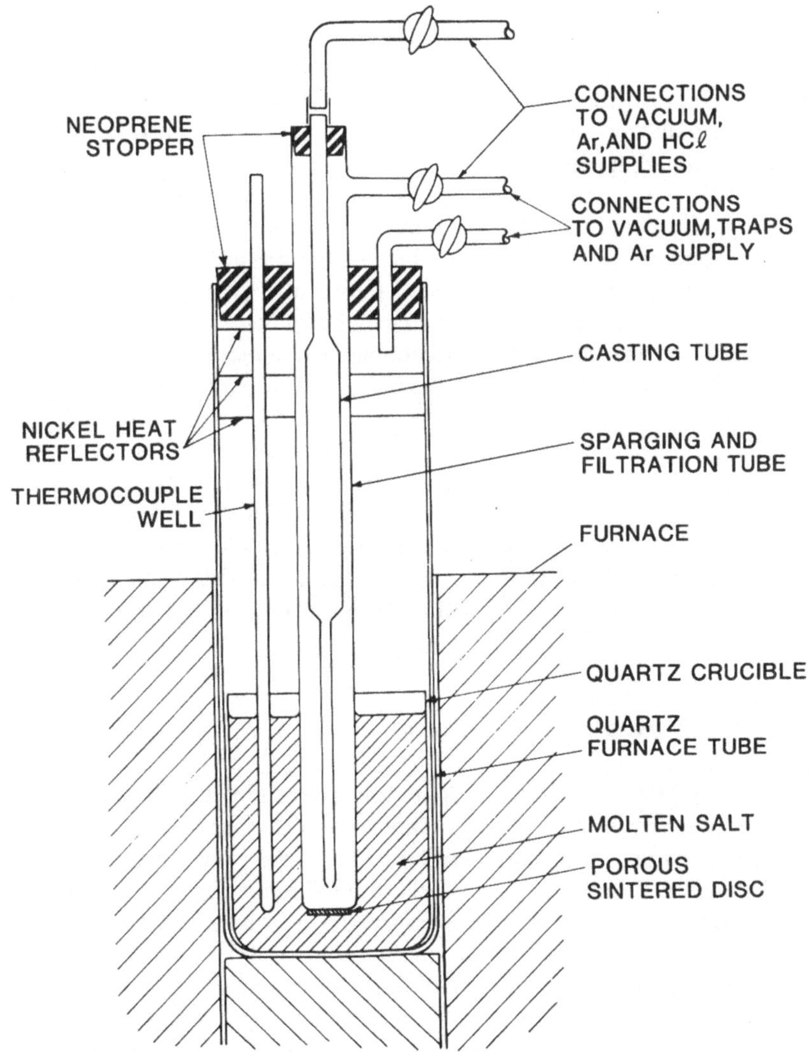

FIGURE 9. Apparatus for purification of molten salts by sparging and filtration.

chloride during melting and for 30 min after melting. The molten salt was sparged briefly with argon to remove HCl and then filtered to remove black particulate material that was always observed when commercially supplied "analytical reagent" lithium chloride was melted. As an indication of the effectiveness of the treatment, purified lithium chloride could be held in quartz crucibles at 650°C for several days with no observable etching, while the same salt that had only been vacuum-dried etched quartz severely within a few hours. After a series of experiments showed that thermal arrest temperatures of these salts were the same when measured during sparging with argon and with hydrogen chloride, the remainder of the observations were made during sparging with hydrogen chloride to prevent buildup of any oxygenated species.

Others have used equally rigorous drying and hydrohalogen sparging techniques in the preparation of lithium chloride/potassium chloride eutectic for electrochemical studies.(91)* Recent attempts at Los Alamos to remove oxygen-containing species from kilogram quantities of calcium chloride/10% calcium oxide mixtures by hydrochlorination at 850°C show great promise. Hydrogen chloride appears to be more effective than phosgene for this purpose. In another study,(92) it was reported that chlorine was more effective than hydrogen chloride for preparing pure lithium/potassium chloride eutectic for electrochemical studies when the salt mixture was treated at 450°C, but that the treatment with chlorine was not effective at 740°C. When halogens are used for this purpose, care must be exercised to avoid unwanted higher oxidation states of actinides because these have more volatile halides. Care must also be exercised to avoid container materials that will react and contribute cationic impurities to the molten salts.

6. Metal Preparation and Purification

The actinide metals are prepared by active metal reduction of oxides or salts and by electrorefining. Three general methods of active metal reduction are used: (1) reduction by metal vapor in a vacuum system, (2) reduction in a closed pressure vessel ("bomb" reduction), and (3) reduction in a closed vessel at ambient pressure. As normally performed for high yields, the reductions give almost no separation of the actinide from metallic impurities, although the third method listed is capable of being operated in such a fashion as to give a low yield and some separation from impurities. Metals may, in some instances, be further purified by a process called molten salt extraction wherein the molten metal is equilibrated with a molten salt and the metal is purified by transfer of impurities to the salt phase.

6.1. Reduction by Metal Vapor

With the exception of uranium and thorium, the first reductions(93) of all the actinides to form the metals were by reduction of oxides or salts by active metal vapors using microtechniques. These techniques are still used for preparation of actinium and the transcurium elements.(94) The selection

* See also: S. H. White in Vol. 1 of this series.

of the reductant to be used for these reactions is based on the free energies of formation of the compounds involved, the vapor pressures of the reactants and products, and the melting points of the reactants and products. The free energies of formation of the candidate oxides at 725°C are listed in Table III. Free energies of formation of fluorides and chlorides of these elements at 725°C are listed in Table VI. These tabulations can be used as a guide in selection of the actinide compound and its reductant. For instance, actinium trichloride was reduced by potassium vapor,[(3)] while actinium oxide would not have been reduced by this metal. As was pointed out in Section 4 of this chapter, lithium vapor can be used for reducing fluorides of nonvolatile curium with the resulting lithium fluoride volatilized away from the products. More volatile californium, however, may be lost when reduced by the same technique.[(94)] Particular care must be exercised when working with microgram quantities of materials at high temperatures lest the entire sample be lost by alloying with the support or by vaporization during the experiment.

6.2. *Pressure Vessel or "Bomb" Reduction*

In application of the pressure vessel or "bomb" reduction technique in preparation of actinide metals, a compound (usually oxide or halide) of the actinide is mixed with a metal (usually an alkali metal or alkaline-earth

TABLE VI. Approximate Free Energies of Formation of Selected Fluorides and Chlorides at 725°C (Stated in kcal/g atom of Halogen)[a]

Halide	$-\Delta F$	Halide	$-\Delta F$	Halide	$-\Delta F$
CaF_2	125	ZrF_4	93	$CeCl_3$	66
BaF_2	123	NpF_4	90	$AcCl_3$	63
LiF	122	AlF_3	89	$PaCl_3$	63
LaF_3	121	PuF_4	88	$PuCl_3$	59
CeF_3	120	SiF_4	84	$ThCl_3$	59
NaF	112	TaF_5	75	$MgCl_2$	58
AcF_3	112	UF_6	69	$ZrCl_2$	56
PaF_3	111	ZnF_2	68	$NpCl_3$	55
MgF_2	111	HF	65	UCl_3	53
KF	110	PuF_6	65	$ThCl_4$	53
AmF_3	110	$BaCl_2$	84	UCl_4	46
PuF_3	107	KCl	81	$AlCl_3$	46
NpF_3	102	LiCl	79	$ZnCl_2$	35
ThF_4	101	$CaCl_2$	78	$TaCl_2$	28
ThF_3	100	NaCl	76	$SiCl_4$	28
UF_3	95	$LaCl_3$	67	HCl	24
UF_4	94	$AmCl_3$	67		

[a] Reference 17.

metal) that will react to form a compound that has a more negative free energy of formation than the actinide compound has. This technique is a very important method of preparation of actinides from their compounds, on both commercial and laboratory scales. The choice of actinide compound and reductant is based in part on the free energies of formation listed in Tables III and VI. Also important are the melting points, phase relationships, viscosities, and vapor pressures of the feed salts put into the charge and the product salts that form the slag. The latter must be separated from the metal product. The bomb consists of a metal container (usually steel) that provides strength for pressure containment and of a refractory liner (such as magnesium oxide, calcium fluoride, or fused dolomite) which resists chemical attack. In addition, the liner furnishes both thermal and chemical barriers to attack on the metal container. Excess reductant (about 25% above the theoretical amount required) is added to the mixture of salt and reductant to increase the yield of metal product. This excess is needed to ensure good contact with the compound being reduced (no stirring is provided other than that produced by the reaction itself) and to react with traces of impurities such as water and oxygen. In many instances a "booster" is added to the reactants. The booster is an element or compound that reacts just before or during the early stages of the actinide reduction. The mechanism of the booster in improving reduction yields is a subject of some controversy. One theory is that energy released from the booster reaction simply heats the products of reaction to a higher temperature, at which the slag and metal are less viscous and flow more rapidly to allow better consolidation of the metal product. Another theory is that products of the booster reaction produce a slag with a lower melting point, which provides more time for the metal to coalesce before the slag freezes. A third theory is that the booster triggers the reduction reaction at a lower overall system temperature than would be the case without it, so that the maximum system temperature during reduction is lower, and less spattering and less reaction with the bomb liner are experienced, resulting in higher yields. Any one, two, or none of these effects may come into play during a given reduction reaction. Many questions about the behavior of molten salt/metal systems during bomb reduction remain unanswered. It appears that performance of such reductions is an art rather than a science.

The actinide metals commonly produced by bomb reduction procedures are thorium, uranium, neptunium, plutonium, and americium. Other actinide metals are usually produced by application of microtechniques because of the limited quantities of the elements that are available or because of radiation problems encountered in handling larger quantities.

Bomb reduction on the 100-kg scale is used commercially for thorium and uranium metal production, although both metals have been produced commercially by reduction at ambient pressure in open containers. One method of producing thorium on the 100-kg scale is reduction of thorium tetrafluoride by 25% excess calcium. Zinc chloride equal to 10% of the weight of the fluoride is added as a booster. The pressure vessel is steel with a fused dolomite liner. The charged and sealed bomb is put into a furnace preheated to 660°C. The slag is reported[95] to be $CaCl_2/CaF_2/ZnF_2$ and the metal phase an alloy of thorium containing 4%–7% zinc. After mechanical separation of the metal and salt phases, the zinc is distillled away from the thorium. This reduction reaction must be conducted by the bomb technique to prevent escape of zinc and zinc chloride, whose boiling points are exceeded at the time of the reaction.

A similar bomb reduction method is used commercially for production of uranium on the 100-kg scale. Uranium tetrafluoride is blended with 5% excess magnesium and loaded into a pressure vessel that has a fused dolomite liner. The sealed bomb is put into a gas-fired furnace preheated to 600°C. The heat of reaction is sufficient to completely melt the magnesium chloride and uranium products so that they coalesce, making mechanical separation of the cooled products feasible. The pressure vessel is required in this case because of the relatively high vapor pressures of magnesium and magnesium chloride.

For the experimenter in the field of molten salt studies it is probably more practical to obtain thorium and natural uranium metal and salts from a laboratory that produces these items commercially than to set up equipment to do such operations as bomb reduction. On the other hand, there remains much research to be done before the chemical mechanisms of boosted bomb reduction are understood. Because many fewer health, safety, and security problems are encountered in work with thorium and natural uranium than are encountered in work with other actinides, these two elements are good candidates for work on such problems.

A much smaller scale of bomb reduction is used in preparation of plutonium, neptunium, americium, and highly enriched uranium. Because of the high levels of radioactivity and the criticality constraints, bomb reductions of these elements are restricted to the 1- or 2-kg scale, with the exception of americium, whose reduction has been limited to 25-g batches because of the gamma and neutron radiation levels produced in americium tetrafluoride. The laboratory scale (as contrasted to the commercial scale) bomb reduction technique developed rapidly during the period 1943–1946 and has evolved less rapidly since then. Figure 10 contains sketches of the pressure vessels used up to 1945[96] compared with one of a pressure vessel currently used for preparing 2-kg batches of plutonium. The basic procedure for bomb reduction of the tetrafluorides has not changed greatly from the earlier preparations to the present time. The pressure vessels are steel and the crucibles used as liners are low-density magnesium oxide. The tetrafluoride is mixed with iodine (iodine-to-metal mole ratio is 0.15–0.30) and

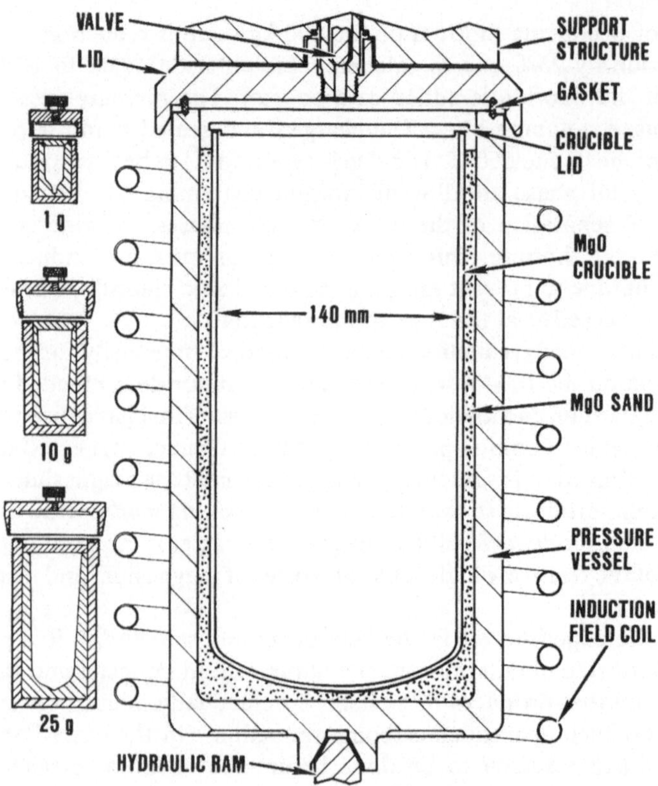

FIGURE 10. Containers used for production of actinide metals by "bomb" reduction of halides. All four containers are shown on the same scale. The three small containers were used before 1945.

granular calcium in an amount that is 25% in excess of that required to react with the fluoride and iodine. This mixture is loaded into the pressure vessel, a magnesia lid is placed on the crucible, and the pressure vessel lid is firmly sealed to the body, using a copper gasket. The bomb is next evacuated and backfilled with argon to a pressure significantly above ambient for a leak check. The pressure is then reduced to ambient and the connecting orifice is closed. The integrity of the seal is very important, since high pressures (greater than 100 psi) are generated at the instant of reaction. If the argon escapes through a leak at the gasket at that instant, it is followed by a mixture of volatile calcium, iodine, and magnesium (from the magnesia crucible) that instantly reacts to enlarge the leak and release large quantities of actinide to the surroundings with explosive force. It can be seen from Fig. 10 that the small pressure vessels had lids that were held on by threads. A threaded plug closed a hole in the lid. In this design, after the bomb was loaded and sealed, the plug was removed and replaced by a threaded tube for the evacuation and argon-filling procedure. The tube was then removed and replaced by the plug before the bomb was heated. The lid of the modern large-scale bomb is held on the assembly by a hydraulic ram. A valve for gas transfer control is mounted permanently in the lid. The bomb is heated by induction-heating methods by direct coupling of the induction field with the wall of the pressure vessel. Reaction of the mixture is detected by monitoring the drop in neutron emission from the fluorine alpha-neutron reaction that diminishes when intimate mixing of the actinide and fluorine is disrupted. Heating is discontinued when the mixture reacts if the charge is greater than 100 g, but for smaller-scale reductions heating must be continued to temperatures about 1000°C to attain good separation of metal and salt. After reaction, the bomb is cooled and the dense metal regulus is mechanically separated from the slag.

6.3. Ambient Pressure Reduction of Actinide Salts

The distinction between bomb reduction and ambient pressure reduction of salts to form metals is often questionable. The major difference in the equipment is that the bomb reduction pressure vessel is always of sturdy construction and is sealed during the reaction, while ambient pressure reductions may be performed in fragile closed containers or even in open containers.

Ambient pressure reductions utilize reactions that have either lower total energy release or a lower rate of energy release governed by the rate of combination of reactants. No booster is necessary, although a small "ignitor" may be used. Neither volatile salts nor volatile metals are heated to high temperatures (above their boiling points) during these reactions.

6.3.1. Reductions by Active Metals

The procedures of commercial production of thorium and uranium at ambient pressure can hardly be distinguished from the bomb reductions discussed in Section 6.2. Natural uranium is produced on the 100-kg scale by igniting a mixture of uranium tetrafluoride and excess calcium by adding a burning pellet of potassium nitrate/lactose (no booster is used).[97] Large-scale reductions of thorium tetrachloride by magnesium produce a thorium/magnesium alloy from which the magnesium is separated by distillation. Since these salt/metal systems oxidize badly at high temperatures in air, provisions are made to blanket the reactants and products with an inert gas until they are cooled.

Research- and development-scale (10 g of actinide) reductions may be conducted in a very simple manner. An extensive series of experiments that resulted in the development of the "pyroredox" process[98] for purifying plutonium was conducted using the simple apparatus shown by Fig. 11. The apparatus consisted of a quartz furnace tube closed at the top by a neoprene stopper through which passed a quartz thermocouple well, a tube for evacuating the tube and backfilling with argon, and a quartz tube for reductant introduction. The granular reductant was contained in a rotatable arm of the latter tube so that it could be introduced gradually. The crucibles used for containing salts to be reduced were quartz, magnesia/10% titania, and tantalum. The salt to be reduced was placed in a crucible that had been previously baked out *in vacuo.* The loaded crucible was quickly put into the furnace tube, which was evacuated briefly and then filled with argon. The rotatable sidearm was removed, loaded with reductant, and returned to the postion shown in Fig. 11. The furnace tube was again evacuated and put into the furnace during continued evacuation. Evacuation was continued during heating until the salt melted. The apparatus was then filled with argon while heating continued to the desired reaction temperature (up to 850°C), at which time the side arm was rotated to gradually introduce the reductant. The salt studied most extensively was a plutonium

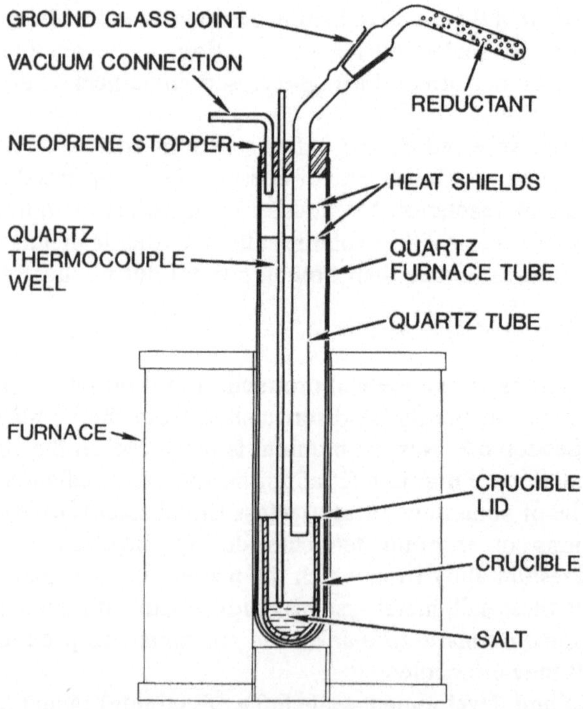

FIGURE 11. Apparatus used for research and development studies of reduction of plutonium chloride.

chloride/sodium chloride mixture. Severe spattering was observed if calcium addition was too rapid, but the total charge of other reductants such as lanthanum and cerium did not produce spattering even if combined with the salt before heating. Thus, the use of the reductant addition tube is unnecessary for some reductants. This apparatus was used to study successfully the conversion of plutonium metal to plutonium chloride/alkali chloride combinations and reduction of the salt by several metals to form plutonium metal and alloys.

Some of the plutonium reactions studied in the simple quartz apparatus described above have been scaled up to the level of hundreds of grams.[99] The processing is carried out in equally simple, although more sturdy, containers. The design of a general purpose apparatus is shown in Fig. 12. The same general design has been used in a variety of sizes, the largest having a crucible diameter of about 77 mm and a plutonium capacity for metal/salt reactions to about 1 kg, the practical capacity for each reaction depending on the energy release during reaction. No reactant introduction tube has been provided, although addition of such a device would be rather easy. The stirrer is fabricated from tantalum or tantalum/10% tungsten.

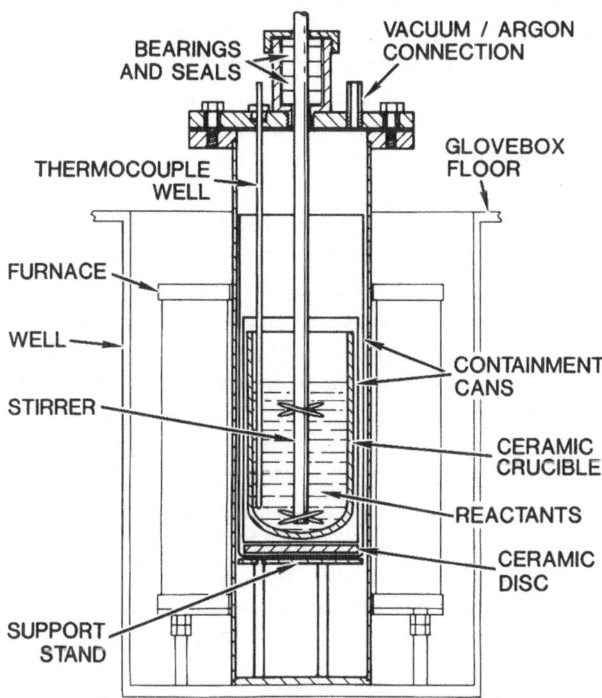

FIGURE 12. Apparatus for ambient pressure oxidation and reduction reactions of actinides in molten salt systems.

The thermocouple well is a closed-end nickel tube inside a tantalum well. This combination resists oxidation on the inside and attack by the salt on the outside. Triple containment (in addition to the primary crucible) is provided to prevent release of molten plutonium to the air environment of the glovebox enclosure. The secondary containment is a tantalum can, which is nested inside a stainless steel container. The furnace tube is an oxidation-resistant alloy such as Inconel or Hastelloy. A variety of supports, baffles, and insulators are used to reduce temperature gradients near the end of the furnace. The bottom end of the furnace tube extends beyond the end of the resistance-heated furnace to a cooler region so that molten plutonium reaching this point would immediately freeze and not alloy with the metal of the tube. Ceramic crucibles often leak small amounts of molten salts and occasionally leak small amounts of molten metal, but the tantalum secondary containment has never been breached.

In operation of the apparatus, the reactants and diluent salt are put into the crucible at room temperature. The crucible is then put into the nested cans of the furnace tube, and the lid of the furnace tube (carrying the thermocouple well and the stirrer in a raised position) is bolted into place. The tube is evacuated, checked for leaks, then filled and pressurized with argon. The furnace is heated to the melting point of the salt. The thermocouple well and stirrer are pushed down into the molten salt and the desired program of heating and stirring is completed. The stirrer and the thermocouple well are then retracted to some level above the surface of the molten salt before the salt is cooled to the solidification temperature. The cooled products are separated mechanically.

The apparatus of Fig. 12 has been used for several different metal/salt reactions. Plutonium metal is reacted with zinc chloride containing potassium chloride as a diluent to lower the melting point of the plutonium trichloride which is formed. The reaction is exothermic, but only limited spattering is observed. This reaction is performed in a tantalum crucible from which the products can be removed mechanically. It has also been shown that this reaction proceeds satisfactorily in quartz containers.[99] The plutonium chloride/potassium chloride is then reduced by calcium metal in a vitrified magnesia crucible in the same apparatus to form plutonium metal. Another reaction conducted in a vitrified magnesia crucible in this apparatus is reduction of plutonium dioxide by granular calcium metal in molten calcium chloride. The amount of calcium chloride is chosen so that the composition of the calcium chloride/calcium oxide product has a relatively low melting point compared to the usual maximum temperature of 875°C.

Another reduction procedure that does not have as its goal the preparation of a specific metal, but rather the removal of all actinides from a given batch of salt, is "salt stripping." The impetus for use of such a procedure is peculiar to the actinides because of the discard limits and great expense of disposal of radioactive wastes. Examination of the free energies of formation of the chlorides listed in Table VI leads one to predict that actinides could be stripped from salts such as calcium chloride or sodium chloride/potassium chloride by calcium metal. One would also predict that sodium could also be reduced, but the uncertainties of the values listed in Table VI do not justify a definite prediction of the extent of sodium reduction that will be observed. Sodium reduction can be restricted by limiting the amount of calcium added. The salt-stripping process and the apparatus shown by Fig. 12 have been used successfully to strip plutonium and americium from chloride processing salts.

6.3.2. Electrolysis

Thorium,[100] uranium,[101] and plutonium[102] metals have been prepared by electrolysis of molten salts on various scales of operation up to multikilogram quantities. The most common procedure is electrolysis of the actinide chloride or fluoride in a mixed-chloride bath contained in a graphite crucible anode. The metal is deposited on a molybdenum cathode. Molten salt bath compositions included sodium chloride/potassium chloride and sodium chloride/calcium chloride. Thorium and uranium have been added as $KThF_5$ and KUF_5 prepared by aqueous means. All three actinides have been introduced as pure anhydrous chlorides and fluorides. The cell design must provide for removal of chlorine gas produced at the anode so that it does not contact the metal deposit on the cathode. The metals are deposited as dendrites, which must be leached free of salt by aqueous means to recover the possibly pyrophoric metal powders. Significant purification from active metals, such as the rare earths, may be achieved by this method of metal production, but it has not become a popular production process.

6.4. Pyrochemical Purification of Actinide Metals

The pyrochemical purifications discussed here are applicable to removal of impurities when the feed material is a metal and when the desired product form is metal of higher purity. These methods are capable of virtually complete removal of certain groups of impurities from actinide metals.

6.4.1. Electrorefining

Molten salt electrorefining is a process whereby an impure metal is oxidized at an anode, transported through a molten salt electrolyte, and deposited as a pure metal ($>99.95\%$) at a cathode. The cross section of a cell that has been used to produce high-purity plutonium on a multikilogram scale[103] is shown by Fig. 13. The container is a cylindrical vitrified magnesia crucible with a smaller anode "cup" in the center. It is provided with a vitrified magnesia stirrer that stirs both the impure molten plutonium anode and the sodium chloride/potassium chloride/plutonium chloride electrolyte. Electrical contact to the anode pool is provided by a magnesia-insulated tungsten rod that dips into the pool. Pure liquid plutonium deposits on the cylindical tungsten cathode and drips into the ring-shaped space beneath it. The cell is operated at 740°C in a furnace tube with nested containers very similar to the ones shown in Fig. 12. Loading, closing, evacuation, argon filling, and stirrer, electrode, and thermocouple well manipulations are performed very much as were described for that furnace. Polarization effects are minimized by the stirring of the anode and electrolyte and by providing a large cathode surface area. The cell current is automatically interrupted momentarily at preset time intervals to measure the back electromotive force (EMF), which is an indication of polarization and/or excessive impurity concentration in the anode, and the current is discontinued if this EMF rises above a preselected value. This procedure is necessary if the high purity of the product is to be maintained.

The higher melting points of the other actinide metals probably make application of electrorefining, as described above, impractical for all actinides other than plutonium and neptunium. Good purification of high-

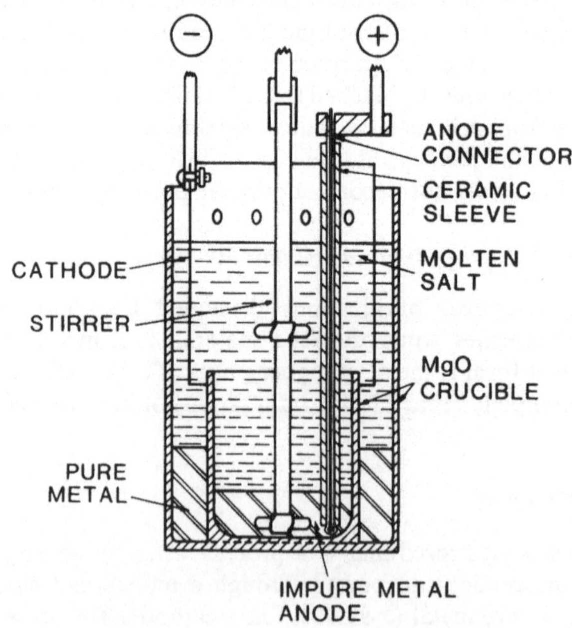

FIGURE 13. Molten salt electrorefining cell.

melting metals can be attained in a molten salt electrolyte system using an anode, which is an alloy of the high-melting metal dissolved in a low-melting metal such as cadmium or bismuth. The pure metal deposits on the cathode as a solid that may be recovered by dissolving away the salt electrolyte or by melting the metal so that the salt floats to the top.

6.4.2. Molten Salt Extraction

The process referred to as molten salt extraction here has been called halide slagging, chloride slagging, and molten salt/metal equilibration.[104–106] It consists simply of agitating molten salt and molten metal together in a container until an equilibrium distribution of the elements of interest is established. The system is then cooled and the metal and salt phases are separated mechanically. The apparatus used is the same as that shown in Fig. 12. The crucible and stirrer materials are selected for the combination of physical and chemical properties that best satisfy the purposes of the procedure. For instance, ceramic crucibles have greater probability of participating in chemical reactions than do tantalum crucibles, but the reaction products generally can be more easily removed mechanically from ceramic crucibles.

The processing application of molten salt extraction most often encountered is removal of small amounts of active metal impurities from selected actinide metals. The capabilities of the technique may be predicted by use of the free energies of formation of halides listed in Table VI. Elements having halides with more negative free energies of formation than those of the major component of the metal phase should be extracted into a salt phase containing an excess of the major metal halide. For example, it would be predicted as a first approximation that rare earths, actinium, protactinium, americium, plutonium, neptunium, and thorium would be extracted from uranium metal equilibrated with a molten fluoride eutectic containing a few mole percent of uranium fluoride. This process might be impractical, however, because the high melting point of uranium would require a high operating temperature, which might not be tolerated by any available compatible container material. One way to circumvent the high temperature problem is to dissolve the metal in a low-melting metal such as cadmium during processing and to distill away the volatile cadmium after extraction is completed. Predictions of extraction behavior become even more uncertain when the alloying technique is used because one must deal with not only the uncertainties in the free energy of formation values but also activity coefficients in the molten salt and in the alloy.

An important application of molten salt extraction is the removal of americium from plutonium.[43,44] Kilogram amounts of plutonium containing (typically) 3000 ppm of americium are contacted with low-melting potassium chloride/sodium chloride/plutonium trichloride in a magnesia crucible in the apparatus shown by Fig. 12. The plutonium halide is either added as plutonium tetrafluoride or trichloride, or generated from the metal by addition of magnesium chloride. Approximately 90% of the americium is transferred from the metal to the salt phase.

During development of the americium extraction process, an interesting sampling technique was developed[104] for sampling molten salt and metal phases to determine the americium distribution. Multiple metal samples were taken by lowering tantalum dippers individually into the metal phase to retrieve samples at different times and temperatures without cooling the furnace. Similarly, multiple salt samples were taken by lowering nickel capsules having a sintered nickel filter in the bottom end (the top end was sealed) into the salt and pressurizing the space above the salt to force molten salt through the filter into the capsule. Salt and metal samples were withdrawn into the cooler upper part of the furnace to solidify them immediately after sampling. All samples were removed from the cool apparatus for analysis after termination of the experiment.

Measurements of distributions of actinides between molten magnesium chloride and zinc/magnesium alloys were made in a somewhat similar fashion with the aim of developing molten salt extraction methods of separation of actinides from each other and from fission products.[105-108]

7. Physical Properties Measurements

New types of instruments useful in making physical property measurements of molten salt systems are becoming available at an increasing rate. New methods of data acquisition and analysis are being developed and implemented. No attempt has been made by this author to survey catalogs of scientific supply companies to obtain descriptions of new physical-properties-measuring equipment that might succeed (but have not been proved) in measurements on actinide salts. Some of the techniques discussed here may have been applied as much as four decades ago. This approach may make the discussion seem outdated, but one defense of the approach is that the techniques actually succeeded in generating reliable data in almost all cases.

7.1. Melting Points and Phase Relationships

It was pointed out in Section 1 and Table I of this chapter that one compilation[2] included phase diagrams of approximately 76 actinide oxide systems and 77 actinide salt systems. No attempt will be made to reference all of those systems here. Brief descriptions will be given of techniques and container materials used in typical studies.

One of the most comprehensive programs for determination of phase relationships in molten salts ever undertaken was the study of binary and ternary fluoride and chloride systems at Oak Ridge, in support of the Molten Salt Reactor experiment.[6] The halides studied were primarily halides of uranium[109,110] and/or thorium[111,112] mixed with halides of alkali and alkaline-earth metals. For thermal analysis, the fluoride salts were typically contained in nickel or graphite crucibles with as many as four crucibles in a single furnace tube.[109] The molten salts were stirred by means of nickel stirrers whose vertical shafts passed through close-fitting graphite sleeves in the lid of the assembly. A slight positive pressure of helium was maintained inside the furnace tube to prevent reaction of the contents with air. Salt temperatures were measured by means of chromel/alumel thermocouples in nickel thermocouple wells and recorded by strip chart recorders. Typical samples weighed 50 g and cooling rates were 3–4°C/min. Relatively little use was made of differential thermal analysis. Phases present in quenched samples were identified by examination with a polarizing microscope and x-ray diffraction. As many as 20–30 quenched samples were prepared at one time by loading the crushed powders into a thin-walled nickel tube 0.10 in. diameter by 5–6 in. long. The tube was crimped at about 0.2-in. intervals while being loaded, and the ends of the tube were finally sealed to protect the samples. Tubes were then heated in a furnace with a thermal gradient and dropped into an oil-quenching bath. The tube was retrieved and samples recovered after the individual compartments were cut open.

Rather different techniques were developed at Los Alamos for study of phase relationships of plutonium trichloride in binary combinations with alkali and alkaline-earth chlorides.[90,113–115] Much greater effort was expended to remove traces of water and oxygen species from the salts and to prevent contamination by these species during observations. The purification method used was described in Section 5.3.2 and the apparatus is shown in Fig. 9. The apparatus used for thermal analysis and differential thermal analysis is shown in Fig. 14. The apparatus consisted of a 50-mm-diam, 38-cm-deep quartz furnace tube, closed at the top by a neoprene stopper which supported two 16-mm-diam, 45-cm-deep quartz tubes. One of the quartz tubes contained the salt being observed, while the second contained either a similar amount of lithium chloride/potassium chloride or an empty ceramic crucible with a similar heat capacity and insulating effect. These smaller tubes were closed at the top by neoprene stoppers that were penetrated by quartz tubes that acted as thermocouple wells, gas sparging tubes, evacuation ports, and salt addition tubes. The salts were sparged with anhydrous hydrogen chloride during measurements (it was demonstrated by test measurements that thermal arrest temperatures were the same

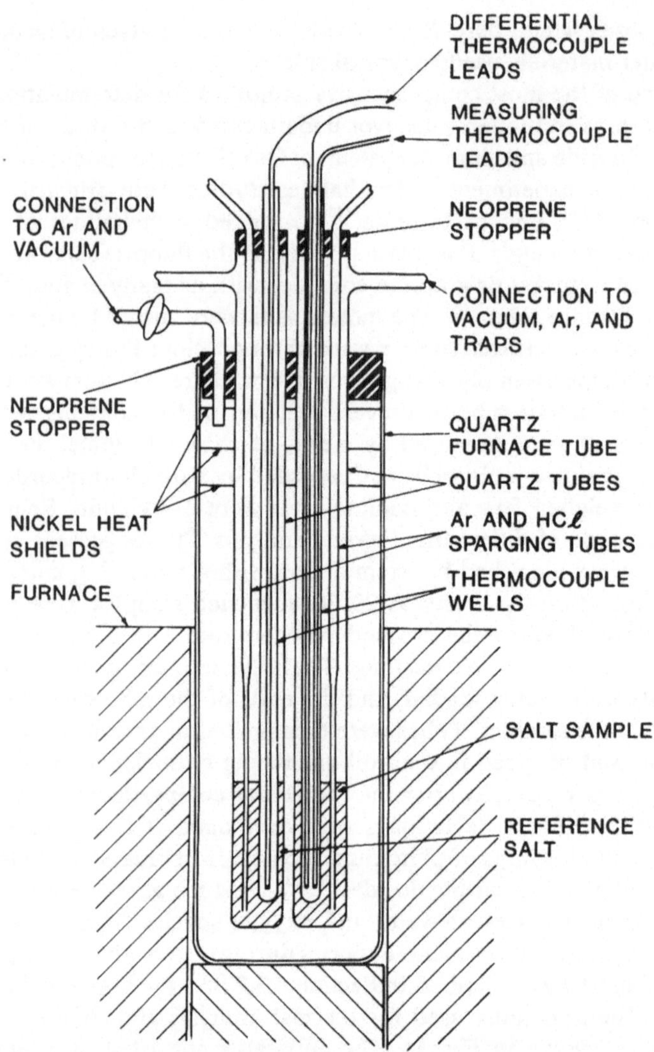

FIGURE 14. **Apparatus for differential thermal analysis of molten salts.**

for hydrogen-chloride-sparged and argon-sparged salts) to give thorough mixing during freezing and to prevent formation of traces of oxychloride. No etching of the quartz was observed even after several days of operation in the temperature range of 450–650°C with hydrogen chloride sparging. Inadequately dried salts containing a high concentration of lithium chloride etched quartz severely within a few hours at 650°C with argon sparging. Temperatures measured by chromel/alumel thermocouples and temperature differences between the two salt containers as measured by opposed chromel/alumel thermocouples were recorded by synchronized strip chart recorders. Differential thermal analysis indications were extremely helpful. Sample sizes were about 20 g and linear heating rates were varied from 1 to 8°C/min. Species present in various mixtures were identified by microscopy and x-ray diffraction techniques. In one of the systems, part of the liquidus curve was too steep to be reliably determined by differential thermal analysis. Chemical analyses of samples taken through a sintered tantalum filter were used to determine liquidus compositions at selected temperatures in this system.[(85)]

Another class of techniques has been used for determination of phase relationships in actinide oxyanion systems. These compounds are often prepared by mixing the actinide oxide with oxides of tungsten or molybdenum and an alkali metal oxide, sealing the mixtures in platinum tubes, and annealing them at elevated temperatures. The products of annealing are studied by application of the techniques of microscopy, x-ray diffraction, absorption spectra, and thermal analysis to determine phase diagrams of molybdates and tungstates. Only limited descriptions of the techniques have been published.[(89,116–118)]

A simple differential thermal analysis apparatus[(119)] using optical sensors has been used for determining phase relationships of the more refractory compounds of actinides in the temperature range of 800–3200°C. A schematic of the apparatus is shown in Fig. 15. An induction heating unit is used to supply power to heat the furnace susceptor, which acts as an oven for heating the crucible containing the sample. The power supply can be programmed to increase or decrease its output at selected rates. A simple optical system delivers light from the incandescent furnace to an optical pyrometer and to two photodiodes. Light emitted from the blackbody hole in the sample container is focused on one diode, while light from a nearby part of the oven is focused on the other. Electrical signals proportional to temperature and to the difference in temperatures are amplified and recorded on a dual trace recorder. The manual optical pyrometer is used for calibration of the recorder temperature trace during each heating and cooling cycle. Five materials whose melting points are well known (copper at 1083°C, platinum at 1770°C, rhodium at 1960°C, alumina at 2050°C, and iridium at 2440°C) were used to check calibration of the system.

7.2. Other Physical Properties of Actinide Salts

7.2.1. Vapor Pressure

Numerous techniques have been applied in determination of vapor pressures of solid and liquid actinides and their salts. The general methods

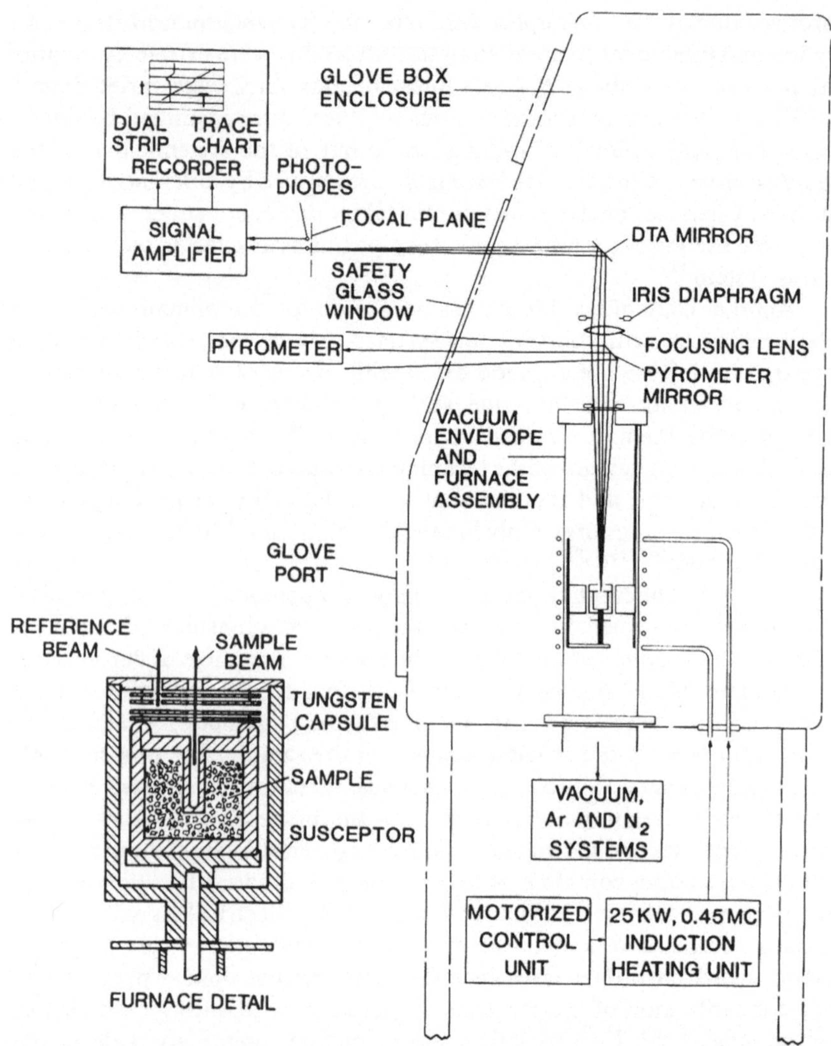

FIGURE 15. Apparatus for differential thermal analysis of actinide compounds at temperatures up to 3200°C.

used are direct-reading gauge, boiling at reduced pressure, transpiration, and Knudsen effusion. In many of the measurements on actinides, the apparatus was set up as in observations of nonactinides except for being partially or totally enclosed in a hood or glovebox. Support electronics such as amplifiers and power supplies are often separated from the sensing units for glovebox work, even though the supplier of a commercially available unit may package the entire apparatus in a single cabinet. Special feedthroughs may be required to connect the sensor, located inside the glovebox, to the amplifier outside the glovebox, but this extra effort may pay rich dividends in reduced maintenance problems of the electronics units.

The simplest apparatus used for vapor pressure measurements is the direct-reading gauge. The gauge may be of the Bourdon type, a diaphragm gauge whose output is read electronically by change in strain or capacitance, a null-diaphragm gauge with automatic or manual balancing gauge, or a null-type sickle gauge with manual balancing. All of these types of direct-reading gauges have been used in measurement of the relatively high (at ambient temperatures) vapor pressures of the hexafluorides of uranium, plutonium, and neptunium.[120–122] The major constraint is that the gauge or null diaphragm must be held at a higher temperature than the sample to avoid condensation in the sensing element. An additional complicating factor in the study of plutonium hexafluoride is its thermal and radiolytic instability. Errors are introduced in the measurement because of buildup of fluorine pressure, and the sensor is contaminated because of deposition of plutonium on all interior surfaces.

Another method which is useful for determining vapor pressures at high pressures (above about 25 torr) and at temperatures harmful to the direct-reading gauges described in the preceding paragraph is observation of boiling points under a reduced pressure of inert gas.[123] The salt is heated in a reservoir of a system containing an inert gas at a pressure higher than the vapor pressure at the temperature of interest. The reservoir is held at the given temperature as the inert gas is very slowly pumped out of the system. Temperature variations of the liquid are observed with a very sensitive null instrument or differential thermocouple system. When the vapor pressure is reached, boiling will be initiated and a small drop in temperature will be observed. The valve to the vacuum system is quickly closed (manually or automatically by actuation of a solenoid valve) and the pressure and temperature are noted. This procedure is repeated for all temperatures of interest. To avoid "bumping" of the liquid, a very slow flow of inert gas can be introduced by means of a capillary or porous tube dipping into the liquid. A variation of this method, called the "quasistatic" method,[124] utilizes a heated salt reservoir that has two restricted tubing connections at its top, a rather limited free volume inside the reservoir, and a relatively large liquid surface. A significant length of the connecting tubes must be inside the furnace to reduce condensation of the sample. One of these tubes leads to a manometer or sensitive pressure gauge and the other leads to a vacuum system through a needle valve. At some point outside the furnace the two tubes are connected by a sensitive differential manometer. In a manner similar to the operation of the reduced-pressure, boiling-point apparatus, the system is filled with inert gas, heated to the desired temperature, and small aliquots of gas are removed. The differential manometer will show temporary differences in pressure as the aliquot of gas is removed but will return to the null point until the pressure reaches the vapor pressure of the salt. At that time the differential manometer will show a "permanent" differential pressure, and the pressure indicated by the gauge is taken as the vapor pressure at the temperature of the liquid. It is claimed that this method is more sensitive than the reduced-pressure, boiling-point method for vapor pressures below about 30 torr. The quasi-static and boiling-point methods have been used for measuring vapor pressures of uranium tetrafluoride,(125) thorium tetrafluoride,(126) and other salts(127) of interest to the Molten Salt Reactor program.

The transpiration or flow method of measurement of vapor pressures probably requires the simplest equipment of any method. It consists simply of flowing a carrier gas over the surface of the sample contained in a "boat" or crucible in a (generally horizontal) furnace tube. It is important that the carrier gas flow be slow enough that saturation is achieved and rapid enough that back diffusion and sample loss are not allowed. The amount of sample transported must be measured accurately, either by weight loss or complete recovery of the condensed product. This method was used for determination of vapor pressures of uranium tetrachloride(128) and of thermodynamic quantities for plutonium tetrachloride,(129) which exists only under a chlorine atmosphere at elevated temperatures.

The Knudsen effusion method has been widely applied in vapor pressure studies of actinide compounds and elements. Tantalum and tungsten have been used for effusion ovens heated by either induction or resistance methods in high vacuum. The effusion orifices have been made by a variety of machining techniques applied to thin sheets or larger blocks of metal. The ovens have been used both with and without inner cups. The inner cups have been tungsten, tantalum, or ceramic, with and without chemical vapor-deposited tungsten coatings. The greatest advantage of use of the inner cup is limitation of creep by the liquid samples. Creep may lead to change of orifice diameter and cause sealing of the lid on the oven so that it cannot be reused. If used, the inner cup should have a sharp lip at the top to prevent or reduce creep of the liquid to the outside of the cup. The oven and its heater should be designed so that the orifice and lid operate at temperatures slightly higher than those at the bottom of the crucible so the sample does not redeposit on the lid and orifice.

The vapor pressure is determined from calculations based on the rate of effusion from the orifice. This is, in theory, most simply determined by measuring the weight loss produced by heating the oven and contents for a specific time at a specific temperature. However, in practice, there are much more rapid methods for measuring rates of effusion. One of these methods is measurement of the amount of sample deposited on a target within precisely controlled geometrical constraints. Several targets are loaded into a holder in the vacuum system. A target-changing mechanism is provided so that several samples can be collected during a run without opening the vacuum system or cooling the furnace. The target holder is often cooled by liquid nitrogen or chilled water to ensure that the condensation coefficient is unity.

Radioactive materials are uniquely suited for vapor pressure studies by the target collection method, since submicrogram quantities of samples collected on targets can be measured quantitatively by application of counting methods. Inherent in this method is the assumption that the identity of effusing species is known. This uncertainty can be overcome by addition of a mass spectrometer to the system to identify species. A mass spectrometer can be used to determine vapor pressures by ion current measurement without a target collection system if an elaborate calibration scheme involving multiplier gain, isotopic abundance, ionization cross sections, and threshold energies is used.[130]

Vapor pressures for plutonium halides, plutonium oxide, and americium metal determined by the target collection and counting method were published in 1950.[131–133] The same method, but quite different apparatus, has been used for determination of vapor pressures of the fluorides of plutonium and americium.[134] Excellent discussions of the technique are given in articles describing vapor pressure determinations for several actinide metals by the Knudsen effusion technique.[135–137]

7.2.2. *Surface Tension and Density*

The two most popular techniques for measurement of surface tension of molten salts can be modified slightly to measure density in the same apparatus. It is estimated that over 80% of the measurements of surface tension are by the maximum gas bubble pressure technique.[138–140] In this technique, a capillary tube having a precisely formed tip is immersed to an accurately determined depth in the molten salt and pressurized with a very slow gas flow to form bubbles at the rate of five or less per minute. The inside diameter of the capillary is on the order of 1 mm and must be very nearly a perfect circle. The end must be flat and perpendicular to the axis of the capillary. There is some difference of opinion as to whether the edge of the orifice should be a "knife edge" or a "small flat." These capillaries have been made of precious metals and refractory oxides such as beryllium oxide. The surface tension is calculated from the depth of immersion, the inside diameter of the tube, the maximum pressure observed during bubble growth, and density of the liquid. The density of the liquid can be measured by noting the change in maximum pressure with an accurately measured change in depth of immersion of a single capillary, or by difference in maximum pressures for two capillaries with accurately known differences in depth of immersion in the same liquid. This technique appears to be most popular in the USSR, where many measurements[141-143] have been made with uranium chlorides in binary and ternary molten salts used in uranium electrowinning and electrorefining cells.

The second most popular method[144] for measurement of surface tension and density in a single apparatus is based on a slight modification of the Archimedes float bob. The bob commonly used to measure density of a liquid by buoyancy is modified by adding an accurately machined rodlike protrusion on its bottom. The density of the molten salt can then be determined by measuring the buoyant force on the bob in the usual fashion. The container is lowered and the discontinuity of the indicated weight noted at the instant of breaking contact between the bob and the liquid. The preferred geometry of the bob is a double cone (base to base), with a rod about 5 mm long by 1 mm diameter extending from the base. The rod must be accurately machined with the face perpendicular to the axis, and the diameter must be known accurately. The material from which the bob is machined must be resistant to corrosion by the molten salts to be observed and its coefficient of thermal expansion must be known. The Archimedean float technique has been used at Oak Ridge for measurement of densities of molten salts containing thorium tetrafluoride, but surface tensions were not determined in that work.[145]

7.2.3. *Viscosity*

Only a limited number of measurements of viscosities have been made with molten salts containing actinides. In the USSR, viscosities of uranium chlorides in molten chloride mixtures[146] and thorium fluorides in molten fluoride mixtures[147] have been measured in support of uranium extraction and electrorefining programs and in support of their fluoride-based reactor program, respectively. The molten salts were contained in cylindrical crucibles suspended in furnaces by torsion wires. No other experimental details were given except that the method of "torsional vibrations" was used. This apparently is the oscillating container method, which is to be compared with the method of oscillations of an immersed cylinder[148] and to capillary flow methods.[149] It is not clear which of these methods gives the most accurate results. Experiments in the Molten Salt Reactor program at Oak Ridge[150] have used a commercially available coaxial cylinder viscometer for measuring viscosities in molten fluorides. This instrument indicates the torque on the drive shaft of a cylinder immersed in the molten salt when the cylinder is rotated at constant speed on the axis of the cylindrical container. A set of viscosity standards was used to calibrate the apparatus.

8. *Absorption Spectra*

This discussion is limited to measurement techniques of the ultraviolet-visible, and near-infrared absorption spectra of molten actinide salts. The general topic of molten salt spectroscopy is discussed in much greater depth by T. R. Griffiths in Chapter 4 of Volume 2 of this series. Some aspects of the subject are also presented by V. E. Norvell and G. Mamantov in Chapter 7 of Volume 1 of this series. Workers at Oak Ridge National Laboratory developed many of the furnaces and cells for study of uranium and thorium halides, which are essentially free of the problems of radioactivity presented by the other actinides. The techniques they developed are therefore applicable to general discussion of molten salt absorption spectra measurements. It is hoped that duplication is limited in the following, which summarizes some of the techniques pertinent to the chemistry of the actinides.

The early work with the more intense alpha emitters plutonium, neptunium, and americium was performed using essentially "benchtop" techniques. The risks were acceptable for several reasons. The quantities of actinides were small (milligrams) so that escape of a small fraction of a sample would be serious but not catastrophic, although one must remember that one milligram of plutonium is on the order of 1000 "body burdens." The salts were contained in quartz or borosilicate glass cells as a primary containment with the spectrophotometer furnace as a secondary containment. The spectrophotometers were used in laboratories with favorable conditions of air flow to protect personnel from airborne radioactive particles. These conditions made possible many successful studies of highly radioactive molten salts.

The earliest[151,152] studies of uranium, plutonium, and neptunium in molten salts were performed with a Beckman DU spectrophotometer in which the cell compartment was replaced by a cooling block and a furnace block, which could be heated to 240°C by recirculated silicone oil. The solvents used were lithium nitrate/potassium nitrate eutectic (mp 132°C) and pyridinium chloride (mp 144°C).

The next step in the evolution of this technique was introduction of a higher-temperature (700°C) furnace[153,154] and interchanging of the positions of the light source and detector of the Beckman DU so that only a small portion of the thermal radiation from the furnace passed through the monochromator to the detector. The molten salt solvent list was then expanded to include lithium chloride/potassium chloride, which was purified by drying *in vacuo* and sparging with hydrogen chloride gas. The liquid eutectic was then dripped into carbon tetrachloride to form pellets. Lithium chloride/potassium chloride prepared in this way did not etch quartz optical cells or the attached quartz reservoirs that were filled, evacuated, and sealed for observation of the spectra.

A somewhat different approach to modification of a Beckman DU spectrophotometer for observation of plutonium in molten salt systems was taken by the author at Los Alamos. As is illustrated in Fig. 16, the light source was removed from the monochromator case and attached to a furnace containing the optical cells inside a glovebox. Light from the source was transmitted through the furnace containing the optical cells, a simple lens system in a brass tube, a quartz window in the brass tube at the plane of the glovebox wall, the monochromator, and finally through the original cell compartment to the detector. The integrity of the alpha enclosure was maintained by sealing a flange of the brass tube to the glovebox wall. This technique allowed keeping the fragile plutonium-containing optical cells in the glovebox enclosure. Other techniques have been developed at Los Alamos to accomplish the same objective. The simplest of these is the use of wells (approximately 15 cm square by 30 cm deep, with appropriate windows sealed to the sides) attached to the floor of the glovebox in a location such that the cell compartment of the spectrometer can be positioned to enclose it. Heaters and optical cells are placed in the well with their optical path in alignment with the windows.

In more recent years, the Beckman DU spectrophotometer has been replaced at Argonne National Laboratory, Oak Ridge National Laboratory, and Hanford Laboratory by the double-beam (Varian) Cary model 14 H

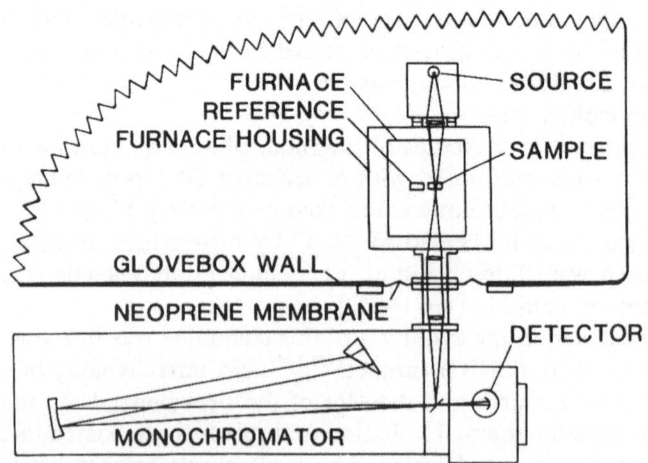

FIGURE 16. Spectrophotometer used for studies of molten plutonium salts.

spectrophotometer, which has a lamp-sample-monochromator-detector arrangement. This arrangement virtually eliminates interference from furnace thermal radiation. A variety of furnace designs[155-158] have been developed, one of which permits operation at temperatures up to 1450°C. Silica optical cells have continued to be the most popular for molten salts except fluorides and metal/salt systems, which are incompatible with silica. Windowless cells, such as the "captive liquid" cells discussed by Griffiths and by Norvell and Mamantov, were developed for these corrosive liquids. Other windowless cells[159,160] that have been used to support corrosive liquids are platinum screens, loops, spirals, and tube segments. The major fault of most of these windowless cells is that the effective path length cannot be accurately determined. A graphite cell equipped with diamond windows[161] has been developed for observation of fluorides. The diamond windows (nominally 5 × 5 × 1 mm) are very expensive and darken at an unacceptable rate at 800°C. The windows are not required to fit tightly, since molten fluorides do not wet graphite and, in the absence of a large pressure differential, will not run out through a small orifice because of surface tension effects.

Other systems for studies of absorption spectra of actinides have utilized commercial monochromators with supporting optics, choppers, light sources, and detectors set up on optical tables. Examples[11,25,162] of this type of system were discussed briefly in Section 4 (Microtechniques). A new concept in commercial spectrophotometers,[27] which may offer the versatility needed for observations of molten actinide salts, is being sold by a new company called Guided Wave, Inc., of Rancho Cordova, California. This instrument uses fiber optic light pipes to transmit light to and from the sample, which may be located at some distance (more than 10 m) from the spectrophotometer. Other developments in computerized data acquisition and analysis and more intense broad-spectrum light sources should make possible more rapid and accurate measurement of absorption spectra with even smaller furnaces and samples than have been required in the past.

9. *Electrochemistry*

Studies of the electrochemistry of actinide molten salts have been rather limited in extent. The applications of electrorefining and electrolysis of thorium, uranium, and plutonium in molten salt systems on the multikilogram scale have already been discussed in Section 6.4.1. These processes seem to have been developed with very little basic research.

The techniques applied in studies of electrochemistry of actinide molten salts are not significantly different from those applied in study of nonactinides. The general techniques have been described in many discussions[163-166] much more comprehensively than in this short treatment. Gloveboxes are required for study of the actinides (with the exception of uranium and thorium), but these enclosures are not unique to the actinides. Traces of oxygen and water interfere so severely with electrochemical studies that glovebox enclosures have been used for studies of hygroscopic non-actinide salts. Cells for studies of actinide systems employ the usual materials such as quartz, borosilicate glass, tantalum, and tungsten. The very reactive molten metals cannot be allowed direct contact with quartz or glass and must be contained by tantalum, tungsten, molybdenum, beryllium oxide, thorium oxide, or other nonreactive material. The fluorides are often studied in containers made of platinum, graphite, nickel, or molybdenum and the oxides of thorium, beryllium, or magnesium.

Only conductance and EMF studies will be reported here. The conductance measurements in recent years have been made almost exclusively on fluorides that may be used in molten salt reactors.[16] The EMF measurements have been made on only a small number of the actinides in chloride systems.

9.1. Conductance Measurements

Almost all of the conductance measurements of actinide salts that are reported in the technical journals were made in the USSR. The reports present data but give very little information concerning the techniques used in making the measurements. The technique most often referenced is the capillary cell method[140,167-170] described by Janz. Because of the use of the plural "capillaries," one may infer that the type of capillary technique used is that which employs two capillary tubes (Fig. 17). The capillaries shown here have inside diameters of about 1 mm and a length of about 25 mm in the portion that dips into the molten salt. The upper portion has a larger diameter to accommodate electrical connectors. When in use, the bottom of this upper portion must be positioned so as to be at a level just beneath the surface of the salt in its crucible. For accurate measurements the capillaries must be held rigidly in a position that is reproducible. Electrical contact is maintained by rods or tubes made of precious or refractory metals. The cell constant is determined by observation of aqueous or molten salt standards.[171] The best descriptions of this type of cell recommend that the capillaries be fabricated from single-crystal magnesium oxide, but they may also be made of glass, quartz, boron nitride and polycrystalline magnesium oxide, or beryllium oxide. Alternating current frequencies in the range of 1-50 kHz have been used for measurements.

Typical studies which have been reported are (1) thorium tetrafluoride with lithium and sodium fluorides using magnesia capillaries and nickel container and electrodes up to 950°C[169]; (2) uranium, thorium, beryllium,

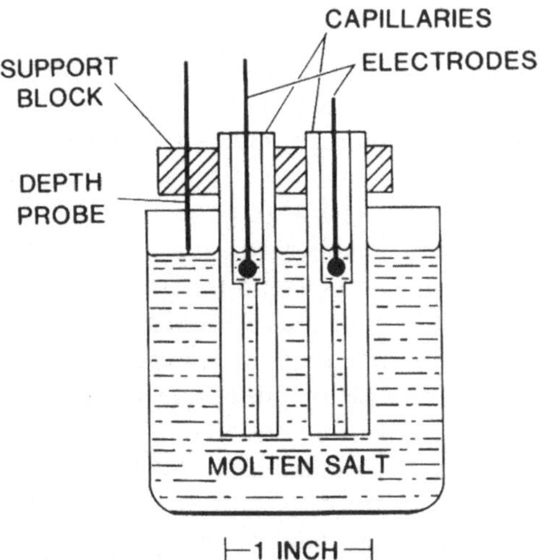

FIGURE 17. Capillary cell used for measuring conductance of molten salts.

and lithium ternary and quaternary fluoride systems using beryllia capillaries and molybdenum electrodes up to 1000°C[170]; and (3) uranium and sodium fluoride using polycrystalline magnesia capillaries up to 1170°C.[172]

9.2. Electromotive Force Measurements

The reported EMF measurements made on actinide molten salt systems have had as their objectives the determination of thermodynamic quantities such as free energies of formation and activities and the elucidation of electrode processes and measurement of current efficiencies. Most of the EMF studies at Los Alamos have had as their objective determination of free energies of formation of plutonium compounds. The free energy of formation of plutonium trichloride was determined by use of the cells[173,174]

$$Pu(liq)/PuCl_3\text{-}MCl(liq)/Cl_2(g) \tag{3}$$

where M was sodium and potassium. The molten plutonium was contained in a porous thoria crucible and chlorine gas was introduced through a graphite electrode. Another study of plutonium trichloride was made by use of the cell[175]

$$Pu\text{-}PuCl_3(s)/BaCl_2(s)/NaCl\text{-}AgCl,Ag(s) \tag{4}$$

Cells employing lithium chloride/potassium chloride eutectic as electrolyte were used to study PuN, Pu_2C_3, $PuC_{(1-x)}$, $PuRu_2$, and $PuFe_2$.[175] One of the unique techniques developed during this work[176] was formation of a "temporary" liquid plutonium electrode by electrodeposition of plutonium from the electrolyte onto a tungsten microelectrode, measuring the EMF, and immediately stripping off the plutonium by electrolysis. This technique alleviates the active metal-container corrosion problem. It was also extended to the uses of a controlled potential to generate a concentration gradient near the electrode surface.[177] This gradient is then allowed to relax under open circuit conditions and the potential versus time curves are treated theoretically to calculate parameters of interest in describing the system. The method was applied in determination of thermodynamic properties of binary Laves phase compounds of plutonium with iron, ruthenium, and osmium.[178]

Among the many electrochemical studies conducted by Inman at Imperial College (London) are EMF studies of uranium and thorium in molten chloride systems. During studies of the cell[179,180]

$$U/UCl_3,\ LiCl\text{-}KCl//LiCl\text{-}KCl,\ AgCl/Ag \tag{5}$$

and the analogous thorium cell,[181] activities in the salt phase were determined and mechanisms of electrode reactions were elucidated. Unique techniques used in this work include preparation and addition of hygroscopic salts to the cell, without exposure to the atmosphere, and a silver/silver chloride reference electrode enclosed by a quartz diaphragm.

Two series of EMF studies of molten actinide salts have been reported from Argonne National Laboratory. In one of these[182] the cell was

$$U/UCl_3,\ LiCl\text{-}KCl//LiCl\text{-}KCl,\ AgCl/Ag \tag{6}$$

Silver chloride/silver electrodes were formed *in situ* by electrolysis using a silver wire anode and auxiliary platinum cathode. In the course of this work it was noted that there was slow corrosion of uranium wire in molten lithium/potassium chloride eutectic contained in Pyrex, while corrosion did not occur when the container was sapphire.

The other work reported from Argonne was determination of thermodynamic quantities of a series of actinide metal alloys. A typical cell used in this series was

$$U/(s)/UCl_3,\ LiCl\text{-}KCl(liq)/U\text{-}Cd(liq) \tag{7}$$

where the uranium electrode was a solid rod and the uranium/cadmium was a liquid contained in an alumina or tantalum crucible. The alloys studied were uranium/cadmium,[183] plutonium/zinc,[184] plutonium/cadmium,[185] and neptunium/cadmium.[186] The free energies of formation and other thermodynamic quantities of several intermetallic compounds were calculated from the data. Activity coefficients of the actinides in these alloys were calculated. It may be speculated that these will be valuable in prediction of separations from impurities by use of such alloys as anodes during electrorefining.

10. Summary

It has been pointed out that the study of actinides in molten salt systems is interesting and challenging. Many areas remain to be explored. The techniques are in many ways similar to those used in studies of other active metal elements such as the alkaline earths and rare earths, but new and modified techniques have of necessity been introduced to meet the challenge of intense radioactivity. Many of the challenges in this field are regulatory in nature—i.e., there are regulations concerning security, accountability, personnel safety, and environmental protection imposed by a large number of regulatory agencies—but it should be recognized that these regulatory agencies and their parent governments all around the world (rather than private corporations) have supported work with the actinides. Indeed, many of the techniques now applied to work with nonactinides in private industries were developed in laboratories established to study actinides.

References

1. National Research Council Committee on Nuclear and Alternative Energy Systems, *Energy in Transition, 1985–2010,* National Academy of Sciences, Washington, D.C., pp. 128–260, W. H. Freeman and Co., San Francisco (1980).

2a. E. M. Levin, C. R. Robbins, and H. F. McNurdie, *Phase Diagrams for Ceramists* (M. K. Reser, ed.), The American Ceramic Society Inc., Columbus, Ohio (1964).

2b. E. M. Levin, C. R. Robbins, and H. F. McMurdie, *Phase Diagrams for Ceramists, 1969 Supplement* (M. K. Reser, ed.), The American Ceramic Society, Inc., Columbus, Ohio (1969).

2c. E. M. Levin and H. F. McMurdie, *Phase Diagrams for Ceramists, 1975 Supplement* (M. K. Reser, ed.), The American Ceramic Society, Inc., Columbus, Ohio (1975).

2d. R. S. Roth, T. Negas, and L. P. Cook, *Phase Diagrams for Ceramists, Volume IV* (G. S. Smith, ed.), The American Ceramic Society, Inc., Columbus, Ohio (1981).

3. J. D. Farr, A. L. Giorgi, M. G. Bowman, and R. K. Money, *J. Inorg. Nucl. Chem.* **18,** 42–47 (1961).

4. D. A. Collins, J. Hillary, J. S. Nairn, and G. M. Phillips, *J. Inorg. Nucl. Chem.* **24,** 441–459 (1962).

5. D. C. Stewart, *Handling Radioactivity,* Wiley, New York (1981).

6. D. C. Stewart, in: *Technique of Inorganic Chemistry,* Vol. III (H. B. Jonassen and A. Weissberger, eds.), pp. 167–175, Wiley, New York (1963).

7. IAEA Advisory Group, *Manual on Safety Aspects of the Design and Equipment of Hot Laboratories, Safety Series No. 30,* International Atomic Energy Agency, Vienna (1981).

8. Seventeen papers in Session C-18, Part II, in: *Proceedings of the Second United Nations International Conference on the Peaceful Uses of Atomic Energy,* Vol. 17, *Processing Irradiated Fuels and Radioactive Materials,* pp. 352–494, United Nations, Geneva (1958).

9. P. J. Peterson, R. L. Thomas, and J. L. Green, in: *Proceedings of the Second United Nations International Conference on the Peaceful Uses of Atomic Energy*, Vol. 17, *Processing Irradiated Fuels and Radioactive Materials*, pp. 664–667, United Nations, Geneva (1958).
10. G. N. Walton (ed. board chmn.), *Gloveboxes and Shielded Cells for Handling Radioactive Materials*, Proceedings of the Symposium on Glovebox Design and Operation, Harwell, February 19-21, 1957, sponsored by UKAEA, Academic Press, New York (1958).
11. G. D. Calkins and P. Schall, in: *Reactor Handbook*, Second Edition, Vol. 1, *Materials* (C. R. Tipton, ed.), pp. 74–83, Interscience, New York (1960).
12. How radiation affects materials, a special report, *Nucleonics* **14**(9), 53–88 (1956).
13. W. R. Grimes and D. R. Cuneo, in: *Reactor Handbook*, Second Edition, Vol. 1, *Materials* (C. R. Tipton, ed.), pp. 425–476, Interscience, New York (1960).
14. W. R. Grimes, D. R. Cuneo, F. F. Blankenship, G. W. Keilholtz, H. F. Poppendiek, and M. T. Robinson, in: *Fluid Fuel Reactors* (J. A. Lane, H. G. McPherson, and F. Maslan, eds.), pp. 569–594, Addison-Wesley, Reading, Massachusetts (1958).
15. D. M. Gruen, R. L. McBeth, J. Kooi, and W. T. Carnall, *Ann. N.Y. Acad. Sci.* **79**, 941–949 (1960).
16. J. W. Koger, *Corrosion-NACE 29*, pp. 115–122 (1973).
17. A. Glassner, The Thermochemical Properties of the Oxides, Fluorides, and Chlorides to 2500°K, Argonne National Laboratory Report ANL-5750, U.S. Government Printing Office, Washington, D.C. (1957).
18. D. C. Christensen and L. J. Mullins, in: *Plutonium Chemistry* (W. T. Carnall and G. R. Choppin, eds.), ACS Symposium Series 216, pp. 409–431, American Chemical Society, Washington, D.C. (1983).
19. G. S. Perry, L. G. Macdonald, and S. D. Wilcox, Contamination of Molten Salts by Crucible Materials, HCTD Technical Memorandum No. 1/84, AWRE, Aldermaston, Berkshire (June, 1984).
20. C. E. Baldwin, in: *Actinide Recovery from Waste and Low-Grade Sources* (J. D. Navratil and W. W. Schulz, eds.), pp. 56–60, Harwood, New York (1982).
21. B. B. Cunningham, in: *Submicrogram Experimentation* (N. D. Cheronis, ed.), pp. 69–87, Interscience, New York (1961).
22. J. R. Peterson, in: *Lanthanide and Actinide Chemistry and Spectroscopy*, ACS Symposium Series 131 (N. M. Edelstein,ed.), pp. 221–238, American Chemical Society, Washington, D.C. (1980).
23. J. N. Stevenson and J. R. Peterson, *Microchem. J.* **20**, 213–220 (1975).
24. E. F. Westrum, Jr., and LeRoy Eyring, *J. Am. Chem. Soc.* **73**, 3399–3400 (1951).
25. J. L. Green and B. B. Cunningham, *Inorg. Nucl. Chem. Lett.* **2**, 365–371 (1966).
26. J. P. Young, K. L. Vander Sluis, G. K. Werner, J. R. Peterson, and M. Noe, *J. Inorg. Nucl. Chem.* **37**, 2497–2501 (1975).
27. The Optical Waveguide Spectrum Analyzer, Guided Wave Inc., Rancho Cordova, California (1984).
28. D. B. McWhan, B. B. Cunningham, and J. C. Wallmann, *J. Inorg. Nucl. Chem.* **24**, 1025–1038 (1962).
29. D. R. Stephens, H. D. Stromberg, and E. M. Lilley, *J. Phys. Chem. Solids* **29**, 815–821 (1968).
30. L. Grainger, *Uranium and Thorium*, pp. 46–73, George Newness, London (1958).
31. F. L. Cuthbert, *Thorium Production Technology*, pp. 50–96, Addison-Wesley, Reading, Massachusetts (1958).
32. F. A. Cotton and G. Wilkinson, *Advanced Inorganic Chemistry, A Comprehensive Text, Third Ed.*, pp. 1096–1098, Interscience, New York (1972).
33. D. O. Campbell, in: *Physico-Chimie du Protactinium*, Colloques Internationaux, Orsay, July 2–8, 1965, pp. 209–223, Centre National de la Recherche Scientifique, Paris (1966).

34. H. W. Kirby, in: *Physico-Chimie du Protactinium*, Colloques Internationaux, Orsay, July 2-8, 1968, pp. 283-291, Centre National de la Recherche Scientifique, Paris (1966).
35. L. Grainger, *Uranium and Thorium*, pp. 1-45, George Newness, Ltd., London (1958).
36. J. C. Burger and J. McN. Jardine, in: *Proceedings of the Second United Nations International Conference on the Peaceful Uses of Atomic Energy, Vol. 4, Production of Nuclear Materials and Isotopes*, pp. 3-9, United Nations, Geneva (1958).
37. R. Gelin, H. Mogard, and B. Nelson, in: *Proceedings of the Second United Nations Conference on the Peaceful Uses of Atomic Energy*, Vol. 4, *Production of Nuclear Materials and Isotopes*, pp. 36-39, United Nations, Geneva (1958).
38. W. W. Schulz and G. E. Benedict, *Neptunium-237 Production and Recovery*, USAEC (now DOE) Technical Information Center, Oak Ridge, Tennessee (1972).
39. L. W. Gray, G. A. Burney, T. W. Wilson, and J. M. McKibben, in: *Transplutonium Elements—Production and Recovery*, ACS Symposium Series 161, pp. 223-242, American Chemical Society, Washington, D.C. (1981).
40. P. C. Doto, L. E. Bruns, and W. W. Schulz, in: *Transplutonium Elements—Production and Recovery*, ACS Symposium Series 161, pp. 109-129, American Chemical Society, Washington, D.C. (1981).
41. L. W. Gray, G. A. Burney, T. A. Reilly, T. W. Wilson, and J. M. McKibben, in: *Transplutonium Elements—Production and Recovery*, ACS Symposium Series 161, pp. 93-108, American Chemical Society, Washington, D.C. (1981).
42. H. D. Ramsey, D. G. Clifton, S. W. Hayter, R. A. Penneman, and E. L. Christensen, in: *Transplutonium Elements—Production and Recovery*, ACS Symposium Series 161, pp. 75-91, American Chemical Society, Washington, D.C. (1981).
43. J. B. Knighton, P. G. Hagan, J. D. Navratil, and G. H. Thompson, in: *Transplutonium Elements—Production and Recovery*, ACS Symposium Series 161, pp. 53-74, American Chemical Society, Washington, D.C. (1981).
44. M. S. Coops, J. B. Knighton, and L. J. Mullins, in: *Plutonium Chemistry* (ACS Symposium Series 216) (W. T. Carnall and G. R. Choppin, eds.), pp. 386-398, American Chemical Society, Washington, D.C. (1983).
45. C. Keller, *The Chemistry of the Transuranium Elements*, p. 532, Verlag Chemie, Weinheim, Germany (1971).
46. T. C. Gorrell, *Trans. Am. Nucl. Soc.* **14**, 343-344 (1971).
47. K. Buijs, F. Maino, W. Muller, J. Reul, and J. Cl. Toussaint, *J. Inorg. Nucl. Chem., Suppl.* **1976**, 209-213 (*Proceedings of the Moscow Symposium on the Chemistry of Transuranium Elements*) (V. I. Spitsyn and J. J. Katz, eds.), Pergamon, New York (1976).
48. L. J. King, J. E. Bigelow, and E. D. Collins, in: *Transplutonium Elements—Production and Recovery*, ACS Symposium Series 161, pp. 133-145, American Chemical Society, Washington, D.C. (1981).
49. F. T. Hagemann, in: *The Actinide Elements* (National Nuclear Energy Series, Div. IV, Vol. 14A) (G. T. Seaborg and J. J. Katz, eds.), pp. 14-44, McGraw-Hill, New York (1954).
50. J. D. Farr, A. L. Giorgi, M. G. Bowman, and R. K. Money, *J. Inorg. Nucl. Chem.* **18**, 42-47 (1961).
51. F. L. Cuthbert, *Thorium Production Technology*, p. 152, Addison-Wesley, Reading, Massachusetts (1958).
52. L. I. Katzin, in: *The Actinide Elements* (National Nuclear Energy Series, Div. IV, Vol. 14A) (G. T. Seaborg and J. J. Katz, eds.), p. 82, McGraw-Hill, New York (1954).
53. F. L. Cuthbert, *Thorium Production Technology*, p. 160, Addison-Wesley, Reading, Massachusetts (1958).
54. A. H. Roberson, Preparation of Thorium Chloride from Thorium Oxalate, U.S. Bureau of Mines Circular 101 (1956).
55. W. Fischer, R. Gewehr, and H. Wingchen, *Z. Anorg. Allg. Chem.* **242**, 161-187 (1939).
56. L. B. Asprey, F. H. Kruse, and R. A. Penneman, *Inorg. Chem.* **6**, 544-548 (1967).

57. P. A. Sellers, S. Fried, R. E. Elson, and W. H. Zachariasen, *J. Am. Chem. Soc.* **76,** 5935–5938 (1954).
58. L. Stein, *Inorg. Chem.* **3,** 995–1001 (1964).
59. D. Brown and P. J. Jones, *J. Chem. Soc.* (*A*) **1967,** 719–723 (1967).
60. M. E. Hendricks, E. R. Jones, J. A. Stone, and D. G. Karraker, *J. Chem. Phys.* **55,** 2993–2997 (1971).
61. D. Brown and P. J. Jones, *J. Chem. Soc.* (*A*) **1966,** 874–878 (1966).
62. D. Brown, J. Hill, and C. E. F. Rickard, *J. Chem. Soc.* (*A*) **1970,** 476–480 (1970).
63. D. Brown and P. J. Jones, *J. Chem. Soc.* (*A*) **1966,** 262–264 (1970).
64. D. Brown, J. F. Easey, and P. J. Jones, *J. Chem. Soc.* (*A*) **1967,** 1698–1702 (1967).
65. H. R. Hoekstra and J. J. Katz, in: *The Actinide Elements* (National Nuclear Energy Series, Div. 4, Vol. 14A) (G. T. Seaborg and J. J. Katz, eds.), pp. 147–148, McGraw-Hill, New York (1954).
66. J. J. Katz and E. Rabinowitch, *The Chemistry of Uranium—Part I* (National Nuclear Energy Series, Div. VIII, Vol. 5), p. 361 (1951).
67. F. S. Patton, J. M. Googin, and W. L. Griffith, *Enriched Uranium Processing,* pp. 63–64, Pergamon, New York (1963).
68. S. Fried and N. Davidson, *J. Am. Chem. Soc.* **70,** 3539–3547 (1948).
69. L. E. Trevorrow, T. J. Gerding, and M. J. Steindler, *J. Inorg. Nucl. Chem.* **30,** 2671–2677 (1968).
70. J. G. Reavis, K. W. R. Johnson, J. A. Leary, A. N. Morgan, A. E. Ogard, and K. A. Walsh, in: *Extraction and Physical Metallurgy of Plutonium and Its Alloys* (W. D. Wilkinson, ed.), pp. 89–100, Interscience, New York (1960).
71. M. J. Steindler, D. V. Steidl, and R. K. Steunenberg, *Nucl. Sci. Eng.* **6,** 333–340 (1959).
72. F. Hagemann, B. M. Abraham, N. R. Davidson, J. J. Katz, and I. Sheft, in: *The Transuranium Elements Research Papers* (National Nuclear Energy Series, Div. IV, Vol. 14B) (G. T. Seaborg, J. J. Katz, and W. M. Manning, eds.), pp. 957–963, McGraw-Hill, New York (1949).
73. S. Fried, *J. Am. Chem. Soc.* **73,** 416–418 (1951).
74. W. V. Conner, *J. Less-Comm. Metals* **25,** 379–384 (1971).
75. R. D. Baybarz, *J. Inorg. Nucl. Chem.* **35,** 483–487 (1973).
76. J. H. Burns and J. R. Peterson, *Acta Cryst.* (*B*) **26,** 1885–1887 (1970).
77. R. G. Pappalardo, W. T. Carnall, and P. R. Fields, *J. Chem. Phys.* **51,** 1182–1200 (1969).
78. L. B. Asprey, T. K. Keenan, and F. H. Kruse, *Inorg. Chem.* **4,** 985–986 (1965).
79. R. D. Baybarz, L. B. Asprey, C. E. Strouse, and E. Fukushima, *J. Inorg. Nucl. Chem.* **34,** 3427–3431 (1972).
80. L. B. Asprey, T. K. Keenan, and F. H. Kruse, *Inorg. Chem.* **3,** 1137–1140 (1964).
81. W. W. Schulz, *The Chemistry of Americium,* pp. 151–159, TID-26971, Technical Information Center, Energy Research and Development Administration (now DOE), Oak Ridge, Tennessee (1976).
82. L. B. Asprey, F. H. Ellinger, S. Fried, and W. H. Zachariasen, *J. Am. Chem. Soc.* **79,** 5825 (1957).
83. J. H. Simons and L. P. Block, *J. Am. Chem. Soc.* **61,** 2962–2966 (1939).
84. R. L. Farrar, Jr., and E. J. Barber, Some Considerations in the Handling of Fluorine and the Chlorine Fluorides, Report K/ET-252, Enrichment Technology Div., Oak Ridge Gaseous Diffusion Plant, Oak Ridge, Tennessee (1979).
85. J. G. Reavis and J. A. Leary, *J. Inorg. Nucl. Chem.* **28,** 1205–1208 (1966).
86. C. W. Koch and B. B. Cunningham, *J. Am. Chem. Soc.* **76,** 1470 (1954).
87. B. B. Cunningham, in: *The Actinide Elements* (National Nuclear Energy Series, Div. IV, Vol. 14A) (G. T. Seaborg and J. J. Katz, ed.), p. 390, McGraw-Hill, New York (1954).
88. D. Brown and J. F. Easey, *J. Chem. Soc.* (*A*) **1970,** 3378–3381.
89. J. Hauck, *J. Inorg. Nucl. Chem.* **36,** 2291–2298 (1974).

90. C. W. Bjorklund, J. G. Reavis, J. A. Leary, and K. A. Walsh, *J. Phys. Chem.* **63**, 1774–1777 (1959).
91. H. A. Laitinen, W. S. Ferguson, and R. A. Osteryoung, *J. Electrochem. Soc.* **104**, 516–520 (1957).
92. D. L. Maricle and D. N. Hume, *J. Electrochem. Soc.* **107**, 354–356 (1960).
93. S. Fried, E. F. Westrum, H. L. Baumbach, and P. L. Kirk, *J. Inorg. Nucl. Chem.* **5**, 182–189 (1958).
94. J. N. Stevenson and J. R. Peterson, *J. Microchem.* **20**, 213–220 (1975).
95. F. L. Cuthbert, *Thorium Production Technology*, p. 180, Addison-Wesley, Reading, Massachusetts (1958).
96. R. D. Baker, Preparation of Plutonium Metal by the Bomb Method, Los Alamos Scientific Laboratory Report LA-473 (May, 1946).
97. L. Grainger, *Uranium and Thorium*, p. 38, George Newness, London (1958).
98. J. G. Reavis, J. A. Leary, and K. A. Walsh, U.S. Patent 2,886,410 (May 12, 1959).
99. D. C. Christensen and L. J. Mullins, Present Status of Plutonium Metal Production and Purification at Los Alamos—1982, Los Alamos National Laboratory Report LA-9674-MS, pp. 18–21 (June, 1983).
100. J. W. Marden, *Trans. Electrochem. Soc.* **66**, 39–47 (1934).
101. L. Grainger, *Uranium and Thorium*, pp. 62–63, George Newness, London (1958).
102. B. Blumenthal and B. Brodsky, in: *Plutonium 1960* (E. Grison, W. B. H. Lord, and R. D. Fowler, eds.), pp. 171–186, Cleaver-Hume Press, London (1961).
103. L. J. Mullins, A. N. Morgan, S. A. Apgar III, and D. C. Christensen, Six-Kilogram Scale Electrorefining of Plutonium Metal, Los Alamos National Laboratory Report LA-9469-MS (September, 1982).
104. L. J. Mullins, A. J. Beaumont, and J. A. Leary, *J. Inorg. Nucl. Chem.* **30**, 147–156 (1968).
105. R. K. Steunenberg, R. D. Pierce, and L. Burris, in: *Progress in Nuclear Energy, Series III, Process Chemistry* (C. E. Stevenson, E. A. Mason and A. T. Gresky, eds.), Vol. 4, pp. 482–486, Pergamon, New York (1970).
106. J. K. Bates, L. J. Jardine, and M. Krumpelt, in: *Actinide Separations*, ACS Symposium Series 117 (J. D. Navratil and W. W. Schulz, eds.), pp. 207–217, American Chemical Society, Washington, D.C. (1980).
107. J. Johnson, *J. Nucl. Mater.* **51**, 163–177 (1974).
108. I. Johnson, J. B. Knighton, and R. K. Stuenenberg, *Trans. Met. Soc. AIME* **236**, 1241–1246 (1966).
109. C. J. Barton, H. A. Friedman, W. R. Grimes, H. Insley, R. E. Moore, and R. E. Thoma, *J. Am. Ceram. Soc.* **41**, 63–69 (1958).
110. C. J. Barton, W. R. Grimes, H. Insley, R. E. Moore, and R. E. Thoma, *J. Phys. Chem.* **62**, 665–676 (1958).
111. R. E. Thoma, H. Insley, H. A. Friedman, and C. F. Weaver, *J. Phys. Chem.* **64**, 865–870 (1960).
112. C. F. Weaver, R. E. Thoma, H. Insley, and H. A. Friedman, *J. Am. Ceram. Soc.* **43**, 213–218 (1960).
113. K. W. R. Johnson, M. Kahn, and J. A. Leary, *J. Phys. Chem.* **65**, 2226–2229 (1961).
114. R. Benz and R. M. Douglass, *J. Phys. Chem.* **65**, 1461–1463 (1961).
115. R. Benz, M. Kahn, and J. A. Leary, *J. Phys. Chem.* **63**, 1983–1984 (1959).
116. M. Pages and W. Freundlich, *J. Inorg. Nucl. Chem.* **34**, 2797–2801 (1972).
117. V. I. Spitsyn, A. N. Pokrovskii, N. S. Afonskii, and V. K. Trunov, *Dokl. Akad. Nauk. SSSR* **188**, 1065–1068 (1969) (Plenum Publishing Corp., New York, translation, pp. 825–827).
118. N. N. Bushev and V. K. Trunov, *Radiokhimiya* **12**, 411–412 (1970).
119. J. G. Reavis, G. R. Brewer, D. B. Court, and J. W. Schulte, *Proceedings of the Nineteenth Conf. on Remote Systems Technology* (American Nuclear Society), pp. 112–117 (1971).

120. B. Weinstock, E. E. Weaver, and J. G. Malm, *J. Inorg. Nucl. Chem.* **11,** 104–114 (1959).
121. A. E. Florin, I. R. Tannenbaum, and J. F. Lemons, *J. Inorg. Nucl. Chem.* **2,** 368–379 (1956).
122. C. J. Mandleberg, H. K. Rae, R. Hurst, G. Long, D. Davies, and K. E. Francis, *J. Inorg. Nucl. Chem.* **2,** 358–367 (1956).
123. H. S. Young and H. F. Grady, in: *Chemistry of Uranium, Collected Papers,* TID-5290, Book 2 (J. J. Katz and E. Rabinowitch, eds.), pp. 749–756, USAEC Tech. Inform. Serv. Oak Ridge, Tennessee (1958).
124. W. H. Rodebush and A. L. Dixon, *Phys. Rev.* **26,** 851–858 (1925).
125. S. Langer and F. F. Blankenship, *J. Inorg. Nucl. Chem.* **14,** 26–31 (1960).
126. A. J. Darnell and F. J. Keneshea, Jr., *J. Phys. Chem.* **62,** 1143–1145 (1958).
127. S. Cantor, R. F. Newton, W. R. Grimes, and F. F. Blankenship, *J. Phys. Chem.* **62,** 96–99 (1958).
128. O. Johnson, T. Butler, and A. S. Newton, in: *Chemistry of Uranium, Collected Papers,* TID-5290, Book 1 (J. J. Katz and E. Rabinowitch, eds.), pp. 1–28, USAEC, Tech. Inform. Service, Oak Ridge, Tennessee (1958).
129. R. Benz, *J. Inorg. Nucl. Chem.* **24,** 1191–1195 (1962).
130. R. Kent, *High Temp. Sci.* **1,** 169–175 (1969).
131. T. E. Phipps, G. W. Sears, R. L. Seifert, and O. C. Simpson, *J. Chem. Phys.* **18,** 713–723 (1950).
132. T. E. Phipps, G. W. Sears, and O. C. Simpson, *J. Chem. Phys.* **18,** 724–734 (1950).
133. N. D. Erway and O. C. Simpson, *J. Chem. Phys.* **18,** 953–957 (1950).
134. S. C. Carniglia and B. B. Cunningham, *J. Am. Chem. Soc.* **77,** 1451–1453 (1955).
135. J. W. Ward, R. W. Ohse, and R. Reul, *J. Chem. Phys.* **62,** 2366–2372 (1975).
136. J. W. Ward, P. D. Kleinschmidt, R. G. Haire, and D. Brown, in: *Lanthanide and Actinide Chemistry and Spectroscopy* (N. M. Edelstein, ed.), pp. 199–220, ACS Symposium Series 131, American Chemical Society, Washington, D.C. (1980).
137. M. H. Bradbury and R. W. Ohse, *J. Chem. Phys.* **70,** 2310–2314 (1979).
138. Y. Umetsu, M. Kawada, E. Nakamura, and T. Ejima, *J. Jpn Inst. Met.* **37,** 1139 (1973) [English translation **16,** 271–279 (1975)].
139. D. A. Nissen and R. W. Carlsten, *J. Electrochem. Soc.* **121,** 500–505 (1974).
140. G. J. Janz, U. Krerbs, H. F. Siegenthaler, and R. P. T. Tomkins, *J. Phys. Chem. Ref. Data* **1,** 589 (1972).
141. V. N. Desyatnik, S. F. Katyshev, and S. P. Raspopin, *At. Energ.* **42,** 99–103 (1977) (Plenum Publishing Corp., New York, translation, pp. 108–111).
142. V. N. Desyatnik and S. F. Katyshev, *Zh. Fiz. Khim.* **55,** 2888–2892 (1981) (The British Library, Letchworth, Herts, translation, pp. 1641–1644).
143. V. N. Desyatnik and S. F. Katyshev, *Zh. Fiz Khim* **56,** 203–205 (1982) (English translation, pp. 125–127).
144. G. J. Janz and M. R. Lorenz, *Rev. Sci. Instrum.* **31,** 18–22 (1960).
145. S. Cantor, D. G. Hill, and W. T. Ward, *Inorg. Nucl. Chem. Lett.* **2,** 15–18 (1966).
146. S. F. Katyshev, Yu. F. Chervinskii, and V. N. Desyatnik, *At. Energ.* **53,** 108–109 (1982) (Plenum Publishing Corp., New York, translation, pp. 565–566).
147. Yu. F. Chervinskii, V. N. Desyatnik, and A. I. Nechaev, *Zh. Fiz. Khim.* **56,** 1946–1949 (1982) (The British Library, Letchworth, Herts, translation, pp. 1189–1190).
148. F. Thorkelp and H. A. Oye, *J. Phys. E* (*Sci. Instrum.*) **12,** 875–885 (1979).
149. G. J. Janz, U. Krebs, H. F. Siegenthaler, and P. T. Tomkins, *J. Phys. Chem. Ref. Data* **1,** 588 (1972).
150. C. T. Moynihan and S. Cantor, *J. Chem. Phys.* **48,** 115–119 (1968).
151. D. M. Gruen, *J. Inorg. Nucl. Chem.* **4,** 74–76 (1957).
152. D. M. Gruen, P. Graf, and S. Fried, Memoire presente a la Section de Chimie Minerale, XVI Congres International de Chemie Pure et Appliquee, pp. 319–327, Paris (1957).

153. D. M. Gruen and R. L. McBeth, *J. Inorg. Nucl. Chem.* **9**, 290-301 (1959).
154. D. M. Gruen, S. Fried, P. Graf, and R. L. McBeth, *Proceedings of the Second United Nations International Conference on Peaceful Uses of Atomic Energy*, Vol. 28, *Basic Chemistry in Nuclear Energy*, pp. 112-124, United Nations, Geneva (1958).
155. D. M. Gruen, *Q. Rev, Chem. Soc.* **19**, 349-368 (1965).
156. C. R. Boston and G. P. Smith, *Rev. Sci. Instrum.* **36**, 206-208 (1965).
157. C. R. Boston and G. P. Smith, *J. Phys. E (Sci. Instrum.) Ser. 2*, **2**, 543-546. (1969).
158. J. R. Morrey and A. W. Madsen, *Rev. Sci. Instrum.* **32**, 799-801 (1961).
159. J. P. Young and J. C. White, *Anal. Chem.* **31**, 1892-1895 (1959).
160. J. Greenberg and L. J. Hallgren, *Rev. Sci. Instrum.* **31**, 444-445 (1960).
161. L. M. Toth, J. P. Young, and G. P. Smith, *Anal. Chem.* **41**, 683-685 (1969).
162. J. P. Young, R. G. Haire, R. L. Fellows, and J. R. Peterson, *J. Radioanal. Chem.* **43**, 479-488 (1978).
163. V. E. Norvell and G. Mamantov, in: *Molten Salt Techniques*, Vol. 1 (D. G. Lovering and R. J. Gale, eds.), pp, 151-176, Plenum Press, New York (1983).
164. H. A. Laitinen and R. A. Osteryoung, in: *Fused Salts* (B. R. Sundheim, ed.), pp. 255-300, McGraw-Hill, New York (1964).
165. R. W. Laity, in: *Reference Electrodes* (D. J. G. Ives and G. J. Janz, eds.), pp. 524-606, Academic, New York (1961).
166. Lantelme, D. Inman, and D. G. Lovering, in: *Molten Salt Techniques*, Vol. 2 (D. G. Lovering and R. J. Gale, eds.), Plenum Press, New York (1984).
167. G. J. Janz and M. R. Lorenz, *Rev. Sci. Instrum.* **32**, 130-133 (1961).
168. V. N. Desyatnik, A. P. Koverda, G. P. Bystrai, and V. I. Kolontyr, *Zh. Prikl. Khim.* **52**, 316-319 (1979) (Plenum Publishing Corp., New York, translation, pp. 288-290).
169. V. N. Desyatnik, A. P. Koverda, N. N. Kurbatov, and V. V. Bystrov, *At. Energ.* **49**, 129-130 (1980) (Plenum Publishing Corp., New York, translation, pp. 583-585).
170. V. N. Desyatnik, A. P. Koverda, N. N. Kurbatov, and V. V. Bystrov, *At. Energ.* **50**, 141-143 (1981) (Plenum Publishing Corp., New York, translation, pp. 134-136).
171. G. J. Janz, *J. Phys. Chem. Ref. Data* **9**, 791-829 (1980).
172. V. N. Desyatnik, A. P. Koverda, and N. N. Kurbatov, *Zh. Fiz. Khim.* **55**, 3128-3130 (1981) (The British Library, Letchworth, Herts, translation, pp. 1782-1783).
173. R. Benz and J. A. Leary, *J. Phys. Chem.* **65**, 1056-1058 (1961).
174. R. Benz, *J. Phys. Chem.* **65**, 81-84 (1961).
175. G. M. Campbell, L. J. Mullins, and J. A. Leary, in *Proceedings of a Symposium of the Thermodynamics of Nuclear Materials with Emphasis on Solution Systems*, pp. 75-88, IAEA, Vienna, September 4-8, 1967.
176. G. M. Campbell, *J. Phys. Chem.* **73**, 350-355 (1969).
177. G. M. Campbell, *J. Electronanal. Chem.* **47**, 387-393 (1973).
178. G. M. Campbell, in: *Plutonium 1975 and Other Actinides* (H. Blank and R. Linder, eds.), pp. 95-104, North-Holland, Amsterdam (1976).
179. D. Inman, G. J. Hills, L. Young, and J. O'M. Bockris, *Trans. Faraday Soc.* **55**, 1904-1914 (1959).
180. D. Inman and J. O'M. Bockris, *Can. J. Chem.* **39**, 1161-1163 (1961).
181. D. Inman, G. J. Hills, L. Young, and J. O'M. Bockris, *Ann. N.Y. Acad. Sci.* **79**, 803-829 (1960).
182. D. M. Gruen and R. A. Osteryoung, *Ann. N.Y. Acad. Sci.* **79**, 897-907 (1960).
183. I. Johnson and H. M. Feder, *Trans. Met. Soc. AIME* **224**, 468-473 (1962).
184. I. Johnson and M. G. Chasanov, *J. Inorg. Nucl. Chem.* **26**, 2059-2067 (1964).
185. I. Johnson, M. G. Chasanov, and R. M. Yonco, *Trans. Met. Soc. AIME* **233**, 1408-1414 (1965).
186. M. Krumpelt, I. Johnson, and J. J. Heiberger, *Met. Trans.* **5**, 65-70 (1974).

3

Cryolite Systems

Dagfinn Bratland

1. *Introduction*

1.1. *General*

Cryolite is the mineral name of the chemical compound sodium hexafluoroaluminate, Na_3AlF_6. The molar or weight ratio between NaF and AlF_3 in a mixture of the two fluorides is often referred to as the *cryolite ratio* (CR) or *bath ratio* (BR), respectively. In pure cryolite the CR is 3, while the BR is 1.5. Mixtures of NaF and AlF_3 having a higher cryolite ratio than 3 are referred to as basic. Correspondingly, a melt with a $CR < 3$ is termed acidic.

In the fused state cryolite is a powerful solvent of many oxides, including alumina, Al_2O_3. Compounds of the type M_3AlF_6, where M is an alkali metal, may act as solvents for alumina. The solubility increases in the order M = Li, Na, K,[(1)] which is what one would expect since complex formation is generally stabilized by large cations. The solvent power has a maximum around a cryolite ratio of 3 because the availability of *ionic* AlF_6^{3-} complexes enhances the solubility. Melts with a $CR < 3$ contain the more covalent complex AlF_4^-. Therefore, the solubility of alumina in these melts will decrease with decreasing CR, although the amount of Al^{3+} increases.

The reason why Na_3AlF_6 is chosen as a solvent for alumina, rather than K_3AlF_6, is that the deposition potential of NaF is higher than that of KF. The heavier alkali metals are more easily codeposited with aluminum than the lighter ones. In addition, potassium metal intercalates with carbon, and ruins cathode linings made of carbon.

Dagfinn Bratland • Institute of Inorganic Chemistry, Norwegian Institute of Technology, University of Trondheim, N-7034 Trondheim-NTH, Norway.

The solubility of a number of oxides in molten cryolite is given in Table I. The oxides of the main group elements of the periodic system together with the rare earth oxides are generally soluble (with the exception of SnO_2) whereas the transition metal oxides are only sparingly soluble. Its solvent power for alumina has been utilized since the commercial electrowinning of aluminum was started in 1887. In this process, the so-called Hall–Héroult process, alumina is dissolved in a molten fluoride mixture consisting mainly of cryolite, and electrolyzed between a carbon anode and a cathode of liquid aluminum at a temperature of about 970°C.

TABLE I. Solubility of Various Oxides in Molten Cryolite and in Cryolite-Alumina Mixtures at 1.000°C[a]

Oxide	Solubility in cryolite (wt%)	Solubility in 95 wt% Na_3AlF_6 + 5 wt% Al_2O_3 (wt%)
SnO_2	0.01–0.1	—
Cr_2O_3	0.1	0.05
Fe_2O_3	0.1–1	10^{-3}
Co_3O_4	0.2	0.1
NiO	0.2–0.3	0.2
Ta_2O_5	0.4	—
CuO	1	1
V_2O_5	1	0.2
ZnO	1–3	10^{-3}
CdO	1–7	—
Mn_3O_4	2	1
TiO_2	5–7	4
FeO	6	—
BeO	9	6
SiO_2	9	—
MgO	11	7
Al_2O_3	13.5	—
ZrO_2	14	—
CaO	14–16	—
CeO_2	3–16	—
BaO	22	—
Na_2O	14–23	—
La_2O_3	19	19
Sm_2O_3	1–20	—
Nd_2O_3	19–21	—
K_2O	28	—
Pr_6O_{11}	31	—
WO_3	88	86
B_2O_3	Unlimited	Unlimited

[a] The data are taken from Ref. 2, p. 365.

As a mineral, cryolite is known to occur at a limited number of places in the world, the most prominent being Ivigtut on the south-west coast of Greenland. The Ivigtut deposit is the only one in the world of commercial significance. It is still (1985) being mined by *Kryolitselskabet Øresund A/S*, a company partly owned by the Danish state. The ore is concentrated in Denmark at the company's plant on Strandboulevarden in downtown Copenhagen. In 1983, 60,000 tons of ore was mined, its composition being 46% cryolite, with silica, topaz, siderite, and fluorspar as the principal associated minerals. This amounts to only 5% of the world production of cryolite, which is mainly produced synthetically by the reaction of hydrofluoric acid with sodium aluminate from the Bayer process,

$$6HF + 3NaAlO_2 = Na_3AlF_6 + 3H_2O$$

Nevertheless, the natural mineral provides the best material for laboratory research, as synthetic cryolite contains more moisture and hydroxyl than the natural variety and it does not have the exact stoichiometric composition.

1.2. Physical Properties of Cryolite

Pure cryolite is colorless. Natural cryolite is colorless to snow white, transparent to translucent. The name cryolite is derived from the Greek words for "ice stone," and this is due to its resemblance of ice when immersed in water. Cryolite crystallizes with monoclinic symmetry (α-cryolite).[3] At 565°C, a reversible transition takes place to a cubic modification (β-cryolite).[4–6] The transition is accompanied by a 7% density change, evidenced by a characteristic decrepitation. At 880°C, a sharp increase in the electrical conductivity takes place, which indicates a second transformation of the crystal, to γ-cryolite.[5] The solid attains a plastic character, and before it melts its conductivity reaches 1 S cm^{-1}. These facts strongly suggest extensive lattice defect formation; the positions of some of the ions are highly randomized.

The melting point of pure cryolite is 1010°C.[6] Melting is accompanied by a volume expansion of 25%.[7] (The volume expansion is 42% compared to α-cryolite at room temperature.) The phase diagrams of the systems NaF-AlF_3[8] and Na_3AlF_6–Al_2O_3[9] are shown in Figs. 1 and 2.

Some important physical properties of cryolite are listed in Table II.

The vapor phase in equilibrium with molten cryolite consists mainly of $NaAlF_4(g)$ (90%), the remainder being $Na_2AlF_5(g)$ and $NaF(g)$.[18] Since the cryolite ratio of the vapour is lower than it is in the melt, the cryolite ratio of the melt will increase when the melt is left to evaporate. (The melt becomes more basic.) The vapor pressure of the melt at the melting point is between 400 and 500 Pa[11] (between 3 and 3.75 torr). It is essential that

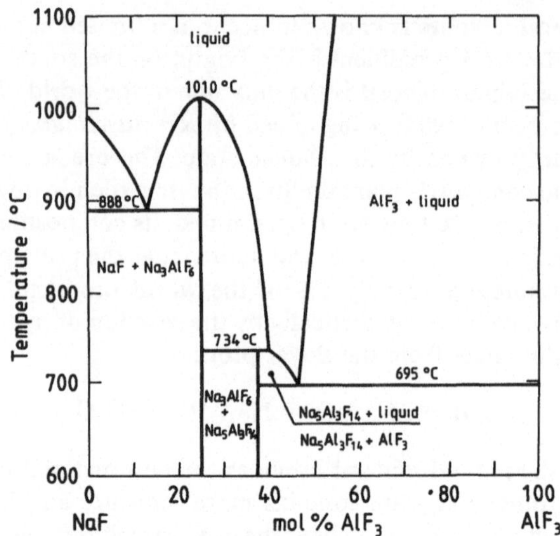

FIGURE 1. Phase diagram of the system $NaF–AlF_3$. From Grjotheim *et al.*[2] Used with permission from Aluminium-Verlag.

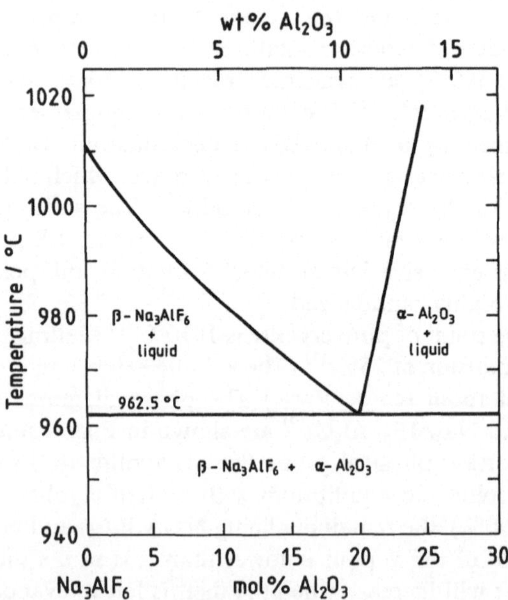

FIGURE 2. Phase diagram of the quasibinary system $Na_3AlF_6–Al_2O_3$. From Grjotheim *et al.*[3] Used with permission from Aluminium-Verlag.

TABLE II. Physical Properties of Cryolite[a]

Melting point (°C)	1010	Ref. 6
Transition temperatures (°C)		
Monoclinic to cubic	565	Refs. 5,
Second order	880	Refs. 5,
Heat of fusion (kJ mol^{-1})	114	Ref. 6
Heat capacity (J mol^{-1} K)	395	Ref. 10
Vapor pressure (Pa)	400–500	Ref. 11
Density (g cm^{-3})		
Monoclinic crystal, 25°C	2.97	Ref. 12
Cubic crystal, 560°C	2.77	Ref. 4
Solid at melting point	2.62	Ref. 5
Liquid at melting point	2.09	Ref. 13
Electrical conductivity (S cm^{-1})	2.83	Ref. 14
Hardness (Mohs)	2.5	Ref. 12
Surface tension (J m^{-2})	133	Ref. 15
Viscosity (mPa × s)	2.29	Ref. 16
Solubility in water (g liter^{-1}) at room temperature	0.4	Ref. 17

[a] All data are given for liquid cryolite at the melting point unless otherwise specified.

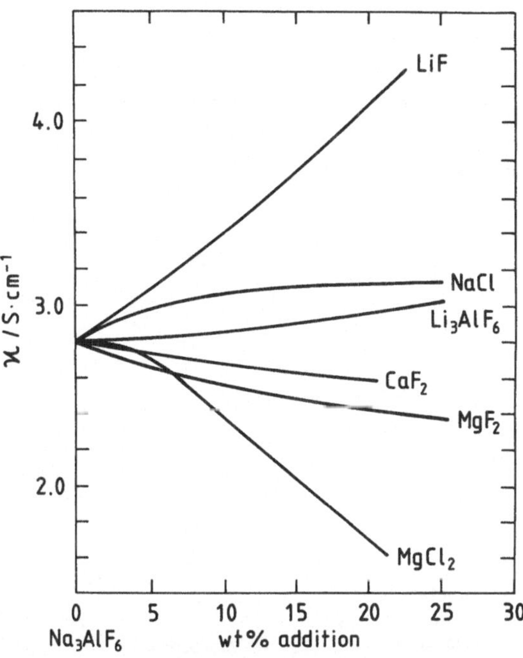

FIGURE 3. The specific conductivity of Na_3AlF_6–MA_x mixtures at 1000°C. From Grjotheim *et al.*[(2)] Used with permission from Aluminium-Verlag.

the duration of an experiment involving molten cryolite is limited as much as possible to minimize composition changes.

The electrical conductivity of molten cryolite is good. The current is mainly transported by the Na^+ ions.[2] The effect of additives of various halides on the conductivity is shown in Fig. 3. Only lithium fluoride, Li_3AlF_6, and sodium chloride have a favorable influence. The effect of additives on the density of molten cryolite is shown in Fig. 4. Generally, the introduction of heavier ions like Ba^{2+} and Ca^{2+} increases the density. The freezing point depression of cryolite for selected additives is shown in Fig. 5. There are two groups: additives that introduce new ionic species, like Ca^{2+}, Li^+, Mg^{2+} and Cl^-, and additives that do not introduce any new species (NaF and AlF_3).

1.3. Uses of Cryolite

Cryolite is used industrially for various purposes. Its electrolytic reduction to aluminum metal has already been outlined.[19]

In the electrolytic refining of aluminum by the Hoopes process, a melt consisting of NaF (25%-30%), AlF_3 (30%-35%), and BaF_2 (33%-38%), is used as the electrolyte between a pure molten aluminum cathode, which floats on top of the high-density bath and an anode of copper-aluminum alloy at the bottom.[19]

In secondary aluminum production (from scrap) cryolite is an important constituent of the flux. It serves to remove nonmetallic impurities like alumina and other oxides, and to promote coalescence of the fused metal. It reacts with metallic magnesium and removes it from the scrap metal, by

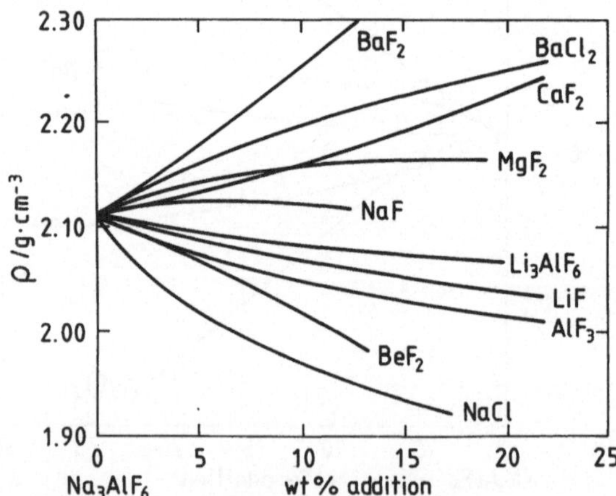

FIGURE 4. Isothermal densities of Na_3AlF_6–MA_x systems at 1000°C. From Grjotheim *et al.*[2] Used with permission from Aluminium-Verlag.

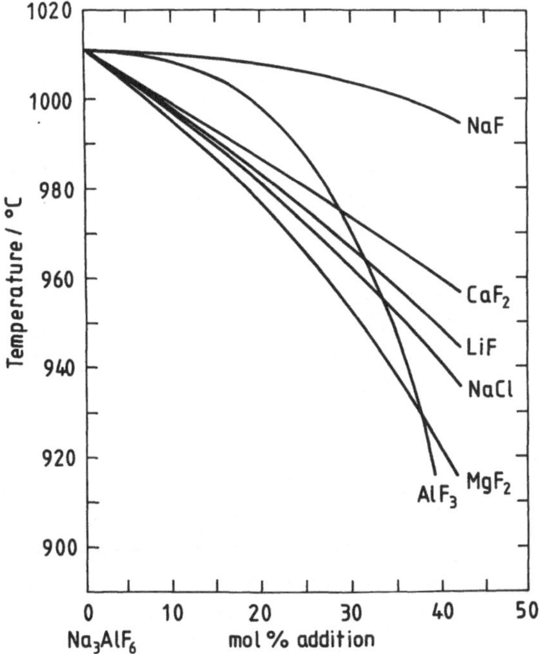

FIGURE 5. The freezing point depression of cryolite for selected additives. From Grjotheim and Welch.[9] Used with permission from Aluminium-Verlag.

an oxidation–reduction reaction,

$$3Mg + 2Na_3AlF_6 = 2Al + 3MgF_2 + 6NaF$$

Cryolite has wide application as flux for soldering aluminum and for welding aluminum and aluminum alloys. It is used as a flux in aluminizing steel, and in the compounding of welding-rod coatings.

Its solvent power for metal oxides is also utilized in the production of other metals, since it removes undesired oxides. It is also used in glass production.

The disadvantage of the great solvent power of molten cryolite is that very few potential container materials are able to resist attack. Refractory metals will oxidize unless the atmosphere is completely free from oxygen because the melt dissolves the protective oxide layer. Only the platinum metals are sufficiently inert to be of any practical use. Usually unalloyed platinum is used to contain molten cryolite in the laboratory. But in some cases alloys are preferred. Graphite may be used as a container material although it should be kept in an atmosphere free of oxygen and water vapor. More details concerning the choice of materials are found in the following section.

2. Container Materials

2.1. Graphite

Ordinary graphite contains clay as binder. This clay consists of metal oxides and is soluble in cryolite, so that care should be taken to find a graphite brand with minimum amounts of clay and other sources of metal oxides.

The porosity of the graphite should be as small as possible, since the molten salt will tend to seep through a porous crucible. In our laboratory, we have made extensive use of the graphite brands produced by Union Carbide Corporation (UCAR). However, not all of their grades are well suited for containing molten cryolite. The grade AGR is too porous, so that the melt seeps through the wall. The preferable UCAR grade has been known by various names: AUC, ECV, CS 49. The grade CS 49 is an especially pure and dense graphite, with a total ash level of 50–200 ppm. Although the internal structure is improved over most other grades due to the fine grain size, one should nevertheless always look out for possible cracks in the material. This graphite comes in rods from $\frac{5}{8}$ to 5 in. diameter. The crucibles are easily machined from rods in any size and shape, within the limits mentioned.

Usually some carbon dust will float on the melt surface in a crucible made from this type of graphite. These are carbon particles shed from the inside of the cell. After an experiment the salt surface has black spots. This does not necessarily contaminate the bulk, but may be harmful for experiments where the surface properties are essential, and for electrode studies. It is common practice to "flush" the new crucible once with molten cryolite. The second time that cryolite is melted in the crucible much less carbon dust will be liberated.

The Carborundum Co. also manufactures a graphite quality which has proved useful as a container material. It has been used for various other purposes, such as electrodes. As the name of the grade implies: Graph-I-tite "G," it is more impervious that most other commercial graphites, although its density is the same as that of the CS 49 graphite, 1.89 g cm^{-3}. Its permeability is ascribed to the manufacturing process, during which the pores are filled and closed with graphite in some undisclosed way. The grain structure is finer than that of CS 49 graphite (maximum grain size 0.2–0.8 mm) and the chances of finding cracks are slim. Its resistance to erosion is superior. The price, accordingly, is about 5 times higher than that of CS 49 graphite, but not as high as pyrolytic graphite. It comes in rods with diameters from $\frac{3}{8}$ to 3 in. Its purity is comparable to the CS 49 graphite, the typical ash content being 400 ppm.

The use of Graph-I-tite "G" is justified by the fact that it is usually possible to carry out a larger number of experiments with each crucible. The Graph-I-tite "G" is poorly wetted by the melt, so it is easier to remove the salt after solidification.

The use of Graph-I-tite "G" as an anode material is recommended since it is consumed more evenly, and it disintegrates much less than does CS 49, so there is less dust formation. Union Carbide also markets pyrolytic graphite, which is manufactured by deposition of carbon on a hot graphite substrate in a vacuum furnace operated from 1900 to 2500°C. The carbon atoms are deposited layer by layer. These hexagonal layers are highly ordered, and are oriented parallel to the surface of the substrate. It is monolithic, free of voids, and gives a material of extremely high density (2.20 g cm^{-3}), which compares to the theoretical value of 2.25 g cm^{-3}. Pyrolytic graphite is available in small plates $\frac{1}{8}$–$\frac{3}{8}$ in. thick and in rods up to $\frac{1}{2}$ in. diam × 6 in. The ash content ranges between 10 and 30 ppm. The small sizes available and the prohibitive price (three orders of magnitude higher than that of CS 49 graphite) makes the material less attractive for all but the most specialized purposes. It has been used in some electrode kinetics studies in molten cryolite.[20,21]

Vitreous carbon (Le Carbone-Lorraine) has also been used as a container material for molten cryolite. The material is produced by carbonization and subsequent thermal treatment of carbonaceous material having strong transverse molecular bonds. As the name implies, this carbon has a glassy appearance, and it does not liberate carbon dust as the grade CS 49 does. It is less dense than the graphites already mentioned, but it has a closed microporosity, and is impervious to gas. The surface is not wetted by molten cryolite. This makes it possible to use the crucible several times if the solidified melt is removed and crushed before remelting. Unless this precaution is taken the crucible may crack. Ready-made crucibles are commercially available. Machining this material is difficult. The price of a crucible is generally 200 times higher than the materials cost of a CS 49 crucible (excluding labor). There are two grades, V10, for temperatures up to 1000°C, and V25, for higher temperatures (up to 2500°C).

Some relevant data are given in Table III.

2.2. Boron Nitride

Boron nitride is an attractive material for containing cryolite melt. Pure boron nitride is a bulky white powder with a true density of 2.25 g cm^{-3}. It does not melt under atmospheric pressure, but sublimes at about 3000°C.

A self-bonded boron nitride body is made by hot pressing the powder in graphite dies at relatively high temperatures under moderate pressure.[26] It is mechanically strong, and its appearance is like ivory. It resembles

TABLE III. Density, Purity, and Price of Some Varieties of Graphite

	Density ($g\ cm^{-3}$)	Ash content (ppm)	Approximate price in Norway 1984 (US$ per cm^3)
Graphite, theoretical	2.25		
Pyrolytic graphite[a]	2.20	10–30	10
Graph-I-tite "G"[b]	1.89	600	0.20
CS 49 = ECV = AUC[c]	1.89	800	0.02
AGR[c]	1.63		
Vitreous carbon V 10[d]	1.50–1.55	400	5
Vitreous carbon V 25[d]	1.50–1.55	400	2

[a] Reference 22.
[b] Reference 23.
[c] Reference 24.
[d] Reference 25.

graphite in many respects: its crystal structure is similar to that of graphite. In machining and lubricating properties, it resembles graphite. It can be readily machined, drilled, threaded, or cut with a band saw. The material is ready for use after machining without any further treatment. Also, like graphite, it shows directionalism in its properties. This indicates partial orientation of the platelike crystals with respect to the direction of the forming pressure applied. One such property is thermal expansion, which is comparable in magnitude to that of graphite.

Although boron nitride resembles graphite in many respects, the two materials differ markedly from each other in electrical characteristics. The volume resistivity of dry hot pressed boron nitride is as high as $10^{15}\ \Omega$ cm at 25°C. It drops rapidly when the temperature increases, but it is still $1 \times 10^9\ \Omega$ cm at 1000°C. Some physical properties of boron nitride are given in Table IV.

Boron nitride is available in various qualities from several manufacturers. Usually the material contains mostly boric oxide as a binder, which is soluble in molten cryolite. This grade of boron nitride accordingly has a limited lifetime in molten cryolite, and the presence of dissolved boric oxide in the melt may interfere with the investigations. The presence of boric oxide also causes the material to absorb water when stored in air, which is detrimental.

In other grades the boric oxide is bonded with calcium oxide, which allows the amount of oxide in the binder to be reduced. This is an advantage for work in cryolite melts. This type of boron nitride absorbs much less water than the previous one. Various grades of boron nitride are compared in Table IV. The manufacturers quoted are those known to the author.

Thonstad[(31)] studied the behavior of boron nitride in molten cryolite and aluminum. He found that the rate of corrosion in cryolite-alumina

TABLE IV. Properties of Various Grades of Boron Nitride

	Carborundum		Union Carbide		
	A[a]	HP[a]	HBR[b]	HBN[c]	HBC[d]
Boric oxide (wt%)	5.0–8.0	<0.10			
Total oxygen (wt%)	4.0–6.0	1.5–2.5	<3.9	<3.9	<1.0
Calcium (wt%)	<0.20	1.50	>0.8		ND[e]
Density (g cm^{-3})	2.08	1.90	1.90	2.00	>1.90
Electrical resistivity 1000°C (Ω cm)					1×10^9
Hardness	385 (K 100)	205 (K 100)			
Water absorption (%) 70°C	1.1	0.16		0.75	0.15
Thermal expansion $°C^{-1} \times 10^{-6}$	2.8[f]	0.0[f]	1.0[f]	1.2[f]	0.35[f]
	0.9[g]	0.0[g]	2.0[g]	3.2[g]	0.6[g]
Oxidation rate in air (wt loss at 1000°C, mg cm^{-2} hr^{-1})	0.046	0.008			
Maximum operating temp. (°C)					
Vacuum or inert atm.	2775	2775	1800	1800	2000
Oxidizing atm.	985	1200	1000	1000	850

[a] Reference 27.
[b] Reference 28.
[c] Reference 29.
[d] Reference 30.
[e] Not determined.
[f] Parallel to pressing direction.
[g] Perpendicular to pressing direction.

melts at 1010°C was 0.23 mg cm^{-2} h^{-1} for pyrolytic boron nitride and 0.31 mg cm^{-2} h^{-1} for hot-pressed boron nitride. The electrical resistivity of pyrolytic boron nitride was found to change very little after immersion in molten cryolite, whereas the resistivity of the hot-pressed material was drastically decreased after immersion. The latter material is more porous and the melt penetrates the pores.

Pyrolytic boron nitride is an extremely pure product. It is made by Union Carbide in a high-temperature, low-pressure process whereby boron nitride is formed by a reaction in the vapor phase. Objects of almost any shape may be formed, but the resulting material is harder than hot-pressed boron nitride, and it cannot be machined. Since it contains no binder the purity is very high (less than 100 ppm impurities). Pyrolytic boron nitride is extremely expensive. For example, one tube measuring 5 mm diam × 50 mm long costs about US $600 (1983 price).

2.3. Metals

The only metals that are reasonbly resistant to molten cryolite are found among the platinum metals and the ferrous metals. The platinum metals possess very good properties in this respect. Iron and nickel and nickel-based alloys are resistant provided the surrounding atmosphere does not contain any oxidizing agent.

Metal crucibles for containing cryolite melt are usually made of platinum. Platinum is very well wetted by this kind of melt.[32] Care should therefore be taken to avoid overfilling the crucible as this may cause the melt to "creep" out. This is particularly critical when the melt contains alumina. Another consequence of this wetting phenomenon is that the crucible is somewhat contracted, both axially and radially, each time a melt solidifies in it. Gradually, both the diameter of the lower part and the height of the crucible are reduced. Contact with molten cryolite also causes the platinum metal to crystallize. This is easily seen on the surface of the metal. Crystallization weakens the metal, leaving it vulnerable to corrosion at the grain boundaries where the impurities are accumulated. One way to counteract this crystallization is to grind the suface of the crucible regularly with fine sand. A polished crucible is more resistant to attack than a roughened one. Ultimately, however, any crucible will have to be scrapped. Platinum metal may be expensive, close to US $ 20 per gram (1987 price). But since the metal is recycled its value is not lost. The refining of used metal amounts to some 5%-10% of the metal value. (The forming costs will, of course, be added to this.) Therefore, a platinum crucible is a capital investment, and most of its value may be reclaimed when it is not needed any more.

A typical platinum crucible for cryolite melts, measuring 45-50 mm diam × 80 mm high × 0.5 mm thick, amounts to $2000 (1987 price). If it is made slightly conical, say top diameter 50 mm, bottom diameter 45 mm, the solidfied salt is removed more easily after the experiment.

A special tool for renovating the crucible is very useful. The tool is made of Teflon and consists of two parts: a hollow block and a piston. The new crucible fits exactly into the cavity of the block, and the piston fits exactly into the crucible. Any deformation of the crucible will be corrected by pressing it into the tool. If this is done regularly the lifetime of the crucible will be greatly increased.

When the need arises to employ metal objects like rods, wire, plate, etc. in cryolite milieu it may often be advantageous to use alloyed platinum instead of pure platinum. This increases the rigidity and hardness of the metal. Platinum-rhodium alloys are used for thermocouples. The plummet used in surface tension measurements should be made of platinum-iridium. Experience has shown that an alloy consisting of 5% gold and 95% platinum is not wetted by molten cryolite.

One should be aware that the platinum metals tend to crystallize, not only when immersed into the cryolite melt, but even when positioned above the melt, as the very exposure to the vapors is enough to cause deterioration of the metal structure.

Nickel crucibles may be used as containers for molten cryolite for a limited time, provided that oxygen and moisture are excluded. Alloyed nickel may also be used. Electrolytic iron is resistant unless there is any oxidizing agent present.

3. *Preparative Techniques*

3.1. *Commercial Cryolite*

The highest purity natural cryolite is available from Kryolitselskabet Øresund A/S, Copenhagen, Denmark. This material is of the provenance of Ivigtut, Greenland. The quality "hand-picked," which was formerly available, is now out of stock. The current premium quality is "stykkryolit" ("cryolite fragments"). The so-called "stykkryolit II" is produced by optical screening under water of choice fragments of mineral cryolite. The dealer states that the composition is very nearly 100% pure mineral cryolite. According to measurements in our laboratory the melting point of this variety is 1010°C.

"Stykkryolit II" is generally perfectly white, but some fragments have spots or obscurities which are almost imperceptible. In "Stykkryolit I" these fragments have been sorted out.

Cryolite fragments are very stable chemically. To release any adsorbed moisture the cryolite should be heated to 130°C before use. At higher temperatures the reaction

$$2Na_3AlF_6 + 3H_2O = 6NaF + Al_2O_3 + 6HF(g)$$

becomes important. Although the Gibbs energy is positive the reaction will proceed to the right since it is an open system, and HF(g) formed will be continually removed. Cryolite should therefore never be heated in air above 300°C.

3.2. *Preparation of Aluminum Fluoride*

In studies of systems involving cryolite one frequently wishes to vary the cryolite ratio. The difficult part is in providing sufficiently pure aluminum fluoride, since this is not commercially available.

In our laboratory the aluminum fluoride is purified by sublimation of smelting grade aluminum fluoride. This material typically contains 88 w % AlF_3, the remainder being alumina and a small amount of sulfate. One hundred grams of the material is placed in a sinter corundum crucible in a vertical tubular furnace. The crucible, 37 mm i.d. and 220 mm high, is covered by a platinum lid, and located in the furnace in a position where the temperature gradient along its axis is a maximum. The furnace is evacuated and heated until the temperature at the crucible bottom reaches 1000°C. The temperature of the lid is then 750°C. The furnace is kept at this temperature for 48 h. The aluminum fluoride condenses as long needles on the platinum lid, and is easily removed after the furnace has cooled. The yield is about 60%. One sublimation is sufficient to give a material of adequate purity.

Simplified representation of the sublimation furnace is shown in Fig. 6.

3.3. *Purification of Alkali Metal Fluorides*

For very accurate work it may also be necessary to purify the alkali metal fluorides. This is done by recrystallization in a platinum crucible. The

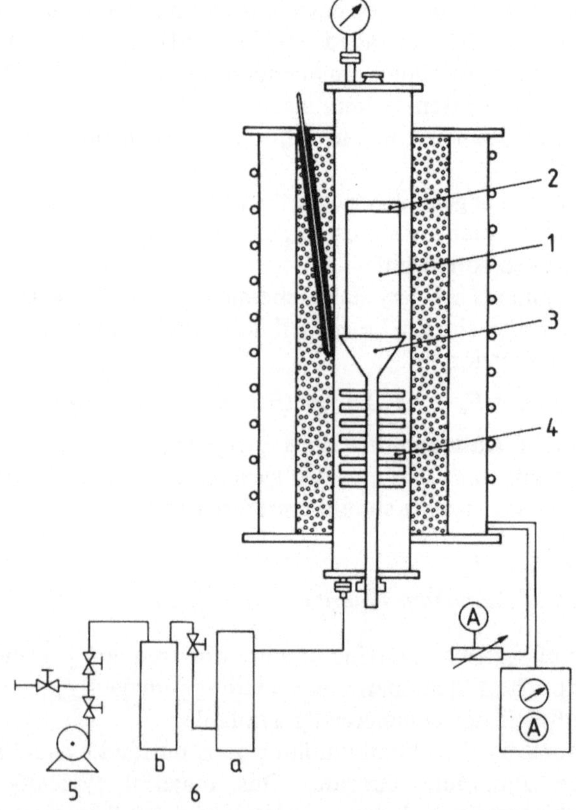

FIGURE 6. Vertical Kanthal-wound furnace for the sublimation of aluminum fluoride. (1) Graphite cell containing crude AlF_3. (2) Platinum lid. (3) Nickel support and support rod. (4) Nickel radiation shields. (5) Rotary pump. (6) Gas purification train: Most of the hydrogen fluoride is absorbed in glass wool (a), the rest condenses in the cool-trap (b), filled with liquid nitrogen.

salt is melted, and cooled very slowly. The purest salt crystallizes on the crucible wall, and the impurities collect in the core of the solid salt. The temperature is controlled by a temperature programmer, and the cooling rate during crystallization is 3-4°C per hour.

The sensitivity of the alkali metal fluorides to moisture increases with the molecular weight. Whereas LiF may be stored in air KF has to be treated with some care, and the heavier fluorides should be kept in a glove-box with dry atmosphere to maintain their purity over a prolonged period of time.

White has also mentioned purification techniques in Chapter 2 of Volume 1 of this series and elsewhere.

4. *The Study of Thermodynamic and Physicochemical Properties*

4.1. *Vapor Pressure*

The vapor pressure over molten cryolite–aluminum oxide mixtures has been measured by the transpiration method and the "boiling-point" method.[18]

4.1.1. *The Transpiration Method*

The transpiration method involves the passing of an inert carrier gas over the condensed sample at constant temperature and total pressure. The flow rate of the gas is constant and small enough to saturate the gas with the vapor, and the vapor is subsequently completely condensed in a cooler area downstream. The vapor pressure is calculated either from the amount of material lost from the sample or from the amount of material condensed, provided the molar mass of the vapor is known.

A simplified drawing of the apparatus is shown in Fig. 7.

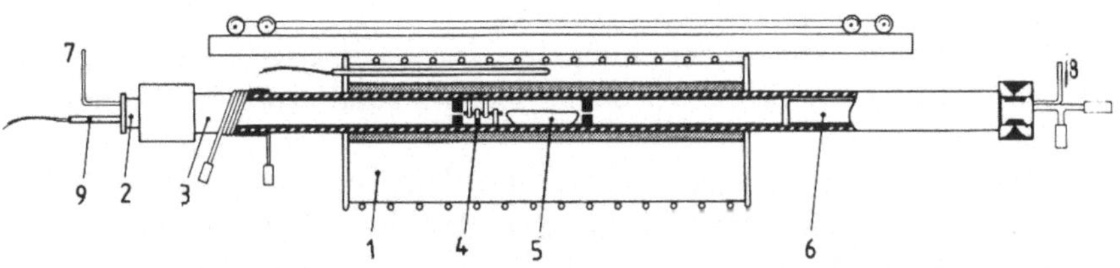

FIGURE 7. Furnace used for measurements of vapor pressure by the transpiration method. (1) Furnace. (2) Steel tube. (3) Inconel tube. (4) Graphite shields. (5) Graphite boat with sample. (6) Cold-finger. (7) Gas entrance. (8) Gas exit. (9) Thermocouple (only partly shown).

4.1.2. The Boiling-Point Method

The boiling point method involves the detection of the point of balance between the inert-gas external pressure and the vapor pressure inside a container. This has been achieved in various ways. One successful approach involves thermogravimetry, with the sample contained in a crucible with a small opening. The crucible, located in the isothermal zone of a tubular furnace, is suspended from a balance, and the apparatus contains an inert gas atmosphere of a pressure which initially is higher than the equilibrium vapor pressure of the system. The inert gas pressure is then lowered slowly. This causes the vapor transport through the opening in the crucible to increase. When the inert-gas pressure becomes smaller than the equilibrium vapor pressure the transport rate increases very rapidly. The rate of mass loss from the sample is recorded as a function of the inert-gas pressure, and this facilitates the determination of the equilibrium vapor pressure. A simplified drawing of the apparatus is shown in Fig. 8.

An alternative procedure of the boiling-point method is to heat the furnace slowly while the inert-gas pressure is kept constant. When the boiling point is reached, the temperature measured by the thermocouple in the melt remains constant even when the furnace temperature continues to increase. This variant of the method is not as practical in studies of cryolite systems, however, because superheating of the vapor and the accurate determination of the boiling temperature present problems.

4.1.3. Other Methods

Other methods for vapor pressure determination like the Knudsen effusion method and high-temperature mass spectrometry, are not applicable with liquid mixtures of cryolite. These methods are only suitable for measuring vapor pressures below 130 Pa. The mixtures in question generally have a vapor pressure up to one order of magnitude higher than that.

4.2. Electrical Conductivity

Molten fluorides have a very high electrical conductivity, besides being very corrosive. Both properties make experimental work in the field a demanding task.

A thorough survey of electrical conductivity measurements in molten fluorides 1927–1967 has been published by Robbins.[33] The main problems fall into three categories:

1. Choice of material for cell and electrodes and their design.

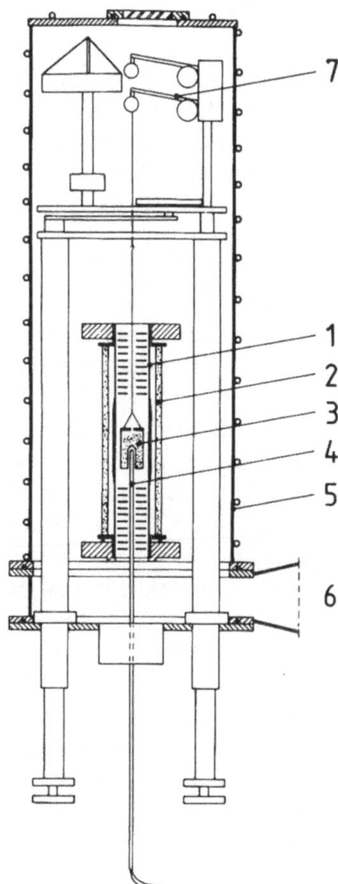

FIGURE 8. Vacuum furnace used for measurements of vapor pressures by the boiling point method. (1) Graphite heater. (2) Graphite radiation shields (with graphite felt). (3) Suspended graphite cell with sample. (4) Thermocouple. (5) Brass bell jar. (6) To diffusion pump. (7) Balance with weight-changing mechanism.

2. Measuring bridges.
3. The frequency dispersion of the measured resistance.

4.2.1. The Cell and Electrodes

Two types of cell have been used with molten cryolite. In the conventional cell, one electrode is a cup and the other electrode dips in the cup. For work in cryolite the electrodes are made of platinum. The cup may have the shape of a hemisphere, or have a hemispherical bottom with cylindrical sides. The other electrode is either simply a rod or it may be a rod whose end has been given the same shape as the cup, as indicated in Fig. 9.

There are at least two ways to use this cell. In the partially filled cell approach, the cup is used as the container of the melt being studied. The cell constant is determined with a definite volume of an electrolyte with known resistivity. Exactly the same volume of the electrolyte under investigation must then be used for the conductivity measurement. This requires accurate knowledge of the density of the molten salt as a function of the temperature. The small volume makes the melt sensitive to accidental contamination.

If the measuring cell is completely immersed in the electrolyte there is no need to know its density. The cell is then kept at a constant depth below the melt surface, and the melt volume is considerably larger. The cup may be provided with a small hole to facilitate its filling when the cell is being immersed in the melt.

The cell constant of the conventional cell is typically of the order of a few tenths cm^{-1}. It is rather easy to construct, but requires quite accurate measuring bridges and accurate determination of lead resistances. The measured resistance is less than 1 Ω. Therefore the accuracy of the results is very sensitive to errors due to temperature gradients and fluctuations, and to polarization.

An alternative to the conventional cell is the so-called capillary cell. A capillary cell has a cell constant of several hundred cm^{-1}. Even with the highly conducting fluoride melts this gives reasonably high resistances to measure, and thus better accuracy. The capillaries have to be made of electrically insulating materials, and that presents a major problem when molten cryolite is involved.

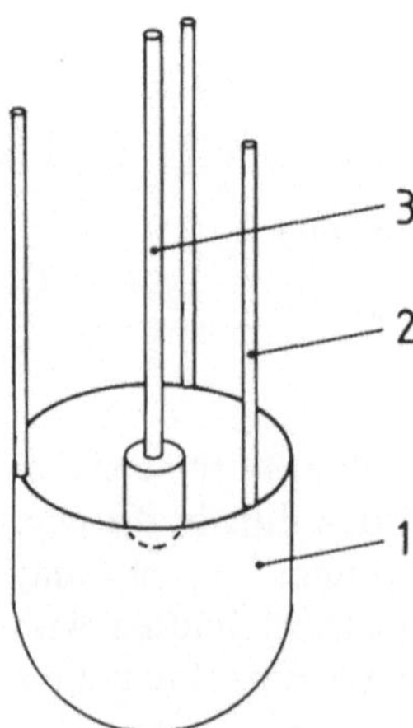

FIGURE 9. Conductivity cell, conventional type. (1) Platinum cup (electrode). (2) Suspending rods (platinum). (3) Platinum electrode.

If the materials problem of the capillary cell could be overcome successfully, therefore, it would be the perfect cell for the purpose.

Yim and Feinleib[34] used a boron nitride cell (Fig. 10). Boron nitride has a very high electrical resistivity (Table IV), and is easy to machine. But hot-pressed boron nitride has a finite porosity, and the pores will absorb molten cryolite. Gradually, then, the cell body becomes a conductor. In Yim and Feinleib's setup a graphite jacket around the cell retarded the rate of melt penetration. (At the present time this problem should be overcome by using pyrolytic boron nitride.) The inside diameter of the capillary was 5 mm, and the capillary was in a vertical position. Inconel was used for the electrode material. The cell was supported by an Inconel carriage, the bottom of which acted as one electrode. The top electrode could be moved vertically. The capillary was filled by immersing the cell block into a crucible containing the melt to be studied. According to Yim and Feinleib boron nitride reacts with platinum at about 1000°C, and the metal becomes brittle. Platinum was also found to stick to the boron nitride after prolonged heating. Platinum consequently has to be discarded as an electrode material for a capillary conductance cell in molten cryolite.

The results obtained by Yim and Feinleib agree well with those which Edwards *et al.*[35] reported. The latter authors used a conventional type cell.

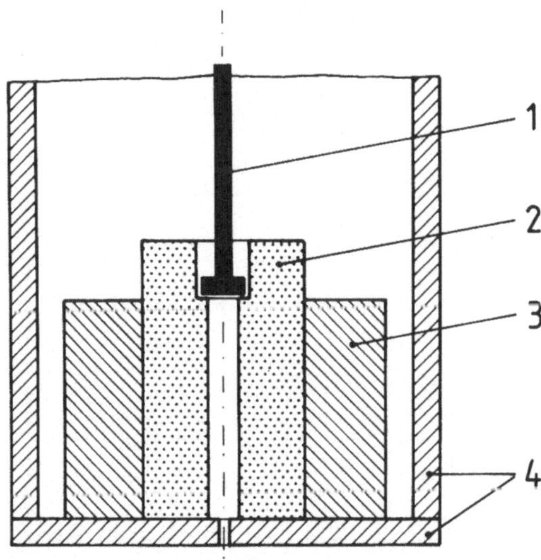

FIGURE 10. Conductivity cell, capillary type. (1) Inconel electrode. (2) Boron nitride capillary. (3) Graphite jacket. (4) Inconel carriage (electrode).

4.2.2. Measuring Bridges

The resistance of the conductivity cells has usually been measured by some form of the classical Wheatstone bridge circuit.[36] Examples are the Jones bridge, the Kelvin bridge, the Thomson bridge, and the Carey–Foster bridge. According to Grjotheim and Matiasovsky[37] a Thomson type bridge is preferred (Fig. 11).

It has been common practice to equate the reading on the variable bridge resistance to the true resistance of the melt. This is only true in dilute aqueous solutions where the resistance is large. In molten salts one has to take the melt/electrode interfacial capacitances into account as well. The problem of determining the true resistance of the melt has been discussed by Robbins[33] and Matiasovsky *et al.*[38,39]

Because of the complex nature of measurements of this kind it appears that some researchers have measured the impedance of the total measuring circuit.[33] Another pitfall is to neglect the resistance of the cell leads when calculating the ohmic resistance of the melt.

4.2.3. Frequency Dependence

Usually the measured cell resistance varies with the applied frequency, f. Common practice is to plot measured resistances versus $f^{-1/2}$, and extrapolate to infinite frequency. This problem has been discussed at length by Winterhager and Werner,[40,41] by Matiasovsky *et al.*,[38,39] and by Robbins.[33]

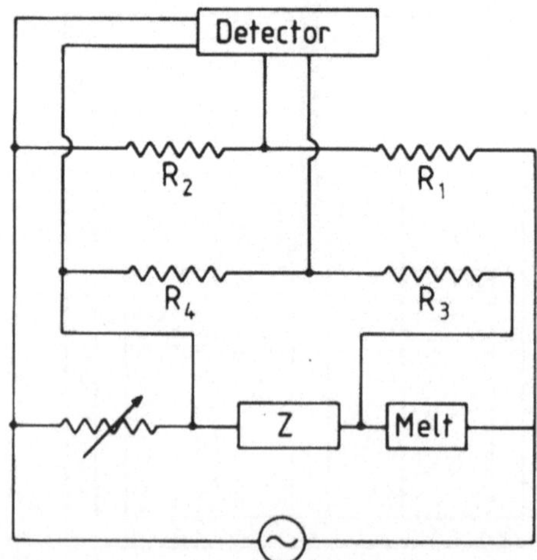

FIGURE 11. Thomson-type bridge for measurements of electrical conductivity in molten salts. R_1–R_4: ohmic resistances. Z: impedance of connections.

4.3. Density

The common method for the determination of the density of fluoride melts is hydrostatic weighing.[13] The sinker, a sphere or a biconical cylinder made of platinum or a platinum alloy, is suspended from a balance by a wire of the same material as the sinker. The volume of the sinker is usually determined by calibration in molten sodium chloride or potassium chloride. These measurements are most accurately and conveniently undertaken by means of a sensitive electronic balance.

4.4. Surface Tension

A survey of various methods applicable for surface tension measurements of molten salts is given by Janz *et al.*[42] Of the methods described, the maximum bubble pressure method, the ring method, and the pin method have found application in studies of molten cryolite.

4.4.1. The Maximum Bubble Pressure Method

The maximum bubble pressure method involves a capillary immersed in the melt. A bubble is formed very slowly at the capillary tip. The maximum pressure in the bubble is determined the moment it bursts.

In their work, Bloom and Burrows[15] used a capillary of platinum/10% rhodium, threaded into a stainless steel tube. The assembly could be moved vertically by means of a vernier rack and pinion. Owing to severe corrosion, the lower portion of the steel tube had to be replaced after every 15 h use. Argon-filled bubbles were formed in the melt, and a manometer with dibutyl phthalate was used to measure the maximum bubble pressure.

4.4.2. Detachment Methods

The ring method and the pin method are rather similar to each other experimentally. They may both be classified as detachment methods. An object in the shape of a ring or a pin is dipped in the melt. When the object then is lifted out of the melt, the apparent weight of the object decreases abruptly the moment it is detached from the melt surface. This weight change is the detachment force, equal to the weight of the liquid column lifted by the object just before detachment. The surface tension may be calculated when the detachment force, the perimeter of the object, and the specific gravity of the liquid is known.

Vajna[43] determined the surface tension in molten cryolite mixtures by means of the ring method. He worked using an open furnace, with a platinum ring and a torsion balance for measuring the detachment force.

Bratland *et al.*[44] used the pin method, with a platinum/10% rhodium sinker, like the one described under specific gravity measurements. The whole apparatus, including the Mettler electronic balance, was kept in a closed space filled with highly purified nitrogen. The results agreed well with those obtained with the maximum bubble pressure method by Bloom and Burrows.[15]

4.5. Viscosity

The viscosity of $NaF-AlF_3$ mixtures has been measured with an oscillation viscometer by Brockner *et al.*[16] The essential part of the viscometer is the torsion pendulum, comprising a platinum/10% iridium cylinder, accurately machined. The cylinder is immersed in the molten cryolite; the damping and the period of the oscillating pendulum is measured. Absolute viscosities are calculated from damping constants determined by a digital method based on time measurements. Calibration is not necessary.

5. Electrodes in Molten Cryolite

5.1. General

Electrodes are necessary in studies of galvanic cells and as reference electrodes in kinetic studies.

Chemical equilibria are studied by galvanic cells. Thermodynamic data, for example activity data, may be derived from such measurements. So far it has proved difficult to provide electrodes that give sufficiently reproducible results.

In kinetic studies where current is passed through an electrolytic cell it is common to include a third electrode that does not carry current. This electrode is referred to as a reference electrode. The potential difference is measured between the reference electrode and the working electrode.

A survey of electrodes used in cryolite melts has been given by Grjotheim *et al.*,[45] by Minh and Redey in Chapter 4 of this volume, and by Lantelme, Inman, and Lovering in Chapter 5 of Volume 2.

5.2. Gas Electrodes

A simple oxygen electrode applicable for investigations in molten cryolite, may consist of a platinum wire inside a sintered corundum tube dipping into the melt (Fig. 12). The wire is flushed with oxygen passing through the alumina tube. This electrode has been used to study the cell

$$Al/Na_3AlF_6 + Al_2O_3/Pt, O_2$$

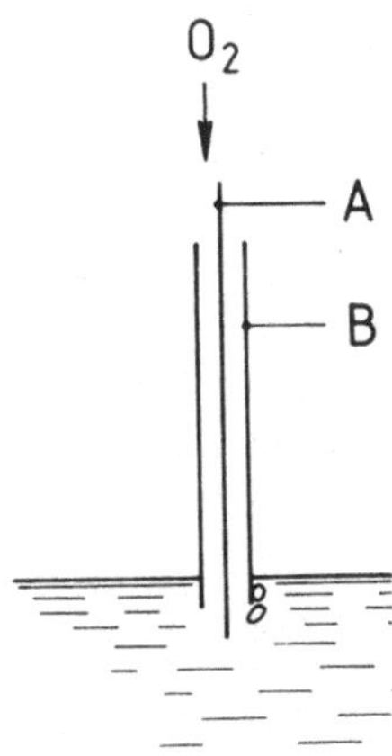

FIGURE 12. Pt, O_2 electrode for use in cryolite-alumina melts. From Grjotheim *et al.*[2] Used with permission from Aluminum-Verlag. (A) Platinum wire. (B) Alumina tube.

When the oxygen electrode compartment is separated from that of the aluminum electrode by means of an alumina diaphragm to prevent transport of dissolved metal from the aluminum electrode to the oxygen electrode, it appears that the electrode behaves reversibly.

A simple C, CO_2 electrode consists of a graphite sleeve fixed to one end of a copper tube (Fig. 13). Carbon dioxide is passed through the tube to provide continuous flushing of the electrode. The graphite electrode is not attacked appreciably by the gas, but some enlargement is observed in the hole through which the gas passes. Graphite is less reactive than other types of carbon in this respect. The reaction between carbon and carbon dioxide contaminates the gas with carbon monoxide.

With a C, CO electrode there should be only limited attack on the carbon by the passage of carbon monoxide, and it has been used for determining the alumina activity.[46]

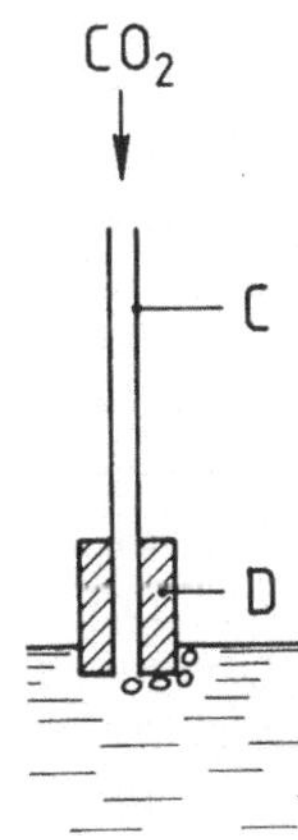

FIGURE 13. C, CO_2 electrode for use in cryolite-alumina melts. From Grjotheim *et al.*[2] Used with permission from Aluminium-Verlag. (C) Copper tube. (D) Graphite electrode.

The Pt, CO_2 electrode appears to be considerably less stable than the C, CO_2 electrode.[45]

A carbon electrode dipping into a cryolite melt generally must be considered as a kind of C(CO, CO_2) electrode when CO and CO_2 are present above the melt and there is no dissolved metal near the electrode, but the potential will not be well defined. Graphite electrodes without any flushing of gas have been used.

The C, CO_2 electrode seems to be the one best suited as a reference electrode.

5.3. *Oxide Electrodes*

The metal oxides, Fe_3O_4, Fe_2O_3, Cr_2O_3, and SnO_2 have been tested as to their performance as electrodes by Rolin and Gallay.[47–49] However, potential measurements of cells of the type

$$W/Na_3AlF_6 + Al_2O_3/Cr_2O_3, \quad \text{Inconel}$$

give unreasonable results. Only SnO_2 seems to offer some promise. It has been studied by Rolin and Ducouret[50] in the cell

$$Al/Na_3AlF_6 + Al_2O_3/SnO_2, \quad Pt$$

which was kept under a protective atmosphere of nitrogen. It was claimed that SnO_2 is chemically stable under oxidizing conditions, and that the emf was stable within ± 5 mV. Stannic oxide is soluble in molten cryolite to some extent (Table I).

5.4. *Metal Electrodes*

Since the melting point of aluminum, 660°C, is far below that of cryolite, the metal cannot be used as an electrode without support. An electrode design recommended by Piontelli[51] has been used by many authors (Fig. 14). The molten metal is contained in a tube made of sintered corundum or boron nitride, closed at one end. The tube has a small hole just above the metal surface, providing contact between the metal and the cryolite melt in the cell. External electrical contact with the electrode is provided by a molybdenum, tantalum, or tungsten wire inside an alumina or boron nitride protective tube, which dips into the molten metal.

Piontelli and Montanelli[52] used a graphite rod coated with electrolytically deposited aluminum. Rey[53] used a rod of ZrB_2 similarly coated. Graphite should not be used as a container for the aluminum electrode, because a short-circuited galvanic cell will then be set up. The stability of the aluminum electrode as a reference electrode is generally within 3–5 mV, which is slightly better than for the C, CO_2 electrode.

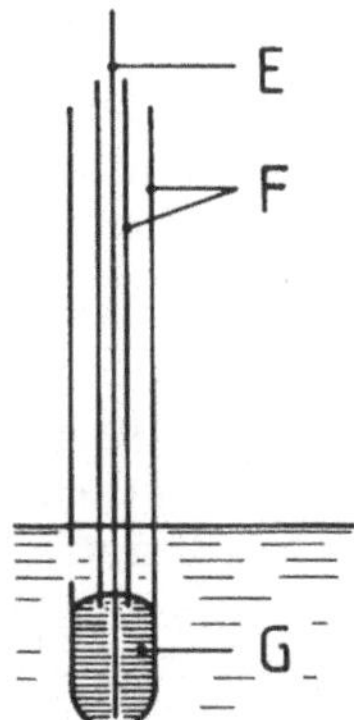

FIGURE 14. Aluminum electrode for use in cryolite-alumina melts. From Grjotheim *et al.*[2] Used with permission from Aluminium-Verlag. (E) Molybdenum wire. (F) Alumina or boron nitride tube. (G) Aluminum.

A solid aluminum electrode can be realized only with an aluminum alloy, provided that the alloying component is more noble than aluminum. The activity of aluminum will generally be less than unity. Unfortunately, the composition of the alloy will tend to change during contact with molten cryolite since aluminum is soluble in the melt.

One such alloy, iron–aluminum, has been investigated by Rolin and Gallay.[47]

Solid electrodes have been made by depositing aluminum on a substrate metal, for example, by applying a cathodic current to a metal rod or wire in a cryolite melt. Tungsten seems to be the best choice of substrate, because it does not seem to alloy appreciably with aluminum.

A sodium electrode is conceivable only as an alloy owing to its volatility. The lead–sodium electrode was first used in cryolite melts by Feinleib and Porter.[54]

5.5. Conclusion

No electrode is universally applicable as a reference electrode in cryolite melts. The C, CO_2 and the SnO_2 electrodes seem to be the best choices under oxidizing conditions. Under reducing conditions, the aluminum electrode is probably the most suitable one.

6. Spectroscopy in Molten Cryolite

6.1. General

Spectroscopic studies of molten cryolite are scarce owing to the obvious materials problems. The cell to contain the melt has to be transparent to the light of the pertinent wavelengths, and at the same time resistant to the liquid to be investigated.

Two different cell designs have been used for molten fluorides. One is a *windowless cell.* The other is a cell with *diamond windows.* Griffiths has given additional design information in Chapter 4 of Volume 2 of this series.

6.2. The Windowless Cell

The term "windowless" is somewhat unfortunate, since the light obviously does not pass through the cell wall. The cell has also been referred to as a "captive liquid" cell, which seems to be a more apt description, but the former term is widely used in the literature. The windowless cell must be made of a material that is wetted by the contained liquid. The light passes through the cell through two holes in the cell wall, and if the dimensions of the holes are right, the liquid will be held captive in the cell through surface tension forces. The proper hole size depends on the surface tension of the liquid.

The windowless cell was first tried by Young[(55)] for ultraviolet-visible spectroscopy of molten fluorides. The same idea was utilized by Solomons *et al.*[(56)] in the first published investigation of Raman spectra in molten cryolite. Their cell was made of boron nitride. Two horizontal channels in the cell, intersecting at a right angle, contained the melt. One channel provided the path of the laser beam and the other the outlet for the scattered light. The melt was kept from running out of the cell by the capillary action of the melt in retaining holes in the "ceiling" covering the cell cavity. The level of the melt was at the level of the top of these retainer holes. The spectrum of molten cryolite at 1030°C was determined by laser excitation using the 694.3-nm radiation from a ruby laser. The results are questionable, however, for reasons stated below.

Gilbert *et al.*[(57)] reinvestigated the spectrum of molten cryolite, and discussed the difficulties related to obtaining Raman spectra at very high temperatures.[(58)] They point out that the main source of error is the strong background due to blackbody radiation. Therefore, the apparatus has to be designed in such a manner that the spectrometer does not see any surface that may emit strong blackbody radiation. This radiation is much more intense in the red than in the blue spectral region. In the experiments of Solomons *et al.*,[(56)] where a ruby laser was used, the blackbody radiation, accordingly, interfered strongly with the Raman emission. Gilbert *et al.*[(57,58)] used the 488.0-nm line of the argon laser. Their cell was similar to that of Young.[(55)] It was made of graphite, and measured 10 mm o.d. × 40 mm long. To prevent the spectrometer from "seeing" the cell wall, which emits thermal radiation, the cell was provided with a window (a hole) facing the window used for the Raman light (Fig. 15). These windows were 3 mm diam. The laser beam passed through two 1.5-mm holes in the cell wall in a direction perpendicular to the direction of the Raman light. The furnace and the rest

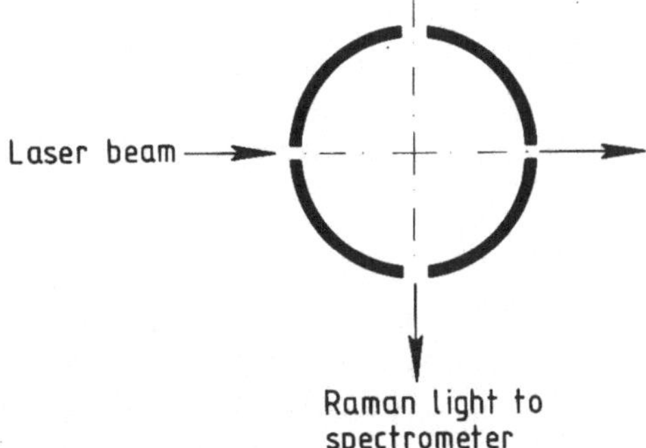

FIGURE 15. The paths of light through a "windowless" Raman cell after Gilbert *et al.*[57,58]

of the apparatus were provided with windows matching the cell windows to allow uninterrupted passage of the laser beam, and prevent the spectrometer from sensing thermal radiation.

The experimental procedure described by Gilbert *et al.*[57,58] involved the addition of the exact amount of salt needed to fill the cell. In previous work[55,56] the superfluous liquid was allowed to run out of the cell. The new procedure was motivated by the fact that the addition of AlF_3 decreases the surface tension drastically, making it more difficult to retain the melt in the cell. When no molten cryolite leaves the cell it is possible to enclose the cell in a silica tube, which is an advantage.

The spectra of molten cryolite obtained by Gilbert *et al.* seem more reliable than the previous ones, particularly because the information they render about the ionic species of the melt is more in harmony with those obtained from indirect methods.

The windowless cell has the following shortcomings[59]:

1. The light passes through a curved gas-liquid interface. The curvature makes it difficult to measure the path length accurately, especially when it is short.

2. Light transmission across the interface is not rectilinear, so the cells must be placed at the focus of the spectrometer beam.

3. The evaporation of the sample is more extensive from a windowless cell than from a closed cell. This may cause the composition of the melt to vary with time.

4. The windowless cell cannot be agitated without risking loss of sample.

Some cells have been tried where the light is transmitted vertically through the horizontal liquid surface. The light then leaves the cell either through apertures below the liquid level[60] or by reflection from a mirror in the bottom of the melt. The latter solution was suggested by Cocks *et al.*,[61] who tried a tungsten mirror for the purpose. There are several disadvantages of such a solution. One is that the light has to pass through the surface of the melt. Impurities collect preferentially in the surface layer and at the bottom, and thus will interfere with the spectral observations.

6.3. The Diamond-Windowed Cell

A diamond-windowed cell for spectrometry of molten fluorides was first tested by Cocks *et al.*,[61] and later by Toth *et al.*[59] Diamond is not attacked by molten fluorides. It has very good optical transmission properties over the near infrared, visible, and ultraviolet spectral ranges. No other material combines these two essential properties, but the price is extremely high, especially for diamond windows cut to precise dimensions.

The sealing between the windows and the cell is the critical point of the cell design. Cocks *et al.*[61] made the cell of copper, since it is soft enough to serve as a gasket material. Toth *et al.*,[59] on the other hand, took advantage of the fact that graphite is generally poorly wetted by molten fluorides. Although the seal between the windows and the cell body is not gas-tight, it is adequate to prevent leakage of melt around the windows.

A similar graphite cell with diamond windows was used by Ratkje and Rytter[62] for the study of molten cryolite and other related fluorides. The cylindrical graphite cell was positioned horizontally and covered by a boron nitride sleeve to protect the surrounding silica tube from being corroded. The cell had two diamond windows, separated by an angle of 90°, the window admitting the laser beam facing downwards. Since there was only one window for the Raman light, the spectrometer would pick up blackbody radiation, as pointed out by Gilbert *et al.*,[58] and, consequently, give poorly defined spectra.

7. Miscellaneous Techniques

7.1. Removing Cryolite

Since molten cryolite wets most container materials it is often difficult to remove it after an experiment. Usually it is helpful to make the container (crucible) slightly conical.

Platinum crucibles are designed to last several experiments. The bulk of the salt may be emptied while molten, but a residue will usually remain stuck to the crucible wall. One recommended method to remove this is to treat the crucible with molten potassium pyrosulfate. This salt melts easily over a laboratory gas burner. Care should be taken not to overheat the melt.

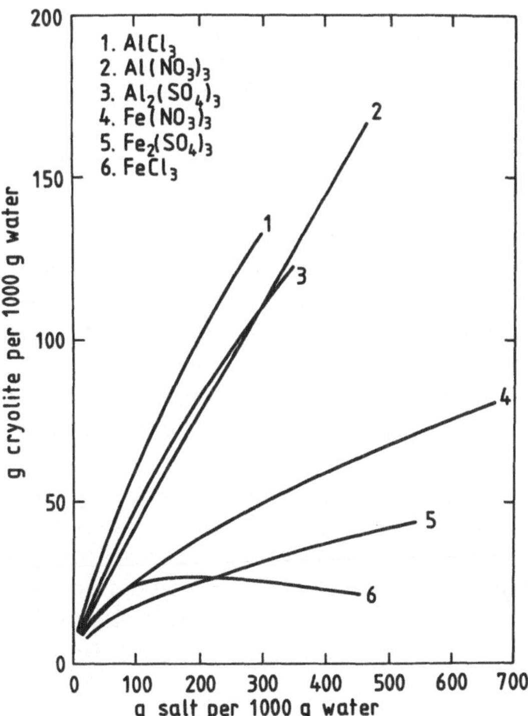

FIGURE 16. The solubility of cryolite in various aqueous salt solutions at 25°C. The data are from Frere.[63]

More effective is a treatment with molten sodium carbonate but this requires a furnace or a large gas burner.

The cleaning is finished in a boiling aqueous solution of aluminum chloride or some other suitable salt. The solubility of cryolite in some aqueous salt solutions at 25°C is shown in Fig. 16.

7.2. Cryolite Mixtures

In studies related to the process metallurgy of aluminum the experiments frequency involve cryolite mixed with alumina; often an aluminum metal phase will be present.

The presence of oxide changes the properties of the melt. The melt wets the container material better, and it takes more effort to clean the crucibles. The wetting of platinum is so drastically improved that the melt creeps up along the walls and out of the crucible. After a period of time the crucible may be practically empty. Impurities of heavy metal ions (lead or bismuth) reduce this wetting.[32]

The presence of aluminum metal necessitates an inert atmosphere above the melt. Aluminum nitride may be formed if nitrogen is present, so it is necessary to work under a noble gas protection. The choice of noble gas is directed by the cost.

The presence of aluminum metal creates a certain vapor pressure of sodium metal due to the reaction

$$Al(l) + 3Na^+ = Al(III) + 3Na(g)$$

which has a standard Gibbs' energy of +120 kJ mol^{-1} at 1300 K. [Al(III) means Al^{3+} complexed.] This may cause sodium metal to condense elsewhere in the apparatus. The equilibrium above is shifted to the right if the cryolite ratio is increased.

Since most metals alloy with platinum this metal is ruled out as a container material for molten aluminum. Graphite is the most likely choice. Molten aluminum wets graphite and carbon so poorly that it tends to contract to a large sphere in the crucible, rather than form a horizontal surface. Boron nitride is not attacked by molten aluminum.

Other possibilities for container material are silicon carbide and metal borides like titanium boride (TiB_2) and zirconium boride (ZrB_2). Silicon carbide and titanium boride are commercially available as crucibles.

Nickel may also be used as container material for cryolite/aluminum melts, but its lifetime may be shorter.

7.3. *Laboratory Furnaces*

The most practical design of a laboratory furnace is the vertical tube furnace. Accurate work depends on the availability of a furnace with a uniform temperature zone. Typically such a furnace has a central zone of 10 cm in the axial direction, where the temperature varies less than ±1°C. This is accomplished by an appropriate design of the furnace and a temperature controller.

The furnace is essentially a refractory tube, surrounded by a heating element, and thermally insulated by an appropriate material (kiesel-guhr, vermicullite). The tube is susceptible to attack by the fluoride vapors, which reduces its lifetime. Generally it does not make any difference what kind of furnace tube is used. Sintered corundum is 5 times as expensive as ordinary porcelain tubes (e.g., Haldenwanger's brand, "Pythagoras"), but the durability of sintered corundum is not so much greater than that of porcelain that its use is justified for that reason.

For less demanding work a smaller tubular furnace is adequate. This furnace may be shorter and wider than the precision furnaces, giving it a potlike design. Covered with a lid it provides a convenient way for rapid melting of salt. Even graphite crucibles may be used for melting cryolite in such a furnace since the airburn of the graphite is rather slow.

Lantelme has described furnace construction in some detail in Chapter 5 of Volume 2 of this series.

8. Conclusion

The experimentalist will find that molten cryolite is not the most agreeable substance to handle. This fact should be rather obvious from the preceding paragraphs of this chapter. Its high melting point, its corrosivity, its emission of troublesome fluorine-containing vapors, its strange ways of wetting solids, all make up a disturbing combination of properties causing potential problems when it comes to designing apparatus and experiments in the laboratory. It should also be borne in mind that whereas all contact between human skin and molten salt is harmful, any contact with molten fluorides is especially so since the fluoride ion has a way of penetrating the tissue much deeper. All experimental work with molten cryolite, therefore, takes more thoughtful planning than most other work of this kind.

References

1. A. I. Beljajew, M. B. Rapoport, and L. A. Firsanova, *Metallurgie des Aluminiums*, Vol. I, pp. 69–70, VEB Verlag Technik, Berlin (1956).
2. K. Grjotheim, C. Krohn, M. Malinovsky, K. Matiasovsky, and J. Thonstad, *Aluminium Electrolysis, 2nd Ed.*, Aluminium-Verlag, Düssseldorf (1982).
3. St. v. Naray-Szabo and K. Sasvari, *Z. Kristallogr.* **99**, 27–31 (1938).
4. E. G. Steward and H. P. Rooksby, *Acta Crystallogr.* **6**, 49–52 (1953).
5. G. J. Landon and A. R. Ubbelohde, *Proc. R. Soc., London* **240A**, 160–172 (1957).
6. Reference 1, p. 85.
7. G. J. Landon and A. R. Ubbelohde, *Trans. Faraday Soc.* **52**, 647–651 (1956).
8. J. L. Holm, *Acta Chem. Scand.* **27**, 1410–1416 (1973).
9. K. Grjotheim and B. J. Welch, *Aluminium Smelter Technology*, p. 30, Aluminium-Verlag GmbH, Düsseldorf (1980).
10. *JANAF Thermochemical Tables, 2nd. ed.*, Natl. Bur. Standards, NSRDS-NBS 37 (1971).
11. Reference 2, pp. 109–111.
12. J. D. Dana, *The System of Mineralogy*, Vol. 2, pp. 124–125, Wiley, New York (1951).
13. J. D. Edwards, C. S. Taylor, L. A. Cosgrove, and A. S. Russell, *J. Electrochem. Soc.* **100**, 508–512 (1953).
14. Reference 2, pp. 153–154.
15. H. Bloom and B. W. Burrows, *Proc. First Australian conf. Electrochem., Sydney, Hobart, Australia, 1963*, pp. 882–888, Pergamon, New York (1965).
16. W. Brockner, K. Tørklep, and H. A. Øye, *Ber. Bunsenges. Phys. Chem.* **83**, 12–19 (1979).
17. *Gmelins Handbuch der Anorganischen Chemie*, 8. Aufl., Syst. nr. 35, Teil B, p. 375, Verlag Chemie, Berlin (1934).
18. H. Kvande, Thermodynamics of the System NaF–AlF_3–Al Studied by Vapour Pressure Measurements, Dr. techn. thesis, University of Trondheim, Norway, 1979.
19. G. A. Wolstenholme, in *Molten Salt Technology* (D. G. Lovering, ed.), Plenum, New York (1982).

20. J. Thonstad, *Electrochim. Acta* **15**, 1569–1580 (1970).
21. J. Thonstad, *Electrochim. Acta* **15**, 1581–1595 (1970).
22. Technical Information, Bull. No. 442-212 II, Union Carbide, New York.
23. Carborundum Speciality Graphite Products, Carborundum, Sanborn, New York, Form C-1241A, 1983.
24. Graphite Speciality, Union Carbide Europe.
25. Graphite and Vitreous Carbon, Le Carbone-Lorraine, Paris, 1981.
26. K. M. Taylor, *Ind. Eng. Chem.* **47**, 2506–2509 (1955).
27. Carborundum, information leaflet, Form A-2005, 1971.
28. Union Carbide Corporation, Technical Information, Bulletin No. 442-205.
29. Union Carbide Corporation, Technical Information, Bulletin No. 442-206.
30. Union Carbide Corporation, Technical Information, Bulletin No. 442-217.
31. J. Thonstad, *Electrochem. Techn.* **6**, 346–349 (1968).
32. T. R. Scott, *Nature* **157**, 480–481 (1946).
33. G. D. Robbins, *J. Electrochem. Soc.* **116**, 813–817 (1969).
34. E. W. Yim and M. Feinleib, *J. Electrochem. Soc.* **104**, 622–626 (1957).
35. J. D. Edwards, C. S. Taylor, A. S. Russell, and L. F. Maranville, *J. Electrochem. Soc.* **99**, 527–535 (1952).
36. Reference 2, p. 153.
37. K. Grjotheim and K. Matiasovsky, *Tidsskr. Kjemi, Bergv., Metallurgi* **26**, 226–231 (1966).
38. K. Matiasovsky, B. Lillebuen, and V. Danek, *Rev. Roumaine Chim.* **16**, 163 (1971).
39. K. Matiasovsky, V. Danek, and B. Lillebuen, *Electrochim. Acta* **17**, 463–469 (1972).
40. H. Winterhager and L. Werner, *Praezizions-Messverfahren zur Bestimmung des elektrischen Leitvermoegens geschmolzener Salze*, Westdeutscher Verlag, Cologne (1956).
41. H. Winterhager and L. Werner, *Bestimmung des elektrischen Leitvermoegens geschmolzener Fluoride*, Westdeutscher Verlag, Cologne (1957).
42. G. J. Janz, J. Wong, and G. R. Lakshminarayanan, *Chem. Instrumentation* **1**, 261–272 (1969).
43. A. Vajna, *Alluminio* **20**, 29–38 (1951).
44. D. Bratland, C. Ferro, and T. Østvold, *Acta Chem. Scand.* **A37**, 487–491 (1983).
45. Reference 2, pp. 194–207.
46. M. M. Vetyukov and N. Van Ban, *Tsvet. Met.* **44**(8), 31 (1971) [*Sov. J. Non-Ferrous Met.* **12**(8), 35 (1971)].
47. M. Rolin and J. J. Gallay, *Bull. Soc. Chim. Fr.* **1960**, 2093–2096.
48. M. Rolin and J. J. Gallay, *Bull. Soc. Chim. Fr.* **1960**, 2096–2100.
49. M. Rolin and J. J. Gallay, *Electrochim. Acta* **7**, 153–178 (1962).
50. M. Rolin and A. Ducouret, *Bull. Soc. Chim. Fr.* **1964**, 790–797.
51. R. Piontelli, *Ann. N. Y. Acad. Sci.* **79**, 1025 (1960).
52. R. Piontelli and G. Montanelli, *Alluminio* **25**, 79 (1956).
53. M. Rey, *Ct. R. Acad. Sci. Paris.* **260**, 5528 (1965).
54. M. Feinleib and B. Porter, *J. Electrochem. Soc.* **103**, 231–236 (1956).
55. J. P. Young, *Anal. Chem.* **36**, 390–392 (1964).
56. C. Solomons, J. H. R. Clarke, and J. O'M. Bockris, *J. Chem. Phys.* **49**, 445–449 (1968).
57. B. Gilbert, G. Mamantov, and G. M. Begun, *J. Chem. Phys.* **62**, 950–955 (1975).
58. B. Gilbert, G. Mamantov, and G. M. Begun, *Appl. Spectrosc.* **29**, 276–278 (1975).
59. L. M. Toth, J. P. Young, and G. P. Smith, *Anal. Chem.* **41**, 683–685 (1969).
60. A. S. Quist, J. B. Bates, and G. E. Boyd, *J. Chem. Phys.* **54**, 4896–4901 (1971).
61. G. C. Cocks, J. B. Schroeder, and C. M. Schwartz, Battelle Memorial Institute Report No. BMI-1185, Columbus, Ohio, February-April 1957.
62. S. K. Ratkje and E. Rytter, *J. Phys. Chem.* **78**, 1499–1502 (1974).
63. F. J. Frere, *J. Am. Chem. Soc.* **58**, 1695–1697 (1936).

4

Reference Electrodes for Molten Electrolytes

Nguyen Quang Minh and Laszlo Redey

1. Introduction

Molten salts have been used in industrial metallurgical processes for many years. The production of aluminum metal by the electrolysis of alumina dissolved in molten cryolite (the Hall–Héroult process) is an example. Since the 1950s considerable attention has been given to the use of molten salts in other applications such as batteries, fuel cells, and nuclear reactors, as well as the processes of catalysis, coal gasification, and electroplating. At the same time, interest in the structure and fundamental properties of molten salts has expanded. Because molten salts are ionic liquids, they are useful for testing ionic theories and models, and they provide a medium for studying various phenomena and processes without the presence of water.

Electrochemical measurements play an important role in determining the properties of molten salts. Besides yielding electrochemical data, they can be used to obtain other useful information. For example, the measurement of equilibrium potentials of galvanic cells is one of the most attractive and direct means of obtaining data on the thermodynamic properties of molten salt systems. Electrode polarization is the common method to investigate corrosion phenomena, while electroanalytical techniques offer convenient ways to monitor the concentration and control the process. Complex formation in molten salts can be investigated by electrochemical transient techniques such as chronopotentiometry and polarography.

In most cases, electrochemical measurements require the use of a reference electrode, and the success of the measurements depends to a large extent on the reliability of that electrode.

Nguyen Quang Minh and Laszlo Redey • Argonne National Laboratory, Chemical Technology Division, Argonne, Illinois 60439.

Several types of reference electrodes are commercially available for use in aqueous systems. However, molten salt reference electrode are always "custom-made;" therefore, the selection of a suitable reference electrode for a molten salt system and establishment of its reliability are major tasks. Unlike aqueous systems, there is a wide range of molten salt systems, ranging from low-temperature melts (e.g., 1-butylpyridinium chloride–aluminum chloride, m.p. $<0°C$) to very high-temperature melts (e.g., cryolite, m.p. 1010°C), and each molten salt system has different chemical and physical behavior. Therefore, it is usually necessary to consider what is the best reference electrode for each molten salt system separately. In addition, many molten salt systems are quite corrosive because of their high melting temperature and high reactivity. As a result, the design of reference electrodes is influenced by the corrosiveness of the molten salt. Thus, a distinct problem of working with molten salt reference electrodes is selection of suitable materials of construction.

This chapter is an attempt to present a comprehensive treatise on the practical aspects of reference electrodes in molten salts. Emphasis is placed on providing details and suggestions on construction and operational characteristics of different reference electrodes; some discussions on theories which pertain to the operation of reference electrodes are also included. The subject of reference electrodes has been reviewed previously in a book chapter by Laity[(1)] and in a monograph by Alabyshev, Lantratov, and Morachevskii.[(2)] These texts, however, appeared in the 1960s and are somewhat dated. Also, detailed information concerning essential operations of reference electrodes is lacking in those texts. Thus, an up-to-date review on the subject is needed. This chapter is also intended as a follow-up of the more general discussion on molten salt reference electrodes in the chapter "Electrochemistry—I" by Lantelme, Inman, and Lovering in Vol. 2 of this series.[(3)] In this chapter, the design, construction, operation, and limitation of various reference electrodes in common molten salt systems are discussed. It is hoped that the chapter will serve as a starting point for both molten salt specialists and nonspecialists in selection of reference electrodes suitable for their experimental investigations, as well as a practical guide on construction and operation of these electrodes.

Although this is primarily a practical text, it is instructive to outline some of the theoretical considerations that are relevant to reference electrodes. A brief review of theoretical principles concerning reference electrodes is thus given in Section 2. It is followed by a general discussion on principles of reference electrode construction and application in Section 3. Detailed descriptions of reference electrodes in various molten salt systems are presented in Section 4.

2. *Theoretical Principles*

This section gives a brief recapitulation of the theoretical principles that are especially important in construction or application of molten salt reference electrodes. For a detailed description on any particular aspect, the reader should consult the numerous books, review articles, and original papers that discuss the theory of electrodes in general and reference electrodes in particular.[1-3] In this section and Section 3 the references are mostly review articles, which provide the reader with citations to original works.

The terminology used in the electrochemical literature and the symbols for these terms are not always consistent. The usage in this chapter—which follows Vetter's definitions[4] and is basically in conformance with the International Union of Pure and Applied Chemistry (IUPAC) recommendations for the electrochemical nomenclature[5]—is summarized in Table I.

2.1. *Electrode Potentials and Potential Scales*

The term "electrode potential" defines a potential difference between the electrode in question and another electrode, which is chosen as a reference. Consequently, all potential measurements in electrochemical systems are performed with a suitable reference electrode. To obtain meaningful results, the reference electrode should be reversible and reproducible, and its potential should remain constant for the duration of the experiment.

There are differences in the requirements for a reference electrode, depending on the type of application. To measure thermodynamic quantities, the numerical value of the reference-electrode potential must be known and precisely measurable relative to a standard, or it must be the standard itself, i.e., defined as zero. In electrode kinetic studies, when the main interest is changes of a test electrode potential, the stability of the reference electrode potential is the primary requirement (for more details of this subject, see Section 3.2 and Table VII).

For the thermodynamic-quantity measurement, the potential of an electrode is determined relatively on a thermodynamically defined potential scale; therefore, it can be compared with the potentials of other electrodes of either the same experiment or those of other studies. The potential scale is calibrated in units of volts; its zero point is defined by the potential of an arbitrarily chosen standard reference electrode system and designated by naming this standard reference electrode. Different potential scales are applicable for different solvent media, e.g., the potential scale of the standard hydrogen electrode for aqueous solutions. Each chemically different molten

TABLE I. Terminology and Symbols

Terms[a]	Remarks	Equilibrium condition[b]	Current load[b]	Symbol
Electrode potential	Generic term; relative to a reference electrode	Y/N	Y/N	ε
Equilibrium electrode potential	Relative to a reference electrode; characterized by an electrode reaction	Y	N	ε
Standard electrode potential	Under standard conditions	Y	N	ε^0
Cell voltage	Generic term	Y/N	Y/N	U
Open circuit (cell) voltage, OCV		Y/N	N	U_{ocv}
Electromotive force, emf (cell potential)	Measured between two electrodes; characterized by a cell reaction; may include a liquid-junction or membrane potential	Y[c]	N	E
Standard electromotive force, standard emf	As above; both electrodes under standard conditions	Y[c]	N	E^0
Electrode polarization, polarization	Generic term; electrode-potential change relative to an open circuit value as a result of current load; excludes IR-voltage drop	N	Y	η
Overvoltage (overpotential)	Electrode-potential change relative to an equilibrium value as a result of current load	N	Y	η
Electric potential	Electrical term		Y/N	ϕ
Electric potential difference	Measured in the same phase		Y/N	$\Delta\phi$
Liquid-junction potential (diffusion potential)	Between two different, miscible electrolytes	N	Y/N	$\Delta\phi_{lj}$

Membrane potential	Used here as potential difference across an ionic-conducting membrane	N	Y/N	$\Delta\phi_{mj}$
IR-voltage drop (IR-drop)	Electrical potential difference between two points of a conductor as a result of a current flow	N	Y	$\Delta\phi_{ir}$
Cell, galvanic cell, electrochemical cell	Generic term	Y/N	Y/N	NA
Active material	Refers to the electrochemically active components of an electrode	NA	NA	NA
Electrolyte	An ionic conductor, includes solvent and solute	NA	NA	NA
Concentration	Generic term for defining relative quantity of a component in any units; in the generic meaning, includes activity	Y/N	Y/N	c
Activity		Y/N	Y/N	a
Activity coefficient		Y/N	Y/N	γ
Molarity	mol/liter solution	Y/N	Y/N	M
Molality	mol/kg solvent	Y/N	Y/N	m
Mole fraction		Y/N	Y/N	N
Ion fraction	of ion i	Y/N	Y/N	N_i
Transference number	of ion i	N	Y	t_i

[a] Parentheses indicate alternative terms not used in this work.
[b] Y = yes; N = no; Y/N = the term applies for both conditions; NA = not applicable.
[c] Here Y means that both electrodes are in equilibrium with their surrounding electrolyte, i.e., charge transfer equilibrium across phase boundaries and local chemical equilibria within phases exist.

salt or salt mixture—as a solvent medium—must be considered a unique system with its own potential scale (for more details, see Section 2.6).

The potential scale provides a tool for comparing potentials (ε) of different electrodes or for assessing material behavior in electrochemical interactions on the basis of the relative positions of the electrode potentials on the scale. When combined with other variables into a two- or three-dimensional coordinate system, the potential scale is useful for the precise description and interpretation of electrochemical events. Examples include the electrochemical equilibrium diagrams (Pourbaix diagrams), various polarization curves, corrosion diagrams, and potentiometric titration curves; these are only a few of the possible combinations.

2.2. Cell Types, Half-Cells, and Standard States

2.2.1. Cell Types

Electrode potential measurement involves an electrochemical cell with, at least, two electrodes. Each electrode has two components: (i) an ionic-conducting phase, the electrolyte, and (ii) an electronic-conducting phase. (The term "electrode" is sometimes restricted to the latter phase.) A summary of cell types is given in Table II.

Examples of cell types shown in Table II are also meant to illustrate cell diagrams and symbols used for cell description. The vertical lines in the cell diagrams indicate the following: solid lines, phase boundary; broken lines, junctions between miscible liquids; and double broken lines, liquid junctions in which the liquid-junction potential is negligible (e.g., Cell 3, if the AgCl concentration is sufficiently low). Names of ionic-conducting membranes are spelled out between vertical dotted lines in cell diagrams, as shown in Cell 10, Section 2.3.2. Since the electrode properties depend on temperature, the cell temperature (T) must be specified. Not shown in these cell diagrams are the actual electrode terminal materials, which may be different from the indicated electrode material (Ag, Pb, or graphite in the examples given in Table II) if inert and do not affect the electrode process. Connections between materials of electrodes and electrode terminals may introduce—often overlooked—thermo-emf problems, which will be discussed in Section 3.1.4. In the cell diagrams, in addition to the temperature, values and units of c and p are specified; usually, the corresponding emf is also indicated (not shown in Table II).

In the cell diagrams, the negative electrode is usually on the left-hand side; however, if the diagram specifies a cell with a test electrode versus a reference electrode, the latter is on the left-hand side. This convention results in the proper sign of the electrode potential (versus the indicated reference electrode), since the cell diagrams are always written so that the sign of the measured emf reflects the right-hand side versus left-hand side polarity.

When the electrolytes of the two electrodes are identical the cell is called "cell without transference," e.g., Cell 1, Table II. When the two electrolytes differ in composition or concentration of their components, the cell is "with transference," e.g., Cell 3, Table II. Then the electromotive force of the cell includes a liquid-junction potential, which is not a thermodynamic quantity, i.e., it is not an equilibrium potential.

The emf of cells without transference is in agreement with the Nernst equation written by using the activities of the components of the corresponding reaction. For a AgCl-formation cell (Cell 2) this equation is based on the reaction

$$Ag + \tfrac{1}{2}Cl_2 \rightleftharpoons AgCl \tag{1}$$

and written as

$$E = E^0 - \frac{RT}{F} \ln \frac{a_{AgCl}}{p_{Cl_2}^{1/2} a_{Ag}} \tag{2}$$

where $a_{Ag} = 1$ since silver is in a pure state. Consequently, cells without transference can be used to measure (i) standard emf or standard electrode potential (e.g., the standard potential of Ag/Ag^+ electrode in Cell 1 if p_{Cl_2} is equal to unity), (ii) activity of an electroactive component (e.g., the activity of AgCl in Cells 2, 5, and 6, where KCl—or more precisely the K^+ ion—is considered electrochemically inert; i.e., it does not take part in the electrode, processes, although it may influence the activity of the electroactive component). Equation (2) is the Nernst equation written in the form that refers to a cell reaction.

Cells with a junction of different electrolytes (the cells with transference and the Daniell cells) are not in thermodynamic equilibrium because they involve a liquid-junction potential. Consequently, the emf value that is calculated by the Nernst equation would not match the measured emf, irrespective of whether the electrodes themselves are in thermodynamic equilibrium with the surrounding electrolyte. However, as will be discussed in Section 2.3, in many cases the junction potential between molten electrolytes is small; therefore, the difference between the measured and the calculated values may be negligible.

Cells 4–7 in Table II are concentration cells; however, in contrast to Cell 4, Cells 5–7 have no liquid junction between the two electrolytes. In these cells, graphite or silver separates the left and the right compartments of the cell. These cell compartments are usually fabricated as two independent cells and connected together by an electronic conductor, as shown in Cell 7. Cells 8 and 9 are examples of the two basic types of the Daniell

TABLE II. Examples of Cell Types

Cell No.	Cell diagram[a]	Cell type[b]
1	$-Ag_{(s)}\|AgCl_{(l)}\|Cl_{2(g)}$(graphite) p_{Cl_2}	FC WOT
2	$Ag_{(s)}\|AgCl_{(l)}, KCl_{(l)}\|Cl_{2(g)}$(graphite) c_{AgCl} p_{Cl_2} at standard pressure	FC WOT
3	$Ag_{(s)}\|AgCl_{(l)}, KCl_{(l)} \vdots\vdots KCl_{(l)}\|Cl_{2(g)}$(graphite) c_{AgCl} p_{Cl_2}	FC WT
4	$Ag_{(s)}\|AgCl_{(l)}, KCl_{(l)} \vdots AgCl_{(l)}, KCl_{(l)}\|Ag_{(s)}$ c_{AgCl} c'_{AgCl} $c_{AgCl} < c'_{AgCl}$	CC WT
5	$Ag_{(s)}\|AgCl_{(l)}, KCl_{(l)}\|Cl_{2(g)}$(graphite) $Cl_{2(g)}\|AgCl_{(l)}, KCl_{(l)}\|Ag_{(s)}$ c_{AgCl} p_{Cl_2} p'_{Cl_2} c'_{AgCl} $c_{AgCl} < c'_{AgCl}$ $p_{Cl_2} = p'_{Cl_2}$	CC WOT
6	(graphite) $Cl_{2(g)}\|AgCl_{(l)}, KCl_{(l)}\|Ag_{(s)}\|AgCl_{(l)}, KCl_{(l)}\|Cl_{2(g)}$(graphite) p_{Cl_2} c_{AgCl} c'_{AgCl} p'_{Cl_2} $p_{Cl_2} < p'_{Cl_2}$ $c_{AgCl} = c'_{AgCl}$	CC WOT

7	$Ag_{(s)} \mid AgCl_{(l)}, KCl_{(l)} \mid Cl_{2(g)}$(graphite) ← electron conductor → (graphite) $Cl_{2(g)} \mid AgCl_{(l)}, KCl_{(l)} \mid Ag_{(s)}$ c_{AgCl} p_{Cl_2} p'_{Cl_2} c'_{AgCl} $c_{AgCl} < c'_{AgCl}$ $p_{Cl_2} = p'_{Cl_2}$	CC WOT
8	$Pb_{(l)} \mid PbCl_{2(l)} \vdots AgCl_{(l)} \mid Ag_{(s)}$	Daniell Cell
9	$Pb_{(l)} \mid PbCl_{2(l)}, KCl_{(l)} \vdots AgCl_{(l)}, KCl_{(l)} \mid Ag_{(s)}$ c_{PbCl_2} c_{AgCl}	Daniell Cell

[a] (s), (l), and (g) indicate solid, liquid, and gas phases. The electronic conducting phase for a gaseous active material is written in parentheses.
[b] FC: formation cell; CC: concentration cell; WT: cell with transference; WOT: cell without transference.

cells having either two pure electroactive salts or two dilute electroactive salts.

Electrode potential measurements with a reference electrode eventually utilize one of the described cell types. Electrodes of these cells are often called half-cells. Thus, either of the half-cells may be chosen as a reference electrode and the other as a test electrode; both must be constructed according to the needs of the experiment. Practically, any electrode system that satisfies the criteria of reversibility, reproducibility, and stability (see Section 3.1) can be refined to a useful reference electrode device. It is important to remember, however, that during the potential measurement the reference electrode works in a cell that is the physical realization of one of the above cell types and the cell usually involves a liquid junction or a membrane. Consequently, the limitation in applicability of the Nernst equation must be considered if a liquid-junction potential is present.

Direct chemical reaction between the constituents of the electrodes must be prevented by proper cell and reference-electrode design because these reactions create nonequilibrium conditions in the cell. For example, the following reactions must be prevented: in Cells 1-3, the direct reaction between chlorine and silver, which produces an unknown increase of AgCl concentration at the surface of the silver electrode (Cells 2 and 3) or which may decrease the partial pressure of chlorine (Cell 1); and in Cells 8 and 9, the reaction between Ag^+ ions and the lead electrode

$$Pb + 2Ag^+ \rightleftharpoons 2Ag + Pb^2Pb^{2+} \tag{3}$$

which renders the activity of lead undetermined on the surface of the electrode by Ag–Pb alloy formation. Direct reactions can be prevented by placing diffusion barriers (porous diaphragm or capillary connection) or an ionic-conducting membrane between the electrolytes. Detailed discussions of these components are given in Sections 2.4 and 3.1.2.

The cell diagram relates to the overall chemical reaction that will take place spontaneously if the active materials are in direct contact or the electrodes of the cell become short-circuited. As a rule, in a spontaneous reaction, the constituents of the positive electrode oxidize that of the negative electrode. The overall cell reaction is the same as the direct spontaneous reaction (supposing that external current source is not connected to the cell), e.g., the reaction in Cells 1-3 is shown by Eq. (1) and in Cells 8 and 9 by Eq. (3).

The emf of a cell can be calculated from thermodynamic data (or vice versa) using Eq. (4) or (5), since for a chemical reaction whose Gibbs' free energy change is known the corresponding cell emf is calculable. The Gibbs' free energy change (ΔG) during the electrochemical reaction is proportional to E according to the following equations:

$$\Delta G = -zFE \tag{4}$$

or

$$\Delta G^0 = -zFE^0 \tag{5}$$

where ΔG is the Gibbs' free energy change, z is the number of electrons involved in the reaction as defined by the overall cell reaction, F is the Faraday constant, and the superscript ^{0} indicates standard conditions. Equation (4) relates to a formation cell in which the electroactive components are not in the standard state, e.g., to the formation of diluted AgCl in Cell 2. At constant volume and pressure

$$\Delta G = \Delta H - T\,\Delta S \tag{6}$$

where ΔH is the enthalpy change in the cell reaction and ΔS is the entropy change. Measurement of the emf (E or E^0) at various temperatures provides means for calculation of the entropy change by

$$E_{T_2} = E_{T_1} + \frac{\Delta E}{\Delta T}(T_2 - T_1) \tag{7}$$

for a temperature range in which $\Delta E/\Delta T$ is constant. In the general case, the entropy change is calculated by

$$dS = zF\frac{dE}{dT} \tag{8}$$

This equation emphasizes the fact that the thermal coefficient of the emf may vary in certain temperature ranges, in addition to the abrupt changes of dE/dT at phase transformations.

Equations (6)–(8) are useful relationships between electrochemical and thermochemical data. However, as mentioned earlier, these calculations result in values that are equal or close to the measured ones only if the measurements are carried out with a cell in which the liquid-junction potential is nil or negligible and if proper care is taken that direct reactions between the active materials of the electrodes are prevented.

Numerical values of ΔH, ΔG, and ΔS can be found in various tabulations.[(6–9)] However, a word of caution is pertinent here: not all tabulations are based on the same reference state, and so values from different calculations may not be consistent or compatible. Therefore, it is better to utilize only a single compilation for related calculations whenever possible. The temperature coefficient of the emf in molten salts is generally in the range of ±0–1.2 mV/K. Pressure dependence of the emf is usually neglected, except when gaseous materials take part in the cell reaction.

2.2.2. Half-Cells, Electrode Potentials

When describing one of the electrodes in a cell, it is convenient to separate the cell reaction into two half-cell reactions, known as electrode reactions. In Cells 1–3 (Table II) the electrode reaction for the silver electrode is

$$Ag^+ + e^- \rightleftharpoons Ag \tag{9}$$

The following types of diagrams are used to describe an electrode (using a molten salt silver reference electrode as an example): Ag/AgCl, KCl or Ag/AgCl, and Ag/Ag^+. The former two forms indicate the composition of the electrode; the latter refers to the actual electrode process [Eq. (9)], since AgCl is highly soluble in molten chlorides.

The form of the Nernst equation that defines the potential of an electrode uses ion activities, e.g.,

$$\varepsilon = \varepsilon^0 + \frac{RT}{zF} \ln \frac{p_{Cl_2}}{a^2_{Cl^-}} \tag{10}$$

where z is the charge number of the chloride ion, ε^0 is the standard electrode potential, a is activity, and p is partial pressure. Although single-ion activities cannot be defined, it is customary to express the electrode potentials in this form to call attention to the ions that take part in the electrode reaction.

An important rule is that increasing the activity of the oxidized form of the electroactive species shifts the electrode potential in the positive direction. Consequently, a positive sign is required in the Nernst equation that defines an electrode potential [e.g., Eq. (10)] if the activities of the oxidized species are written in the numerator.

Formalistically, the emf of a cell is the difference between the Nernstian potentials of the two electrodes of the cell ($E = \varepsilon_1 - \varepsilon_2$). Nevertheless, to calculate either (i) a cell emf or an electrode potential versus a reference electrode from known activities of cell-reaction components or (ii) component activities from a measured emf, the form of the Nernst equation that uses cell-reaction-component activities [e.g., Eq. (2)] should be used.

Knowledge of the form of the electrochemically active species that actually take part in the electrode reaction is important, because the form of the electroactive species—and, consequently, the electrode potential—is affected by (i) redox changes in the charge number of an ion, (ii) complex formation, and (iii) acid–base reactions.

Examples of redox changes of ions in molten salts for iron and oxide ions are

$$Fe^{3+} + e^- \rightleftharpoons Fe^{2+} \quad \text{and} \quad O_2^{2-} + 2e^- \rightleftharpoons 2O^{2-}$$

and an example of a complex-formation reaction for aluminum ion is

$$NaCl + AlCl_3 \rightleftharpoons Na[AlCl_4]$$

Reviews on evidence of complex-ion formation in molten salts and experimental methods used for detecting complex ions are available.[10]

Similarly to the aqueous solutions, electrochemical reactions in molten salts are influenced by the acidity of the medium.[11] Two concepts have been proposed to describe the acid-base properties of molten salts: (i) the oxo-acidity—the Lux-Flood concept considers an oxygen ion acceptor-donor reaction (e.g., $CO_3^{2-} \rightleftharpoons CO_2 + O^{2-}$, where CO_3^{2-} is the oxo-base and CO_2 is the oxo-acid); (ii) the solvo-acidity—the Lewis-acidity concept considers other than oxygen ion acceptor-donor reactions (e.g., $[AlCl_4]^- \rightleftharpoons AlCl_3 + Cl^-$, where the Cl^- ion is an acceptor, $AlCl_3$ is a chloro-acid, and the donor, $[AlCl_4]^-$, is a chloro-base). Acid-base reactions and equilibrium potential versus acidity diagrams in high-temperature ionic melts are reviewed by Tremillon.[12]

2.2.3. Standard States and Standard Potentials

Standard states and standard potentials are defined in two different ways in the electrochemistry of molten salts. In the first, the standard electrode potential (ε_p^0) is defined by a formation cell that involves a single salt. The emf of this cell (e.g., Cell 1 in Table II) is equal to the standard electrode potential of either electrode relative to the other electrode, i.e., the standard electrode potential of the "other" electrode is zero by definition. This is true if the active materials are in the pure state and the gaseous components are at unit pressure.* This standard electrode potential is symbolized by ε_p^0 (p subscript stands for the pure salt condition). This standard state definition is based on the customary practice of assigning an activity coefficient of unity to any ion in its pure salt medium, irrespective of the temperature. The Nernst equation is assumed to be valid even if the electrolyte contains a negligible quantity of an electrochemically inert salt, i.e., the salt mixture is not very far from the pure electrolyte composition (Raoult's law is in effect). This is the case in Cell 2, if KCl is a minor component. Properly measured ε_p^0 agrees with the value calculated from thermochemical measurements [Eqs. (4)–(8)] within the limits of experimental errors of both methods.

The second way to define standard state utilizes the other extreme of the concentration range. For dilute mixtures of the electroactive salt, the standard state is more conveniently defined by the infinitely dilute condition, i.e., by assigning an activity coefficient of unity of the electroactive ion for

* Since the atmosphere is not an accepted SI unit, the bar (= 100 kPa) has been recommended as standard pressure and unit of pressure for electrochemical events.[13] This practice would result in a slight difference between the new standard potential values and the present ones that can be found in compilations published so far on the one-atmosphere standard pressure bases.

this condition (Henry's law is in effect). For example, this approach is applicable for Cell 2 when a minor quantity of AgCl is mixed with KCl, which can be considered electrochemically inert in the given electrode reaction. Then, the standard potential (at any specified temperature) is obtained by extrapolating the experimentally established plot of

$$E - \frac{RT}{zF} \ln N_{AgCl} \quad \text{versus} \quad N_{AgCl}$$

to zero concentration. This procedure is based on the following equality:

$$E^0 = E - \frac{RT}{zF} \ln N_{AgCl}$$

obtained from Eq. (2) if p_{Cl_2} is unity, E is the measured emf of Cell 2, and E^0 is both the standard emf to be determined and also the standard potential of the silver electrode (ε_i^0) for infinitely dilute solution versus the standard chlorine electrode (i.e., defining $\varepsilon_p^0 = 0$ for the chlorine electrode). Although the standard state of the infinitely dilute solution is a hypothetical one, ε_i^0 can be used in the Nernst equation in the usual manner. To satisfy the Nernst equation, the numerical value of ε_i^0 determined by extrapolation to zero ion concentration is set equal to the standard value that refers to unit concentration (expressed in any chemical concentration scale). Under this condition the Nernst equation correctly describes the electrode potential within the limits of the range found for ideal solution behavior. This practice is very convenient at the low end of the concentration scale. A great many experiments have proved that ideal solution behavior exists (i.e., ε_i^0 is reasonably constant) to 0.1 mol% or, for several salt mixtures, even up to 1 mol% concentrations.(14) In standard emf series the ε_i^0 is used(15) (e.g., in Table V, Section 2.6). The ε_i^0, which is often called formal potential, defines a hypothetical standard state in which the electroactive salt has a unit concentration, but in all respects has the thermodynamic properties of the solution at infinite dilution. Similarly, standard potentials ε_i^0 can be measured relative to a metal reference electrode(16) (e.g., relative to an Ag/Ag^+ electrode in Cell 9, if concentrations of the solutes in both half-cells are sufficiently low).

A strict distinction must be made between the two kinds of standard electrode potentials (ε_p^0 and ε_i^0) because they are numerically different, as shown in Table III. Comparison of the differences of ε_p^0 and ε_i^0 for different electrodes or the same electrode in different solvent–salt systems provides useful information about ion interactions in liquid salt mixtures.(14) Activity coefficients, standard partial molal free energies, and entropies of mixing can be calculated from the difference $\varepsilon_p^0 - \varepsilon_i^0$, Table III. Equality of ε_i and ε_p is an indication that the activity coefficient of the chloride salt (when

TABLE III. Standard Potentials of Selected Metal Electrodes versus a Standard Silver Electrode[a]

Electroactive salt	ε_i^0 (V)			ε_p^0 (V)		
	700°C	800°C	900°C	700°C	800°C	900°C
$MnCl_2$	−1.206	−1.190	−1.172	−1.010	−0.981	−0.955
$ZnCl_2$	−0.860	−0.835	−0.810	−0.665	−0.650	−0.630
$CrCl_2$	−0.758	−0.740	−0.728	−0.565	−0.526	−0.505
TlCl	−0.665	—	—	−0.665	—	—
$CdCl_2$	−0.620	−0.580	—	−0.415	−0.367	—
$FeCl_2$	−0.520	−0.510	−0.498	−0.305	−0.292	−0.270
$CrCl_3$	−0.425	−0.385	−0.345	−0.385	−0.380	−0.375
$PbCl_2$	−0.390	−0.376	−0.355	−0.315	−0.286	−0.270
$SnCl_2$	−0.370	−0.354	−0.340	−0.405	−0.435	−0.450
$CoCl_2$	−0.324	−0.300	−0.275	−0.195	−0.155	−0.135
CuCl	−0.260	−0.256	−0.260	−0.140	−0.145	−0.152
$NiCl_2$	−0.140	—	—	−0.095	—	—
AgCl	0.0	0.0	0.0	0.0	0.0	0.0
$CuCl_2$	+0.170	+0.180	+0.192	+0.410	+0.430	+0.450
Cl_2/Cl^-	+0.845	+0.820	+0.795	+0.850	+0.826	+0.794

[a] Data are taken from Flengas and Ingraham.[16] ε_i^0 is the standard electrode potential for infinitely dilute solution in equimolar NaCl-KCl electrolyte. ε_p^0 is the standard electrode potential for pure metal-salt electrolyte, calculated from thermochemical data by Hamer *et al.*[17]

chlorine electrode is used as basis of comparison) is the same in the pure state and in the dilute solution. This procedure requires that both ε_i^0 and ε_p^0 are measured versus the same reference system, e.g., in Cell 1 and Cell 3, where the emf of Cell 1 is ${}_{\text{Ag vs. Cl}_2}\varepsilon_p$ and the emf of Cell 3 is ${}_{\text{Ag vs. Cl}_2}\varepsilon_i$; ε_p^0 can also be determined from thermochemical data by using Eqs. (5)–(7). The differences in $\varepsilon_p^0 - \varepsilon_i^0$ observed for various metal electrodes can be attributed to the change of the activity coefficient over the concentration range.[16] The concentration range within which the activity coefficient of the metal salt is constant can be detected by a concentration cell (Cell 4) using the well-known relationship

$$a = \gamma c \tag{11}$$

and

$$E = \frac{RT}{zF}\ln\frac{c'}{c} = \frac{RT}{zF}\ln\frac{a'}{a} - \frac{RT}{zF}\ln\frac{\gamma'}{\gamma} \tag{12}$$

In this experiment, c' is kept constant and c is varied in the range of interest. Within the range of c in which E obeys the Nernst equation γ' is equal to γ. Detailed discussions of these relationships are available.[16,18] Although Ref. 16 is related to the specific case of metal electrodes in equimolar NaCl-KCl electrolyte, it may be of general interest. An extensive review on formation and concentration cells, ion interactions, and excess free energies of binary salt mixtures has been given.[19]

Selection of the actual conditions of a reference electrode system are immaterial with respect to the potential measurement. However, all the necessary determinants of the electrode system must be precisely specified:

i. Compositions and concentrations of the solvent–salt system.
ii. Composition and concentration of the electroactive component.
iii. Composition of the electronic-conductor phase.

and conditions:

iv. Temperature; and
v. Partial pressure of the gaseous electroactive component(s), if any.

Determinants (i) and (ii) together define the reference electrolyte. For the sake of comparability of different electrodes (e.g., on a potential scale), the conditions of the standard state must be convenient enough that the same standard can be employed for a variety of electrodes.

In the Henry's law range, any chemical concentration units suffice in the Nernst equation; however, the numerical values of the ε_i^0 potentials vary according to the chosen concentration scale (see Table V). The ion fraction (N_i) scale is usually preferred for three reasons. First, N_i is independent of temperature, and the density of the melt is not needed in the calculation. Second, the maximum value of N_i is unity; it would be an inconveniently large number with molarity or molality units when the ε_p type standard is used. Third, although individual ion activities cannot be measured, it is customary in the molten salt electrochemistry to assume that the ion activity is equal to the ion fraction in dilute additive systems, at least in the concentration range where ε_i^0 is measured constant within the limits of the required precision. This practice is referred to as the quasi-thermodynamic treatment of ion activities. N_i is defined as either a cation or an anion fraction according to whether cations or anions are involved in the electrode reaction.

The reference-electrode user often works with reference and test electrodes under experimental conditions that are obviously far from those of either type of standard. Therefore, knowledge of the effects of various conditions on the potential is very important when results of different experiments are to be compared.

2.3. *Significance of the Electrolyte-Junction Potentials*

Two kinds of electrolyte junctions are used in an electrochemical cell between the reference and test electrolytes: either (i) a liquid junction between the two liquid electrolytes, or (ii) a membrane junction, in which a solid (pore-free), ionic-conducting membrane is placed between the two liquids. The first generates a liquid-junction potential, and the second a membrane potential across the junction interphase.

In any electrochemical work that involves measurement of cell voltage and electrode potential, two important factors (among others) must be taken into consideration: whether an electrolyte-junction potential exists, and, if so, how it varies. Careful experimental design, in many cases, minimizes the electrolyte-junction potential or its variation, or at least offers some control of the phenomenon.

2.3.1. Liquid-Junction Potential

A liquid-junction potential usually is included in the measured value of electrode potential. Its contribution, however, cannot be defined unambiguously, since it is not a thermodynamic quantity and may even vary significantly from one experimental condition to another. Therefore, the reference-electrode user must be aware of its significance and of the factors that control its magnitude.

A liquid-junction potential (which is a potential difference between the two sides of a liquid junction) develops where two electrolytes, different in composition or in concentration of one or more of their constituents, come in direct contact. The liquid-junction potential is generated by the diffusion processes that take place in the junction interphase region; hence, an alternative designation is "diffusion potential." The term "interphase" refers to a three-dimensional layer between the two electrolytes and thereby is distinguished from an interface, which is a sharp boundary between two phases. Across the interphase, a gradual concentration change takes place as a result of intermixing and interdiffusion of the different electrolyte components. Since diffusion is an irreversible process, an overall equilibrium cannot be in effect when there is an electrolyte junction in the cell. This is true even when the electrodes themselves are in equilibrium with their surrounding electrolytes.

The contribution of the liquid-junction potential to the cell emf can be estimated by evaluating the transport properties of the species involved in the liquid-junction region. These estimations, although difficult and of limited accuracy, are very valuable. Good estimates of liquid-junction potentials are important in experimental design, interpretation of results, or comparison of electrode potential data measured in identical or different electrolyte systems. The molten salt electrolyte system is defined as a liquid salt or liquid-salt mixture (usually a eutectic) that provides a solvent medium for an electroactive salt, which is a minor component in the solution.

Precise calculation of liquid-junction potentials requires data on transference numbers and activity coefficients of ions in their complex interactions, as well as knowledge of how the junction is formed physically. The latter requirement means that a precise calculation also requires a knowledge of the concentration profiles of ions in the junction region. Since the concentration profiles of the diffusing ions influence the value of the liquid-junction potential, the reference electrode user must be aware that the liquid-junction potential is usually time dependent.

Newman gives a detailed mathematical treatment for assessing the magnitude of liquid-junction potentials and the effects of different types of junctions.[20] Although his examples are given for aqueous cells, similar treatment is applicable for molten salt cells if the respective differences between the aqueous- and molten-salt systems are considered.

Assessments of junction potentials specifically for molten electrolytes are given by Laity,[1] Blander,[14] and Klemm.[21] The liquid-junction potential under isothermal conditions ($\Delta\phi_{1j}$) usually is expressed as

$$\Delta\phi_{1j} = -\frac{RT}{F}\int_{\text{I}}^{\text{II}} \sum_{i} \frac{t_i}{z_i}\, d \ln a_i \tag{13}$$

where t_i is the transference number of ion i in an arbitrarily chosen frame of reference for ionic velocities (the frame of reference, however, must be the same in all experiments to be correlated), z_i is the charge number, a_i is the ion activity, and I and II specify the compositions of the electrolytes on the two sides of the junction. In molten salt experiments, the frame of reference is usually the porous diaphragm or the wall of the cell. Calculations with Eq. (13) require a knowledge of the individual ion activities and the $t_i = f(a_i)$ function. Transference numbers in molten salt mixtures have a strong concentration dependence[22,23] since the concentration of one anion or cation can only be changed at the expense of the other like-charged ion. These values are not available in many cases; therefore, the numerical value of the liquid-junction potential cannot be calculated precisely.

The fully ionized character of most molten salts, however, allows a quasi-thermodynamic approach to be taken to calculate the liquid-junction potential. This approach equates ion activity with ion fraction on the following grounds. The fully ionized molten salts, according to Temkin,[24] are arranged in two interpenetrating quasi-lattices, one for cations and the other for anions; the cations are surrounded by anions and vice versa. Addition of another salt causes foreign cations to mix randomly with the original cations in the cation quasi-lattice without interaction between the foreign and the original cations and disturbing the original lattice order (a similar effect occurs for anions). In this ideal mixture, the activity of a component (a_{CA}) is

$$a_{CA} = N_C N_A \tag{14}$$

where N_C is the cation fraction of cation C^+ and N_A is the anion fraction of anion A^-. If the mixture contains only one anion (A^-) and several cations, then

$$a_{CA} = N_C \tag{15}$$

since N_A is equal to unity. Similar reasoning applies when there is a common cation and several anions.

Using the quasi-thermodynamic concept for solving Eq. (13) (Ref. 1, p. 549), one finds that only small (1–2 mV) liquid-junction potentials occur in molten electrolytes when the difference between the ion fractions of the diffusing ions on the two sides of the junction is small, e.g., the difference of the AgCl concentrations, $\Delta N = N - N'$ in Cell 3 or 4 is less than 0.01 mole fraction. By setting a small ΔN in an experiment, one can measure cell emf or electrode potential (i.e., by measuring electrode potential against a reference electrode) without significant interference of the liquid-junction potential. Furthermore, if N and N' are small, the low concentration range allows one to find the concentration limits of ideality in salt mixtures (see ε_i determination, Section 2.2). Numerous variations of the concentration ratios can be produced to manipulate the emf without violating the requirement for the low concentration ($N_i < 0.01$) of the electroactive ion. Indeed, many examples in the molten salt literature support the validity of this concept and prove that the Nernst equation gives emf values up to concentrations of the electroactive ion as high as $N_i = 0.01$.[16,25–30] This relationship is demonstrated by the usual method of representation of this measurement in Fig. 1.[30] The straightness of the emf versus $\log(N/N')$ line and the correct value of the slope in agreement with the $2.303(RT/zF)$ factor indicate also that the activity coefficient does not change in the observed concentration range, as discussed above in Section 2.2. The favorable conditions of the negligible liquid-junction potentials, however, are limited to salt mixtures that have only one electroactive ion and common counterions, as is the case for Cell 3 or Cell 4, where Ag^+ ion is the electroactive ion and Cl^- is the counterion in both salts in the AgCl–KCl salt mixture.

It is interesting to note that according to Eq. (13) the liquid-junction potential is equal to zero if like-charged ions have equal mobilities[31] and if the activities in Eq. (11) are set equal to mole fractions. This phenomenon provides another explanation of the small liquid-junction potentials observed under certain conditions in molten salts. It has been pointed out[32] that, for a great number of molten salts, the relative mobilities of like-charged ions differ by less than 10% or 15% over a wide range of concentrations and conductivities. For example, small difference in mobility and, indeed,

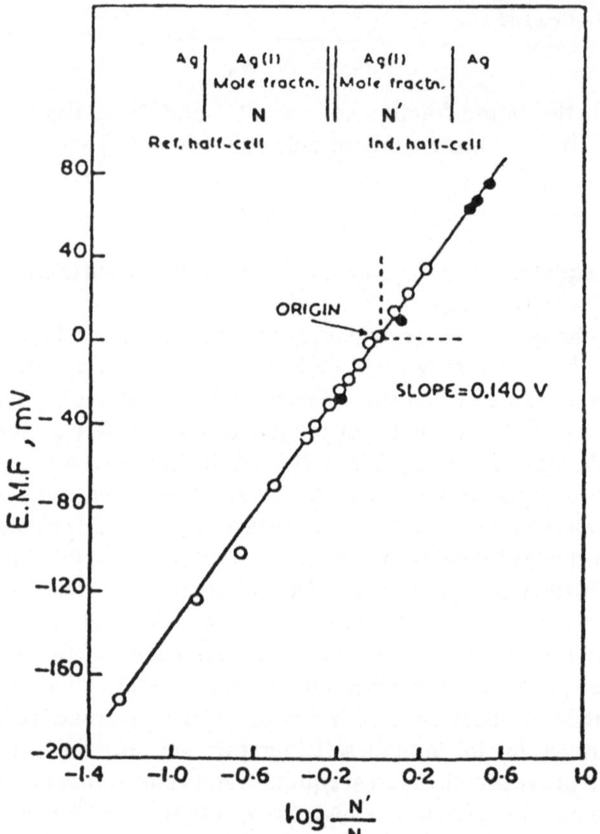

FIGURE 1. **Representation of the Nernstian response of the emf of a concentration cell to various mole-fraction ratios of the silver chloride.[30] (By permission of the publisher.)**

almost negligible liquid-junction potential was observed in CA + C′A type binary salt mixtures, where C and C′ are different cations and can be any of Li^+, Na^+, K^+, and Ag^+ ions in different combinations, and A is Cl^-, Br^-, or NO_3^- ion.[33,34]

Methods used to calculate the magnitude of the liquid-junction potentials in a porous diaphragm were reported by Smirnov and co-workers.[35-39] Their calculations indicated only a few mV (generally less than 10 mV) liquid-junction potentials for several salt systems in the range of 650–1000°C.[35,39] These authors also found that the liquid-junction potential reaches a steady-state value at high temperatures within a few minutes after the contact between the two melts is made across the porous barrier.[36]

Another method for assessing liquid-junction potentials for binary mixtures with a common ion, according to Laity,[1] uses the equation

$$\Delta\phi_{lj} \sim \frac{RT}{F}\frac{\Lambda_{1X}-\Lambda_{2X}}{z_1\Lambda_{1X}-z_2\Lambda_{2X}}\ln\frac{z_1\Lambda_{1X}}{z_2\Lambda_{2X}} \tag{16}$$

where Λ is the equivalent conductivity of a pure salt. Subscript X refers to the common ion, and subscripts 1 and 2 refer to the differing ions of the two salts (1X and 2X); X may be either anion or cation. This rough approximation requires only data on the equivalent conductivities of the pure salts that are involved in the liquid junction formation. It may be used for junctions of molten salts of significantly different chemical composition, Daniell cells (Cells 8 and 9), or cells in which the ion-fraction difference is large. Tabulated equivalent conductivity values of molten salts can be found in several compilations.[40,41] There are indications, however, that values obtained by Eq. (16) are too high and probably represent only the high end of the range obtained from other, more precise calculations [e.g., by using Eq. (13)].

Assessments of various experimental conditions, including aqueous electrolytes, indicate that the liquid-junction potentials are most common in the range of $\Delta\phi_{lj} = 5$–60 mV and usually are considerably smaller in molten salts than in aqueous solutions[1,14,20,22] (refer to Section 2.5 for explanation). This potential range hints at a range of possible error when comparisons are made of electrode potentials or potential scales that have been measured in different electrolytes without considerations of estimates for liquid-junction potentials. Between pure salts, however, substantially higher liquid-junction potentials may exist. For example, between pure AgCl and equimolar NaCl–KCl melts a liquid-junction potential as high as 300 mV was found at 700°C.[16]

The following conclusions can be drawn regarding liquid-junction potentials:

i. Liquid-junction potential is present whenever different electrolytes come into contact.
ii. The effect of the liquid-junction potential on experiments involving reference electrode measurements must be assessed.
iii. In cells with junction of molten salts of closely related chemical composition and under certain conditions—as described above—the liquid-junction potential may be small or negligible.
iv. Whenever possible, the liquid-junction potential should be eliminated by careful experimental procedure and design of the cell and reference electrode.
v. In electrode-kinetic measurements, a liquid-junction potential may be present at the junction of the reference electrolyte and the test

electrolyte, but must be constant. In thermodynamic-quantity measurements, however, it must be made negligible or nil or corrected for.

2.3.2. Membrane Potential

A special type of the electrolyte junction is the liquid$_1$–solid–liquid$_2$ junction, where all these phases are ionic conductors. The solid electrolyte, which is in the form of a thin membrane, is impermeable to the liquids (1 and 2). The ionic-conducting membrane is distinguished from a porous diaphragm. The latter is made of an insulator material, but the pores are permeable by the molten electrolytes and permit direct contact between the liquids. Thus, in a diaphragm a liquid-junction potential develops. The porous diaphragms are representatives of the diffusion barriers (Section 3.1.3), and the liquid-junction potential generated in them can be described according to the previous section.

The ionic-conducting membrane has use in the membrane reference electrodes or, as they are often called, membrane electrodes (an example is shown on the diagram of Cell 10). A membrane electrode is a half-cell consisting of (i) a thin membrane forming a bulb-shape end of a heavy-walled container tube, and (ii) a reference electrode submerged into the reference electrolyte which is contained within the membrane. Membrane electrode systems and designs will be reviewed in Section 4. Membrane electrodes are used in devices that generate reference potential in kinetic investigations, in ion-concentration measurements, and—under carefully controlled conditions—in thermodynamic studies.

In general, the membrane conducts exclusively the cation (or the anion) that is abundant in both electrolytes; the counterions, the anions (or the cations), are immobilized by the rigid network of the crystal structure. On each surface of the membrane, a junction potential is generated. The difference of the two junction potentials is called the "membrane potential." If the conditions are identical on both surfaces, the junction potentials are equal and, consequently, the membrane potential is zero.

Sodium-ion-conducting glass membranes were extensively studied in molten sodium or other alkali salts, for example, in cells built according to the diagram of Cell 10.

$$\overbrace{Ag_{(s)}\,\Big|\,\underset{a_{AgCl}}{AgCl_{(l)}},\ \underbrace{LiCl_{(l)},\ KCl_{(l)}}_{\text{eutectic}}\ \vdots\,\begin{array}{c}\text{Pyrex}\\\text{glass}\end{array}\vdots}^{\text{membrane electrode}}\ \underset{a'_{AgCl}}{AgCl_{(l)}},\ \underbrace{LiCl_{(l)},\ KCl_{(l)}}_{\text{eutectic}}\ Ag_{(s)}$$

Cell 10

The emf of this cell (and other similar cells) can be described by

$$E = \frac{RT}{F} \ln \frac{a'_{Ag^+}}{a_{Ag^+}} - \frac{RT}{F} \int_{a_i}^{a'_i} \sum_i t_i \, d \ln a_i \tag{17}$$

This equation shows that the emf includes two components: (i) the electrode potential difference according to the Nernst equation (the first term), and (ii) the membrane potential contribution, $\Delta\phi_{mj}$ [the second term, which in essence is similar to Eq. (13) used to describe a liquid-junction potential]. Since in this case liquid-junction-type events control the magnitude of the $\Delta\phi_{mj}$, this model is commonly called the liquid-junction theory of the membrane electrodes. Although the electrolytes on the two sides of the membrane in Cell 10 do not contain sodium ion, the membrane works with lithium salts as well as with sodium salts(42); unstable membrane potential has been observed, however, when the membrane is submerged into a sodium-free melt of potassium or cesium salts,(43) indicating the importance of the cation size. But the addition of sodium salt even in a low concentration assures the proper functioning of the membrane. Equation (17) is obeyed in the presence of as low as 2 mol% sodium salt with a sodium-ion conducting glass membrane.(43)

The measured emf of Cell 10 closely follows the Nernst equation, indicating that the membrane potential contribution to the measured emf is small in this arrangement.(42,44,45) Similarly, near-Nernstian response of the emf was found for electrodes with quartz or Vycor membranes, which had Na^+-ion impurities.(46) Also, not too high membrane potential contribution was observed in formation cells. For example, the emf of the formation cell Ag|AgCl : glass : AgCl|Cl_2 was nearly the same as that of Cell 1 in Table II. It is this fact that makes ionic-conducting membranes useful in measurement of formation cells in which direct reaction between the active materials of the electrodes must be prevented. Glass membranes have been studied in Daniell cells (Cell 9).(47,48) It has been pointed out that the material of the membrane and the chemical composition of the molten electrolyte have a great influence on the membrane potential. Details of these findings will be discussed in Section 4.1.2.

Reference electrodes of those metals that form metal solutions with their molten salts (Section 2.5) can be implemented only if they are separated by an ionic-conducting membrane from the rest of the cell. Being a perfect barrier, the membrane separates the electrolytes, preserving their composition for a long time (if chemical stability of the membrane permits). Several reference electrodes made of low-melting metals or alloys have been prepared in this way, and they will be described in Section 4.1.6.

Ionic-conducting membranes possess ion-exchange properties, and ion exchange takes place on the surface of the membrane if cations other than the charge carrier ions of the membrane are also present in the molten electrolyte. The ion exchange increases the membrane potential by adding an additional term to the right-hand side of Eq. (17) (not shown in the equation). This term is described by the ion-exchange model of the membrane potential, which was introduced first for glass in aqueous electrolytes[49,50] and later extended to glass/molten-salt systems.[51-54] The model accounts for the ion exchange on the surfaces of the membrane and the diffusion of the exchanged ion in the immobile anion network. It was found valid for Pyrex glass, e.g., in molten nitrates[51,53] and some other systems[55-59]; but general application of the model for glass in molten salts is limited by the lack of knowledge of the variables involved. The ion-exchange and charge-carrier properties of the glass[60] and the mobilities of various charge-carrier ions[61] vary considerably according to the composition, structure, and impurity level; furthermore, chemical reactions may considerably modify the physicochemical properties of the glass surface.[62] Therefore, it is not surprising that for molten-salt systems a generally valid description of the membrane potential phenomenon that involves ion-exchange is not available.

The membrane potential is equal to zero if the conditions on both faces of the membrane are identical. However, this is seldom the case. In reference-electrode work, therefore, the membrane potential is an unwanted component in the measured emf, except in those cases of ion-concentration measurements when it is the useful signal itself.

Frequently, the two sides of the membrane do not have identical surface properties. If so, a noncalculable asymmetry potential is involved in the membrane potential even if the two liquids are identical. Asymmetry potentials from a negligible value to $\sim$10 mV[51] or higher were reported for various conditions.

Variation of the membrane potential with time is another important effect to be considered. Emf change caused by asymmetry-potential variation and ion-exchange effects may last for hours. The ion-exchange theory indicates that the potential becomes constant as soon as ionic equilibrium is reached on the surface, but it cannot account for the longer-lasting and sometimes detrimental mechanical stress effects and the accompanied potential variations that the exchanging ion may create in the membrane structure.

At present, the membrane-potential phenomenon in molten salts is not very well understood. There are many unknown details of those processes that determine the magnitude of the membrane potential. Thus, when working with membrane reference electrodes, the investigator cannot rely on calculated values, but has to measure or experimentally estimate the membrane potential contribution under respective experimental conditions.

Comprehensive studies and reviews of molten salt membrane electrodes are still awaited. Extensive investigations with significance for reference electrodes have been reported and reviewed only for glass membranes in certain molten salts,[63] and for β- and β''-alumina and stabilized zirconia as key components of sodium/sulfur batteries and high-temperature oxygen sensors, respectively. So far, only sporadic studies have been carried out on new ionic-conductor materials that are offered by the advancement of material science. The authors believe that this interesting field of electrochemistry deserves more attention in the future.

2.4. Types of Reference Electrodes

Reference electrodes are usually classified on the basis of (i) their electrochemical working principle or (ii) their method of application.

For the first type of classification, electrodes are distinguished as reference electrodes of the first kind, of the second kind, and of the third kind.

(1) Electrodes of the First Kind. The potential of an electrode of the first kind is controlled by the activity of the ion that takes part in the electron-transfer process. For example, the potential of an Ag/Ag^+ electrode is set by the activity of the silver ion in an electrolyte that contains a highly soluble silver salt.

(2) Electrodes of the Second Kind. The potential of this electrode is determined by the activity of an anion through the solubility product of a sparingly soluble salt of the electrode metal. In principle, the electrodes of the second kind are preferred as reference electrodes, because of their better reproducibility and stability for potential measurements. For molten electrolytes, however, finding a suitable system is more difficult, mainly because the solubilities of most inorganic compounds in molten salts are too high to satisfy the working principle of this type of electrode. Nevertheless, $Pt|PtO|O^{2-}$ [64] and $Ni|Ni_3S_2|S^{2-}$ [65] reference electrodes of the second kind in alkali-metal chloride melts and $Ag|AgCl|Cl^-$ and $Ag|AgBr|Br^-$ systems[66,67] in alkali-metal nitrate melts have been used successfully.

(3) Electrodes of the Third Kind. The potentials of these electrodes respond, through two equilibria, to the activity of a cation different from the one involved in the primary electrode process. An example of this electrode type, used for monitoring Cd^{2+} ion concentration, is a Pd/PdO, CdO/Cd^{2+} half-cell in alkali nitrate melt.[68] Detailed explanations for the three kinds of electrodes can be found in Refs. 3 and 69.

For the second type of classification (method of their application), reference electrodes are classified as (a) half-cell reference electrodes (reference half-cells), (b) inner reference electrodes, and (c) quasi-reference electrodes.

(a) Half-Cell Reference Electrodes. In these devices the reference electrode itself is physically separated from the bulk electrolyte of the study cell by the reference electrode housing, which is made of some insulator material. Only a small opening, usually plugged up with a porous material forming a diffusion barrier, provides an electrolytic junction between the two electrolyte compartments. A special form of the half-cell reference electrodes is the membrane reference electrode. In the membrane reference electrodes an ionic-conducting membrane substitutes the diffusion barrier. The main concern during the application of the half-cell reference electrodes is to maintain a constant concentration of the reference electrolyte in order to ensure potential stability. Principles of their construction will be described in Section 3.1.

(b) Inner Reference Electrodes. In contrast to the half-cell reference electrodes, inner reference electrodes are electrodes immersed directly into the electrolyte of the study cell. The concentration of the electrolyte must be known precisely. This type of electrode finds application in electroanalytical techniques as a concentration sensor to measure activity of the electroactive component of the bulk (test) electrolyte. In this application, the potential of the inner electrode is compared to a half-cell reference electrode. Membrane electrodes have an inner reference electrode in their reference electrolyte, e.g., the left-hand side electrode in Cell 10.

(c) Quasi-Reference Electrode. The working principle of the quasi-reference electrode is based on the observation that some metals (for example, Pt[70,71]) dipped directly into the melt can act as an electrode having a potential that is maintained by some species of the electrolyte in an often unknown electrode process. The potential of this electrode is stable as long as this process is invariant. If this process is stable during the required length of the experiment, as has been observed in several cases, the metal may function as a reference electrode. According to an interesting hypothesis, metal ions formed in the melt by corrosion of the metal of the quasi-reference electrode may strongly adsorb on the electrode surface and give rise to a more or less defined adsorption potential.[72] In general, quasi-reference electrodes are not very reliable devices because they are very sensitive to little changes of the composition or impurity content of the electrolyte. However, in one reported case a Pt quasi-reference electrode was found very stable in fluoride melts even for over a period of months.[73] Quasi-reference electrodes have application in electrode kinetics and other studies if the test electrode potential need not be known on a thermodynamically defined potential scale and only changes in potential are a matter of interest. Various metals such as Pt, Ni, Mo, W, stainless steel, and also graphite have been used as quasi-reference electrodes in voltammetric and chronopotentiometric investigations in molten salts.[74–79]

2.5. *Special Features of Molten Salts and Their Importance for Molten-Salt Reference Electrodes*

In regard to reference electrode design and application, molten salts differ from aqueous electrolytes in at least two important ways: most of them are in a fully ionized state (the ionic liquids) and they dissolve metals in a unique way.

The structural order of molten salts[24,80-83] is basically different from that of aqueous and nonaqueous electrolyte solutions, which characteristically contain hydrated/solvated ions in a matrix of neutral solvent molecules. In molten salts, we cannot distinguish solvent species in this sense. A distinction between the salt-mixture components is made on the basis of their electrochemically active or inactive behavior (i.e., participating or not in an electrode reaction). In this sense, the inactive salt or a mixture of the inactive salts is often called the "solvent" or "solvent system."

The structural differences between the aqueous/nonaqueous electrolytes and the molten salts lead to several important consequences regarding the physical, chemical, and electrochemical properties of these media,[84] which in turn affect the properties of reference electrodes. At this point, however, only those properties that relate to the liquid-junction potential and to the metal-dissolving capability and stability of the electrolyte will be briefly discussed since these are the most important differences for reference electrode design and application.

The junction potential for molten salts generally is smaller than that for aqueous and nonaqueous solutions that are made with only partially ionized solvent. This observation can be explained in the following way. In contrast to electrolytes that contain hydrated/solvated ions and neutral solvent molecules, the content of any one cation or anion in molten-salt junctions can only be changed at the expense of others of like charge. Therefore, the charge concentration—the number of cations or anions per unit volume—may vary only to a small extent across the junction interphase. Consequently, those forces that create the junction potential by separating the positive and negative ions of the diffusing salt in the solvent matrix have limited effects in the interphases between the molten salts.

Another consequence of the fully ionized structure is that the KCl-electrolyte bridge, which is so useful in aqueous electrolytes to make the junction potential negligible, does not provide the same effect in molten salts.

The generally wide potential-stability range (the high decomposition voltage) of the molten salts is another consequence of the absence of a molecular solvent (water or an organic solvent), which is less stable electrochemically than are most of the molten salt solvent systems. For example, molten salt systems provide work media at the potential of the lithium, sodium, aluminum, or fluorine electrodes, while water would react with these elements. The stability window of a molten salt, which is equal to the decomposition voltage of the salt, can be estimated on the basis of the corresponding electromotive-force series (Section 2.6). Stability limits of various molten salt solvent systems are listed in Table IV.

Many metals can dissolve in their own molten salt. The solubility varies widely, depending on the metal, the salt, and the temperature. Molten halides and sulfides are especially good metal solvents; salts containing anions such as nitrates, sulfates, or phosphates do not form stable metal solutions because they tend to decompose in reactions with the dissolved metal. According to Bredig,[85] these are real solutions of the metal rather than colloidal dispersions, as believed previously. Several reviews[70,86–91] describe the properties of the metal solutions in detail. The exact structure of these solutions, however, is not well understood.[89,90]

Metals are generally soluble only in the salts of that metal, although displacement solubility can also take place. As the following equation shows,

$$Li_{(dissolved)} + KCl_{(l)} \rightleftharpoons LiCl_{(l)} + K_{(vapor)} \tag{18}$$

one metal can produce another one from the cations of the salt in an equilibrium reaction. A detailed discussion on the subject is available.[92] Equilibrium constants of various displacement reactions were compiled by Janz.[93] These equilibrium constants vary with temperature, e.g., for Eq. (18) the equilibrium constant is 0.0273 at 354°C and 0.947 at 550°C.[94] These displacement reactions are very important from the standpoint of the potential and chemical stability of the reference electrodes. The displacement reaction may change the composition of the reference system. The new metal may form an alloy with the electrode metal or evaporate from the cell, e.g., as a result of the reaction in Eq. (18), potassium evaporating from an open Li/LiCl,KCl electrode system gradually changes the composition of the electrolyte.

Solutions of metals in molten salts can be divided into two main groups: (i) metallic- and (ii) nonmetallic-metal solutions. Solutions of the alkali, alkaline-earth, and rare-earth metals belong to the first group, and solutions of the transition metals and the IVB and VB metals of the Periodic Table belong to the second. The division between the two groups is made on the basis of electrical conductivity. The metallic-metal solutions have one or two orders of magnitude higher specific conductivities than that of the corresponding pure liquid salts at the same temperature. The magnitude of this difference depends on the metal, concentration, and temperature.[95] The excess conductivity is electronic. The nonmetallic-metal solutions, which do not show an increased conductivity, form "subhalides" in an unusually low valence state of the dissolved metal.

The high electronic conductivity of the metallic-metal solutions, which will short-circuit the electrodes of the study cell through the electrolyte, and the high chemical reactivity of both types of solutions cause severe difficulties in preparation of reference electrodes with the affected metal. To circumvent these problems, an ionic-conducting membrane that is chemically stable in the given medium is used to separate the critical metal or the metal solution from the salt or the rest of the electrochemical cell. Examples of implementation of this principle will be discussed in Section 4.

In addition to the above-discussed differences, there are some others that relate to the elevated working temperature with molten-salt electrolytes and bear importance for reference electrode work.

Temperature variations have more pronounced effects at the high temperatures of molten salt electrochemical work than they do at ambient temperature. Since the RT/F factor in the Nernst equation is substantially larger in proportion to the absolute temperature, the effect of concentration and/or temperature changes on the potential becomes more critical at higher temperatures.

In contrast to the thermodynamics of the ambient temperature systems, there is no universally accepted standard temperature for defining the standard potential of a high-temperature electrode. Only convenience dictates selection of a base temperature, which is usually in the middle of the investigated temperature range and, consequently, may vary from system to system. However, once an investigator has selected a base temperature for a system, subsequent works on related systems generally try to match that temperature for the sake of comparability of the measured potentials. Plambeck(15) recommended standard temperatures for various liquid solvent–salt systems. These are listed in Table IV.

The exchange current density on metal electrodes in molten salts is generally high. For example, molar exchange current densities have been measured in the range of 0.1–200 A/cm^2 in LiCl–KCl eutectic at 450°C (Ref. 40, p. 340). A recent critical evaluation(96) suggested even higher exchange current densities, which ensure fast electron-transfer kinetics even at low concentrations of the electroactive salt. As a result, (i) the chances for finding a satisfactory reversible reference-electrode system are better in molten salts than in aqueous electrolytes; (ii) electron transfer, which is one of the causes of electrode polarization, usually is of only secondary importance relative to mass-transport-related processes in molten salt electrolytes.

Electrochemical studies with molten salts are done over a wide temperature range. The lower end is defined by the melting point of the electrolyte,(40,97) and the upper end is limited only by the volatilization or decomposition of one or more electrolyte components. (Halogenides evaporate(98,99); salts having oxyanions decompose(100,101) at high temperature.) Overlapping liquidus and stability ranges of various salt systems offer conditions for extensive electrochemical studies from room temperature(102,103) to well beyond 1000°C. Table IV shows working-temperature ranges of several selected molten salt electrolyte systems.

2.6. Standard Electromotive-Force Series of Molten Salts

A standard emf series is a tabulated compilation of standard electrode-potential data organized along a potential scale, which, in turn, is specified by a reference electrode system. Each emf series, however, is valid for the specified solvent electrolyte system and standard reference electrode at a specified temperature only. Critically evaluating the literature of molten salt electrochemistry, Plambeck[15] described 14 electrolyte systems as being the most frequently used media in molten salt electrochemical studies and reported emf series of metal electrodes for each. Characterized by their constituents and compositions, some selected electrolyte systems are shown in Table IV.

As an example of the emf series, Table V shows the standard electrode potentials of some selected electrodes (with respect to the electrode that is indicated by a zero potential value). The last column in the table demonstrates the great diversity in precision of various electrode potential measurements. Switching from one reference electrode basis to another within the same solvent electrolyte system and using the same concentration scale is straightforward. Table V shows the great importance of the unambiguous

TABLE IV. Characteristics of

Composition (mol %)		MFW	Temperature of standardization (°C)	Melting point (°C)	Density (g/cm^3)	Conductivity ($\Omega^{-1}\ cm^{-1}$)
LiCl, KCl eutectic	59, 41	55.59	450	352	1.648 ± 0.001	1.572
NaCl, KCl	50, 50	66.50	727	658	1.5630 ± 0.003	2.418 ± 0.005
$MgCl_2$, KCl	32.5, 67.5	81.27	475	~410	1.630 ± 0.02	1.01
$AlCl_3$, NaCl eutectic	63[b], 37	105.63	175	~170	1.640 ± 0.001	0.239 ± 0.004
LiF, NaF, KF eutectic	46.5, 11.5, 42.0	41.28	500	459		
Na_3AlF_6		209.76	—	1010	2.001 at 1100°C	3.052 at 1100°C
Na_3AlF_6 Al_2O_3 saturated			1000	—	2.035 ± 0.005	2.762

use of the different concentration scales (compare the numerical values of the potential of the same electrode in different columns).

Although each emf series is valid in its strict sense only for the defined electrolyte system, a less precise comparison is permissible only if relatively small variations exist in electrolyte compositions (e.g., variations of the LiCl : KCl ratio). Differences in the constituents of the solvent melt cause larger differences in the standard potentials than the composition variations do. This effect can be assessed by comparing ε^0 data sets of the LiCl-KCl eutectic and that of the NaCl-KCl electrolyte systems in Table VI. This difference in constituents of the solvent melt causes 20-30 mV differences depending on the metal being examined. Potentials measured in greatly different electrolyte systems cannot be compared unambiguously, although a great many efforts have been invested in this topic.[92,105]

An emf series is valid only for the specified temperature. Since the temperature coefficients of the emf of a cell built with various electrode

Selected Solvent Electrolyte Systems[a]

Approximate upper working temperature limit (°C)	Useful potential range (V) and limiting processes	Reference electrode	Usual cell container material
550 vaporization	+1.033, Cl_2 evolution −2.543, Li deposition	Ag/Ag^+	Pyrex glass, high-density alumina
~1000	+0.838, Cl_2 evolution −2.372, alkali metal deposition	Ag/Ag^+	Quartz, high-density alumina
1075 vaporization	+0.756, Cl_2 evolution −1.921, Mg deposition	Ag/Ag^+	Pyrex glass (for a few hours), high-density alumina
~400, $AlCl_3$ vaporization	2.1, Cl_2 evolution 0, Al deposition	$Al/[AlCl_4]^-$	Pyrex glass, Teflon
900	~4	Ni/Ni^{2+}	BN, glassy carbon, graphite
1200-1300 vaporization	~2 V on C, anode effect 0, Al deposition	Al/Al^{3+}	BN, graphite
1200-1300 vaporization	2.1, O^{2-} to O_2 on Pt 1.1, to CO_2 on graphite 0, Al deposition	Al/Al^{3+}	BN, graphite, alumina

(*cont.*)

TABLE IV (cont.).

Composition (mol %)		MFW	Temperature of standardization (°C)	Melting point (°C)	Density (g/cm^3)	Conductivity ($\Omega^{-1}\,cm^{-1}$)
$AlCl_3$, BPC[c]	63 37	147.56	—	~−30	1.33 at 25°C	$6 \cdot 10^{-3}$
$LiNO_3$, KNO_3 eutectic	43 57	87.28	177	132	1.973 ± 0.002	0.223
$NaNO_3$, KNO_3	50 50	93.05	250	228	1.947 ± 0.010	0.510 ± 0.003
$LiCO_3$, $NaCO_3$	50 50	74.97	550	~497	2.029 ± 0.003 (eutectic)	1.73 ± 0.03
Li_2SO_4, K_2SO_4 eutectic	80 20	122.81	675	535	2.090 ± 0.002	1.282
NaOH, KOH eutectic	51 49	47.89	227	185	1.81 ± 0.062	1.40
$NaPO_3$, KPO_3	50 50	110.2	700	567	2.231	Not available

[a] Table is based on Plambeck's compilation.(15) MFW is the mean formula weight of the solvent electrolyte system. Temperatures of standardization are recommended by Plambeck, most of the previous experiments were carried out at or close to these temperatures. Density and conductivity are given at the temperature of standardization. Useful potential ranges are indicated by the standard potentials of the evolving elements versus the reference electrode indicated in the next column when given with mV precision; approximate ranges indicated with ~ symbol show voltage span not defined on the potential scale of the reference electrode. Indicated container materials were used in—usually short—laboratory experiments; also other

pairs may differ considerably (they are generally in the 0 to ±1.2 mV/K range) and the value of the coefficient may also show variation in different temperature ranges (e.g., as a result of phase transformation), the sequence of the electrodes in the series may change with the temperature. This phenomenon is very important to consider when electrode-material stability for the required chemical environment is assessed at various temperatures. Hamer *et al.* reported emf values calculated from thermochemical data for metal chloride(17) and other halogenide(106) formation cells for a wide temperature range, up to 1500°C.

Approximate upper working temperature limit (°C)	Useful potential range (V) and limiting processes	Reference electrode	Usual cell container material
	~2.1, oxidation of org. cation 0, Al deposition	$Al/[AlCl_4]^-$	Pyrex
227 anion decomposition	+0.5, $NO_3^- \rightarrow NO_2 + \frac{1}{2}O_2 + e^-$ below −1.30, $NO_3^- + 2e^- \rightarrow NO_2^- + O^{2-}$	Ag/Ag^+	Pyrex glass
528 (600)	+0.45; same as above −1.80	Ag/Ag^+	Pyrex glass
~600 decomposes to CO_2 + metal oxide	+0.1 to 0.2 V, $CO_3 \rightarrow CO_2 + O_2 + 2e^-$ −1.8 to −2.0 V, $CO_3 + 4e^- \rightarrow C + 3O^{2-}$	CO_2, O_2 ref. el.	Pt, Au Porcelain, alumina
~750	+0.7, $SO_4^{2-} \rightarrow SO_3 + \frac{1}{2}O_2 + 2e^-$ ~ −2.0, $SO_4^{2-} + 2e^- \rightarrow SO_3^{2-} + O^{2-}$	Ag/Ag^+	Pyrex glass Pt, Au
500	~1.4 range, depending on the basicity of the melt Anodic: O_2^{2-}, O_2 Cathodic: H_2, alkali metal	(Na/Na^+)	Teflon, glassy carbon, Ni
>1000	~1.0, $2PO_3^- + \frac{1}{2}O_2 \rightarrow P_2O_5 + 2e^-$ $LiPO_3^- + 5e^- \rightarrow P + 3PO_4^{3-}$	Ag/Ag^+	Pt

materials are available. More information about reference electrodes and materials of construction used in various electrolyte systems can be found in Sections 3.2.2 and 4.

[b] Because of acid-base equilibrium, properties of the electrolyte sensitively vary by the Na : Al atom ratio.

[c] BPC stands for *N*-(*n*-butyl)pyridinium chloride. This solvent-electrolyte system is an example of the numerous room-temperature melts; properties of the various systems vary depending on the organic cation and the $AlCl_3$: R-Cl ratio.[102,103]

The emf series is an invaluable tool in various types of electrochemical studies. In reference electrode design and application, it provides (i) a firm theoretical basis for selection of the proper reference electrode that is most compatible with the system to be studied, (ii) information about possible effects of known impurities of the electrolyte or electrode materials, (iii) a good estimate on the stability and corrosion behavior of the test and counter electrodes, as well as construction materials, and (iv) an indication of possible exchange reactions between electrode materials and electrolytes, (e.g., metal displacement reactions in Daniell cells when the electrolyte

TABLE V. Standard Electromotive-Force Series in LiCl-KCl Eutectic at 450°C[a]

Electrode	ε^0_M(Pt) (V)	ε^0_m(Pt) (V)	ε^0_N(Pt) (V)	ε^0_m(Ag) (V)	Precision (V)
Li(I)/Li(0)	−3.304	−3.320	−3.410	−2.593	0.002
Na(I)/Na(0)	−3.25	−3.23	−3.14	−2.50	0.008
Zn(II)/Zn(0)	−1.566	−1.566	−1.566	−0.839	0.002
Ti(IV)/Ti(0)	−1.486	−1.494	−1.539	−0.767	0.05[b]
Cd(II)/Cd(0)	−1.316	−1.316	−1.316	−0.589	0.002
Fe(II)/Fe(0)	−1.172	−1.172	−1.172	−0.455	0.005
Cr(III)/Cr(0)	−1.125	−1.130	−1.160	−0.403	0.01[c]
Pb(II)/Pb(0)	−1.101	−1.101	−1.101	−0.374	0.002
Sn(II)/Sn(0)	−1.082	−1.082	−1.082	−0.355	0.002
S(1), C/S_x^{2-}	−1.008	−1.039	−1.219	−0.312	0.002[d]
Co(II)/Co(0)	−0.991	−0.991	−0.991	−0.264	0.003
Cu(I)/Cu(0)	−0.957	−0.941	−0.851	−0.214	0.004
Ni(II)/Ni(0)	−0.795	−0.795	−0.795	−0.068	0.002
Fe(III)/Fe(0)	−0.753	−0.758	−0.788	−0.031	0.006[c]
Ag(I)/Ag(0)	−0.743	−0.727	−0.637	0.000	0.002
I_2(g)/I^-	−0.207	−0.254	−0.525	+0.473	0.008
Pt(II)/Pt(0)	0.000	0.000	0.000	+0.727	0.002
Cu(II)/Cu(I)	+0.061	+0.045	−0.045	+0.772	0.002
Fe(III)/Fe(II)	+0.086	+0.070	−0.020	+0.797	0.003
Pt(IV)/Pt(II)	+0.142	+0.126	+0.036	+0.763	0.010
Br_2(g)/Br^-	+0.177	+0.130	−0.141	+0.857	0.002
Au(I)/Au(0)	+0.205	+0.221	+0.311	+0.948	0.008
Cl_2(g)/Cl^-	+0.322	+0.306	+0.216	+1.033	0.002

[a] Entries in table were selected from Plambeck's compilation (Ref. 15, p. 68). All electrode potentials are given on the infinitely dilute standard bases (ε^0_i). Subscripts *M*, *m*, and *N* indicate molarity (mol/liter solution), molality (mol/kg solvent), and mol fraction concentration scale values. Symbols in parentheses indicate the reference-electrode system used in a column: (Pt) standard molar or molal Pt/Pt^{2+} reference electrode,[28] (Ag) standard molal Ag/Ag^+ reference electrode.[28]
[b] Precision estimated by Plambeck.
[c] Precision calculated by Plambeck.
[d] Precision extrapolated by Plambeck.

compartments are poorly separated). Examples of these calculations (i–iv) can be found, e.g., in Ref. 17.

3. *Principles of Reference Electrode Construction and Application*

Reference electrodes as measuring tools of electrode potentials are the physical realizations of the theoretical principles discussed in Section 2. Principles of reference electrode construction and application are the same for both aqueous and molten salt systems. Implementation of the principles, however, usually requires different designs and material selection for molten salts. In addition, the reference electrode user faces more technical difficul-

TABLE VI. Comparison of ε_i^0 Standard Potentials of Metal Electrodes in Two Chloride Solvent–Electrolyte System[a]

Electrode system	LiCl-KCl eutectic	NaCl-KCl equimolar
$Mn\|Mn^{2+}$	−1.212	−1.230
$Zn\|Zn^{2+}$	−0.929	−0.930
$Cr\|Cr^{2+}$	−0.788	−0.810
$Cd\|Cd^{2+}$	−0.679	−0.630
$Fe\|Fe^{2+}$	−0.534	−0.560
$Cr\|Cr^{3+}$	−0.523	−0.525
$Pb\|Pb^{2+}$	−0.464	−0.447
$Sn\|Sn^{2+}$	−0.445	−0.410
$Co\|Co^{2+}$	−0.354	−0.372
$Cu\|Cu^{+}$	−0.214	−0.240
$Ag\|Ag^{+}$	0.0	0.0
$Cu\|Cu^{2+}$	+0.189	+0.200
$Cl_2\|Cl^{-}$	+0.853	+0.905

[a] Data for the LiCl-KCl electrolyte were measured at 450°C;[28] for the NaCl-KCl electrolyte, the electrode potentials were measured in the temperature range of 700–900°C (Table III) and extrapolated to 450°C.[16]

ties for high-temperature applications. Contrary to the wide selection and availability of ready-made (off-the-shelf) reference electrodes for aqueous electrolyte applications, reference electrodes for molten-salt work are not commercially available. Consequently, the investigator has to invest considerable time in adaptation (or even invention) and fabrication of the reference electrode, which cannot be done properly without careful design work. The purpose of this section is to provide some guidelines for construction and application of molten salt reference electrodes.

3.1. *Construction and Use of Molten Salt Reference Electrodes*

Selection of a reference electrode for a given task is based mainly upon convenience and compatibility with the chemistry of the system under investigation. In principle, any of the electrode systems listed in the emf series can be used as a reference electrode for a particular investigation. In developing an electrode system into a useful reference-electrode device suitable for a particular measurement, one must be aware of several basic factors:

i. Reversibility of the electrode system (discussed in this section).
ii. Stability and reproducibility of the potential (Section 3.2.1).
iii. Stability of the reference-electrode materials (Section 3.2.2).

iv. Impedance of the reference electrode (Section 3.3.1); and
v. Compatibility of the reference electrode with the type of measurement (discussed in this section).

The degree of emphasis on any of these factors depends on the needs of the measurement in question.

The reversibility, which is ensured by fast electrode kinetics (i.e., high exchange current density), is the basic requirement for a reference electrode system. If the reversibility of an electrode process is not satisfactory, the stability and reproducibility of the potential is questionable, since in this condition the electrode is very sensitive to impurities and current load. If a new electrode system is introduced to the experimentation, the electrode should be tested for reversibility of the potential and for sufficiently fast kinetics.

The reversibility test involves application of the micropolarization test. In this test, the electrode is polarized in an inner-reference electrode type arrangement within a short, usually less than ±5 or ±10 mV potential range in the so-called linear polarization response section of the η vs. i curve. The ohmic resistance of the cell must be low, or the measured polarization value must be corrected for IR-voltage drop. Consequently, this test cannot be performed with a reference electrode that has a high-resistance diffusion barrier. Criteria for the proper functioning of the electrode are (i) straightness of the polarization "curve," which must intersect the origin of the polarization versus current density plot, as shown in Fig. 2a, and (ii) low polarization resistance (η/i, Ω cm^2 of the exposed, geometrical electrode area). A defined value of the polarization resistance, however, cannot be given as a minimum requirement for reversibility; circumstances decide the acceptance level, which is usually not higher than 100 Ω cm^2. Any hysteresis is indicative of irreversibility (Fig. 2b). Reversibility has to be tested at several temperatures in the intended temperature range. The potential of a gas reference electrode must not depend on the rate of the gas flow after steady state and saturation have been reached.

Any counter electrode that has an invariant potential during the test will be satisfactory; it can even be the same electrode system provided its surface area is 10–100 times larger than that of the tested electrode. Potentials measured versus an electrode of known electrochemical properties during slow thermal cycling provide means for calculation of the temperature coefficient of the emf (Fig. 2c). Furthermore, a supposed cell reaction can be confirmed if the agreement between the experimental data and the calculated corresponding thermodynamic quantities is satisfactory.

Adequate reversibility ensures fast and perfect recovery of the potential after a short, inadvertent current load (e.g., caused by a momentary short circuit between the cell electrodes that has not changed the reference

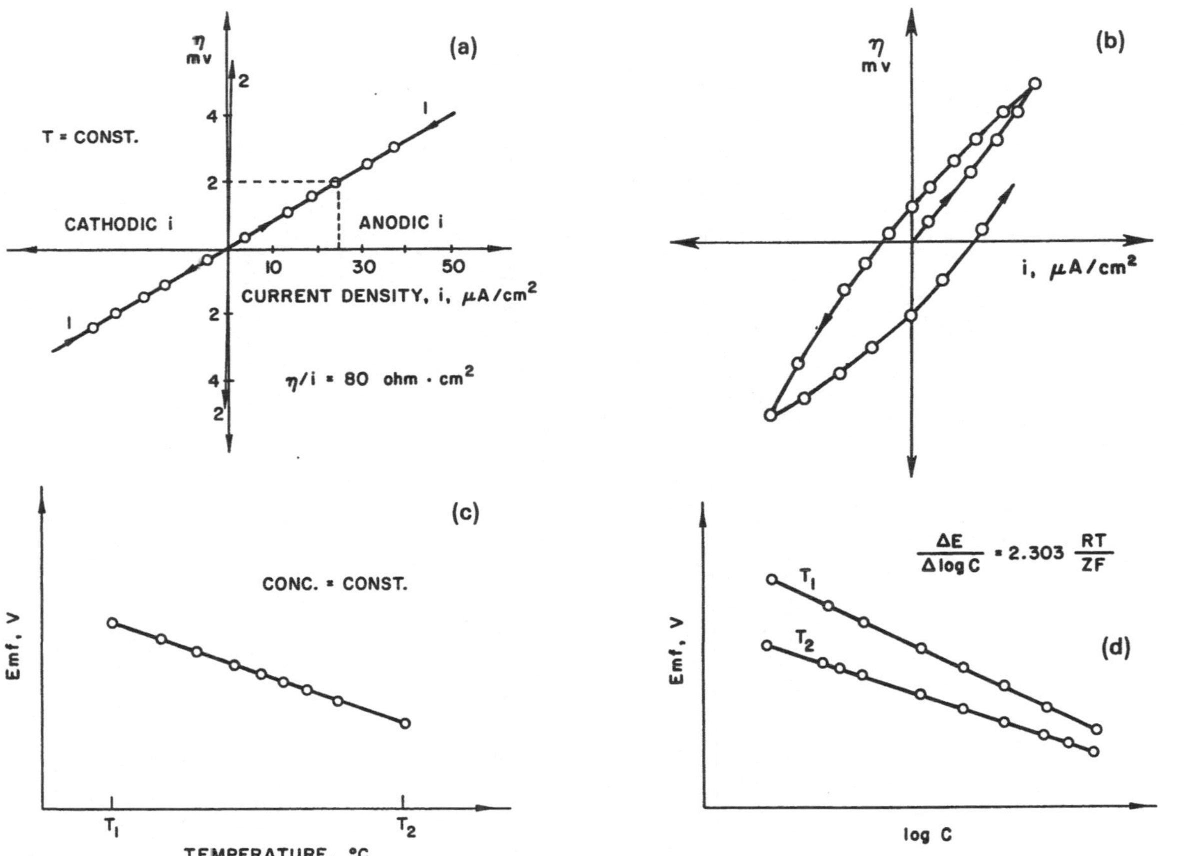

FIGURE 2. (a) Plot obtained in a micropolarization test. 1, Indication of reversibility; 2, Very high polarization resistance indicates irreversibility or inertness of the electrode. (b) Hysteresis indicates irreversibility. (c) Emf measurements at steady temperatures during thermal cycling provide data for entropy-function calculation [Eq. (8)]. (d) Nernstian response of an electrode.

electrode composition significantly). Application of a short current load and observation of the fast potential recovery may be used as a simple test of proper functioning of the reference electrode.

Using a concentration cell with negligible liquid-junction potential, the thermodynamic equilibrium condition of the selected electrode system can be demonstrated by the Nernstian response of the electrode potential for concentration variation. The Nernstian response requires that the slope of the ε versus $\ln c$ line is equal or close to the numerical value of the RT/zF factor (Fig. 2d). For an example of how the above factors are determined and evaluated numerically, see Ref. 30. This paper (Ref. 30), illustrated with numerous diagrams and tables, describes studies of Cl_2/Cl^-, Ag/Ag^+, and Pt/Pt^{2+} electrodes in KCl–$MgCl_2$ eutectic.

Careful work requires a sufficiently low impurity level in the electrolyte and active materials. Since a generally applicable acceptance level cannot be specified, the investigator has to determine the necessary and adequate purity conditions of a reference electrode system, e.g., by the above-described reversibility test. Water[107–110] presents probably the greatest problem as an impurity, except when used as a desired component of the molten salt system, e.g., in hydroxide melts. It is readily absorbed by the salts from the atmosphere and is very difficult to remove either from the salts used to prepare the electrolyte or from the molten electrolyte. Even heating in vacuum is not sufficient to remove water completely from a salt. Besides water, oxygen and oxide species[111] (except when needed for the study) and heavy-metal ion impurities have the most important disturbing effects in the reference-electrode potential determining processes. The chemical and electrochemical methods used to purify the electrolytes and the voltammetric and other analytical methods of electrolyte impurity level evaluation are reviewed in the first volume of this series and in Refs. 107–111. Availability of high-purity, polarographic or optical grade, custom-made salt mixtures[112,113] offers substantial time saving in experimentation. Another source of impurities is the "inert" atmosphere in the cell or glove box. Techniques for maintaining truly inert atmospheres are described in Ref. 3 and a chapter in this volume.

The compatibility factor includes several aspects. The chemistry of the reference electrode must be compatible with that of the test electrode in an inner-type reference electrode application. In a half-cell-type application, this requirement is not so strict. Nevertheless, the reference electrode should be designed to be as similar as possible to the chemistry of the system under investigation. As a general rule, the best choice is the reference electrode system that is chemically identical to the test electrode. Then, the problems of the liquid junction and the impurity effects are greatly reduced. Selection of the type of the reference electrode depends on the type of measurement. For example, the inner type is preferred in thermodynamic studies, and the half-cell-type equipped with a Luggin capillary (Section 3.1.3) is a must in kinetics investigations. For instrumental reasons, a reference electrode with a potential that is outside the range of the test-electrode potential variation may be preferred. This choice eliminates the inconvenient sign change of the measured value during the experiment.

Useful tabulations of various physical properties of molten salts such as density, surface tension, conductance, viscosity, diffusion, enthalpy of fusion, heat capacity, thermal conductivity, vapor pressure, and cryoscopic constants are available for aiding reference-electrode and experiment design.[6,114–117] Also available is a complete description of the molten salt data and molten salt standard programs of the Molten Salt Datacenter (Rensselaer Polytechnic Institute, Troy, New York 12181, USA) along with a list of the published data compilations including machine-readable molten salts bibliography and eutectics database.[117]

3.1.1. Types of Cell Electrodes

The principles of construction and application are in close relationship with the role that a reference electrode plays in an electrochemical cell. There are three types of electrodes involved in potential measurements: test, reference, and counter electrode. The test electrode (also called the working or indicator electrode) is the one that is the subject of the measurement. The reference electrode is used to measure or control the potential of the test electrode. A third electrode, the counter electrode (often called the auxiliary electrode) is required as an electron source or sink to avoid current load in the reference electrode when the test-electrode potential has to be investigated under current load or controlled by a potentiostat.

There are electrode arrangements in which electrodes may play a double role. A reference electrode may simultaneously serve as the counter electrode, if large surface area and robust construction ensure its negligible polarization in the required current range. This principle may be applicable up to a few milliamperes. In many cases, simultaneous potential measurement of both the test and the counter electrode may become necessary. Then both electrodes become test electrodes and are studied by one or two reference electrodes. With two reference electrodes, simultaneous potential measurements on both "test" electrodes can be implemented; moreover, the IR-voltage drop in the electrolyte between the two reference electrodes can be monitored. This latter advantage is used, for example, in dc-current electrolyte conductivity measurement techniques.

3.1.2. Design Principles of Reference Half-Cells

As mentioned in Section 2.4, the three types of reference-electrode constructions are half-cell, inner, and quasi. Of these, the half-cell reference electrode has the widest variety of applications, but also requires the most laborious construction. The main components of the half-cell reference electrode are (i) an electrochemically active part (the electrode itself) equipped with a terminal which provides metallic connection to the electrical circuit, (ii) an inner or reference electrolyte within the reference-electrode housing, (iii) a reference-electrode housing, which is made of an insulator material to separate the reference electrolyte chemically and electrically from the test electrolyte, and (iv) an electrolyte junction, which provides an ionic contact between the two electrolytes. The electrolyte junction is either a direct junction between the two liquid electrolytes or a membrane junction in membrane electrodes. Most of the problems of construction, after the decision for the chemistry of the reference electrode system has been made, relate to the electrolyte junction and the housing; therefore, these two components will be discussed in detail in this section. Design of the inner or quasi-reference-electrodes does not require further discussions, since they are simply metal wires or plates submerged directly into the test electrolyte. Construction of the membrane electrodes will be described in Section 4.

3.1.3. Design of the Electrolyte Junction

By separating the reference electrode compartment from the rest of the test cell, the reference half-cell has three important features: (i) wide selection for the reference-electrolyte composition, (ii) possibility for use of different gas atmospheres in the reference half-cell and the test-electrode compartments, and (iii) allowance for precise positioning of the electrolyte junction relative to the test electrode, a feature that is very important in kinetic studies. A diffusion barrier must be placed in the electrolyte junction to prevent rapid intermixing of the different electrolytes and to hinder interdiffusion of the electrolyte constituents that may cause an impurity effect in the other compartment. Impurities may shift the potential from an equilibrium value to an undefined, varying mixed potential. Examples of the materials used in diffusion barrier construction are fritted glass, porous ceramics (alumina, beryllia, magnesia, yttria, porcelain, mullite, boron nitride, etc.) or graphite, and asbestos thread compacted into a capillary tube made of high-density ceramics or glass (the latter is usually drawn to a fine capillary with the asbestos thread in it). Various forms of electrolyte junctions and cell designs are reviewed in Ref. 3.

When the test-electrode potential is measured under current load, a pinpoint positioning of the electrolytic junction in the vicinity of the test electrode is of primary importance to minimize the error caused by the IR-voltage drop. To satisfactorily control the placement, the junction is formed as a small opening—usually plugged up with a porous material—in the reference electrode housing. A Haber–Luggin capillary (commonly designated as Luggin capillary or Luggin probe), a slim capillary tube attached to the housing, promotes even better positioning of the junction. A long, fine capillary may serve as a diffusion barrier and eliminate the need for a porous plug. In kinetics studies, some form of the capillary junction is required for properly placing the opening, even if the reference and the test electrolytes are identical and a diffusion barrier is unnecessary.

The potential measured under current load will include a significant IR-voltage drop, if the electrolytic junction point is too far from the test electrode. On the other hand, if the junction point is too close, the insulator wall around the junction (the wall of the capillary) causes a shielding error and throws a "current shadow" on the adjacent areas of the test electrode, resulting in decreased local current density. The rules for minimizing the resultant effect on the potential measurement are outlined elsewhere.[(118)]

The classical works of Piontelli[(119–121)] define the proper ways of placing and shaping the Luggin-capillary tip. Also, a modeling calculation on the capillary tip placement is available.[(122)] In most molten salt experiments, however, these special arrangements cannot be implemented because of a lack of a shapable construction material or because of a fixed, unalterable geometry of the cell. Therefore, molten salt cells usually require special capillary connections. For laboratory experimental cells, which have flexibility in geometry and design, an insulator-block arrangement is very popular. Figure 3a shows the main features of this arrangement in a molten-carbonate fuel cell.[(123)] The insulator block, which envelopes the test electrode, ensures uniform current distribution over the face of the test electrode and provides a properly placed electrolyte channel, a "Luggin capillary" between the reference electrode and the paste cell electrolyte. Figure 3b shows the design details of the gas reference electrode used in this arrangement.

Design and placement of the reference electrodes that are to be used in engineering cells must conform to the fixed geometry of the cell (e.g., of battery cells or industrial electrochemical processing cells). In these applications, the design options are even more limited than in laboratory experimental cells. Straight tube-shape reference electrodes equipped with a capillary junction at the lower end have been used in the testing of molten-electrolyte Li–Al/FeS battery cells.[(124)] Potential measurement during scanning of the reference electrodes at the horizontal and vertical edges of the cell electrodes provided information about the potential and current distribution within the cell (Fig. 4). The long reference electrodes were manipulated through an opening of the cell from the outside of the high-temperature zone. An interesting example presented in Fig. 5 shows how the problem was solved in an aluminum smelter in which even a capillary junction was not applicable.[(125)] An inner-type Al reference electrode was prepared *in situ* by depositing aluminum on the tungsten tip. The rest of the assembly

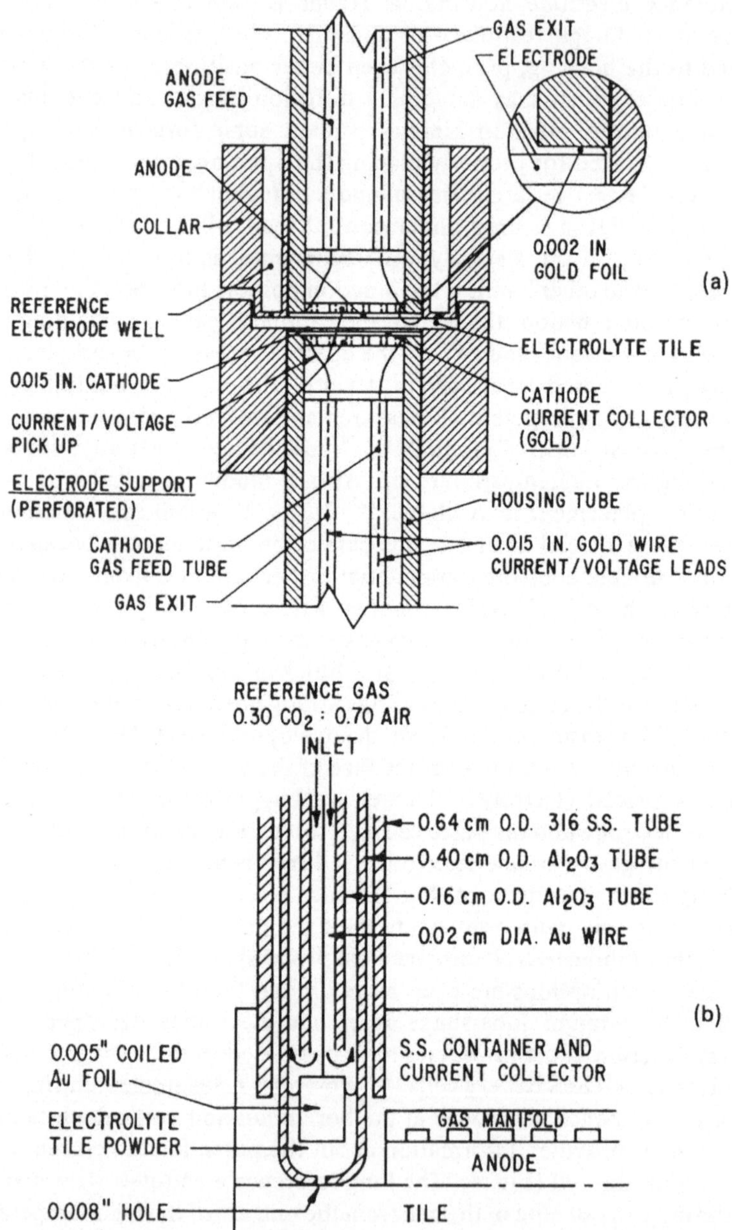

FIGURE 3. **Experimental molten carbonate fuel cell, General Electric Co.[123] (a) Insulator-block design for reference electrode placement. All structural parts made of alumina unless otherwise noted. (b) Gas reference electrode design. (By permission of the publisher.)**

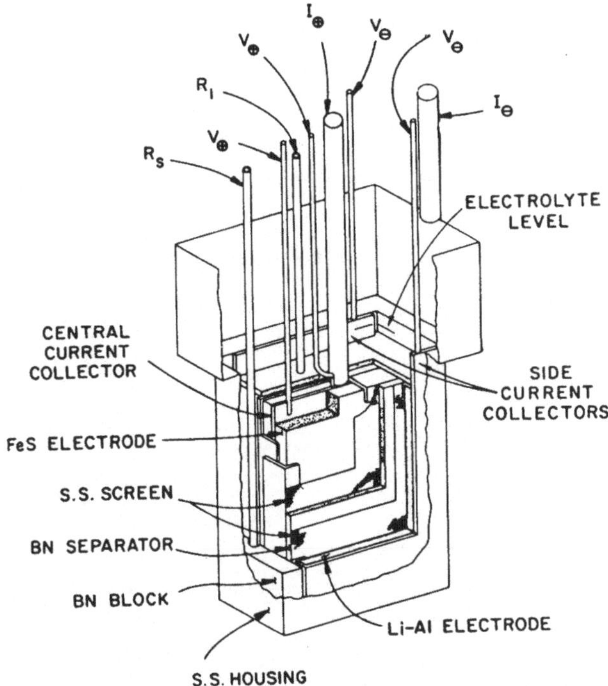

FIGURE 4. Ni/Ni_3S_2 reference electrodes (R_I and R_s) used to measure potential distribution in an Li-Al/FeS cell.[124] V_+ and V_- are voltage leads; I_+ and I_- are current leads. (By permission of the publisher.)

was electrochemically inert and served only as support for the active tip and the voltage lead wire. Instead of a capillary connection, the electrode itself was positioned and scanned in various directions between the molten aluminum cathode and the carbon anode. Although the active tip was not a tiny component, it was small enough relative to the size of the 10,000-A smelter to obtain good spatial resolution of the potential field.

Several conflicting factors must be taken into consideration when constructing a diffusion barrier. A denser and thicker porous plug or a finer and longer Luggin capillary provides a more effective diffusion barrier and prevents significant electrolyte composition changes for a longer time. When an electrolyte-level or gas-pressure difference is present in a cell, a diffusion barrier with high flow resistance is required to minimize an electrolyte flow between compartments of the reference electrode and the test electrolyte. The effectiveness of the diffusion barrier is related to its electrical resistance. Approximately 1–20 kΩ resistance is needed to get a proper diffusion and flow resistance.

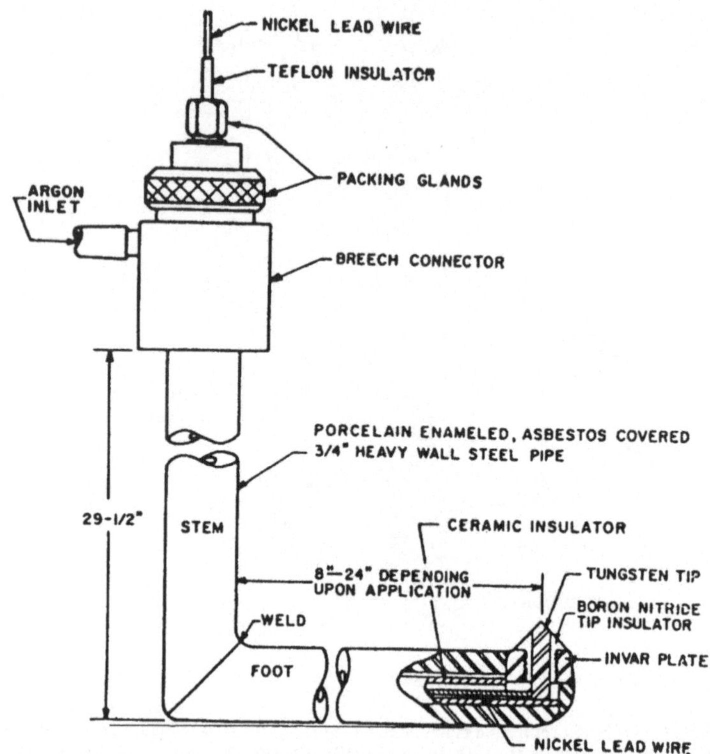

FIGURE 5. An inner-type Al/Al^{3+} reference electrode used as a scanning probe in an aluminum smelter.[(125)] (By permission of the publisher.)

The electrolyte level in the reference electrode compartment must be adjusted to the level of the test electrolyte to eliminate excessive temperature gradients and the danger of the contamination that the latter electrolyte could cause by seeping into the reference half-cell. In room-temperature cells, the reference-electrolyte level is commonly set higher. However, one has to consider that in high-temperature cells this practice may introduce an unknown temperature difference between the electrochemically active parts of the reference electrode and the test electrode. In this case, the higher, protruding reference-electrolyte column is more exposed to cooling effects than the lower, better thermostated test electrolyte.

To prevent diffusion or electrolyte flow, a high resistance is advantageous; however, from the standpoint of the potential measurement or control, exactly the opposite is the case, i.e., a low electrical resistance is desirable. The circumstances for a particular experiment determine whether high or low resistance of the diffusion barrier is more important and what resistance value is a proper trade-off between the two requirements. Further details of this subject will be discussed in Section 3.3.1.

In addition to the porous plug and the Luggin capillary, a third type of the electrolyte junction is the ionic-conducting membrane (discussed in Section 2.3.2). The ionic-conducting membrane is a perfect barrier. However, membrane electrodes are usually bulky; therefore, their positioning for IR-voltage drop "elimination" is problematic unless the membrane electrode is connected to the test cell through a properly placed electrolyte channel (e.g., similarly to the arrangement that is shown in Fig. 3a).

The structure of the diffusion barrier influences the properties of the liquid junction and, hence, the magnitude of the liquid-junction potential. In Ref. 17 on pp. 124–135 and the references cited therein, the reader will find a complete description of the basic types of diffusion processes in a liquid junction and the formulas for estimation of the accompanying junction potentials. Since the formation of a liquid-junction potential is unavoidable if the electrolytes differ, the only way to control it is to keep the conditions as steady as possible within the diffusion barrier and thereby ensure comparability of the potentials that are measured at different times during the experiment. Fortunately, the variation of the liquid-junction potential with time is rather small—even in extreme cases not more than a few millivolts. Nevertheless, in high-precision measurements, this effect has to be considered. Reference electrodes of identical diffusion barriers and well-stabilized diffusion profiles should be used when comparability of the measured potentials is critical.

3.1.4. The Thermo-Emf Signal

Errors in high-temperature potential measurements may occur because of the large difference between the temperature of the signal source and that of the measuring instrument. Schematics of two basic types of cell-component arrangements illustrate below why a careful design is necessary and how this thermo-emf error can be eliminated:

$$\text{Instrument}\,|\,\text{Cu lead}\,|\,M_1\,|\,\text{electrolyte}\,|\,M_2\,|\,\text{Cu lead}\,|\,\text{instrument}$$

$$\qquad TA_1 \qquad TA_2\,TH_1 \qquad TH_2\,TA_3 \qquad TA_4$$

$$\text{if}\quad \begin{aligned} TH_1 &= TH_2 \\ TA_2 &= TA_3 \\ TA_1 &= TA_4 \end{aligned} \quad \text{the thermo-emf} = f(TH_1 - TA_2, M_1, M_2)$$

Cell 11

$$\underset{TA_1}{\text{Instrument}|\text{Cu lead}}|\underset{TA_2\,TH_1}{M_1}|\text{electrolyte}|\underset{TH_2\,TH_3\,TA_3}{M_2|M_1}|\underset{TA_4}{\text{Cu lead}|\text{instrument}}$$

$$\text{if}\quad \begin{array}{l} TH_1 = TH_2 = TH_3 \\ TA_2 = TA_3 \\ TA_1 = TA_4 \end{array} \qquad \text{the thermo-emf} = 0$$

Cell 12

Cells 11 and 12 differ in the way that the electrical leads and the electrode metals (M_1 and M_2) are arranged in the high-temperature (TH) and ambient temperature (TA) zones. In Cell 12 the leads that bridge across the intermediate temperature zone between TH and TA are made of identical metals, while in Cell 11 they are different. Although all interphases and interfaces may be sources of thermo-emf, interfaces at TA_1 and TA_4 do not cause problems, except in measurements requiring microvolt accuracy. On the other hand, the large temperature difference between TA and TH may generate a significant thermo-emf that has to be taken into account. Cells 11 and 12 are different, however, in this regard. The emf of Cell 11 includes a thermo-emf component because the connections of the dissimilar metals (M_1 and M_2) are at different temperatures (TA_2/TA_3 and TH_1/TH_2). This thermo-emf is well defined and accountable. A correction for this emf can be made if necessary, provided that $TA_2 = TA_3$ and $TH_1 = TH_2$. The thermo-emf contribution is generally in the range of 5–20 mV, depending on the metals (M_1 and M_2) and the temperature difference. On the contrary, the thermo-emf contribution in Cell 12 is very elusive if the junction of metals M_1 and M_2 is in the intermediate temperature zone at an unknown TH_3 temperature ($TH_2 > TH_3 > TA_3$). The proper cell design ($TH_1 = TH_2 = TH_3$) eliminates the unknown thermo-emf contribution and actually also excludes this factor from the measured voltage signal. To avoid ambiguity, documentation of high-temperature potential values must define the cell arrangement in the temperature zones and state whether any thermo-emf correction is included in the given values.

Proper care should be taken to ensure that the connections between M_1 or M_2 and the copper leads are placed far enough from the high-temperature zone to avoid uncertain temperatures at TA_1 and TA_2 junctions, which would be caused by radiating heat or gas convection.

When a thermo-emf correction is needed for Cell 11, it is recommended that this quantity be measured directly with the same sort of metal pair that is used in the cell, instead of relying on tabulated values. Experience shows that nominally identical metals may behave differently in a thermocouple, since unknown impurities or alloying agents can significantly influence the value of the thermo-emf.

Temperatures TH_1 and TH_2 must be equal to maintain equilibrium in the test cell. Otherwise, as a result of the thermal gradient, thermal diffusion actuates a gradual change of the electrolyte and electrode composition until a steady-state condition is reached. This process, called the Soret effect, induces potential changes on electrodes M_1 and M_2, resulting in deviations from the equilibrium values. A detailed treatment of the problem (Ref. 17, p. 254) gives guidelines to assess the magnitude of this effect. The thermally induced potential differences are generally in the 0–10 mV range if a large temperature distribution exists within the electrochemical cell. Without proper control, temperature differences may be quite significant. Temperature differences of 10–15 K have been observed in stagnant electrolytes and as high as of 50-60 K in cells not properly thermostated.[126] Stirring of the electrolyte maintains uniform temperature and concentration in the cell (e.g., in the one shown in Fig. 23). Application of heat-distributing metal blocks reduces or eliminates excessive temperature gradients and equally distributes heat flux on the heat-providing surfaces. Properly thermostated cells should not show larger than ± 1 K temperature distributions and variations with time, even for prolonged periods.

An often-overlooked error source is the thermally induced mixed potential, which may cause an error of several millivolts. An example of this elusive phenomenon will be discussed in Section 4.1.2 at the description of the Ag/Ag^+ reference electrode.

A high temperature gradient within the cell causes erroneous readings of electrode temperature if the temperature sensor is placed too far from the test or reference electrode. This error can be significant, considering the high temperature gradients that may exist in high-temperature cells. Incorrect readings of electrode temperature can be eliminated by using special reference electrodes that include a temperature sensor. In one such design (described in Section 4.1.5, Fig. 24) the electronic-conducting phase of the reference electrode simultaneously functions as one of the metal components of the thermocouple. As a result, this device senses the exact electrode temperature under the condition of high temperature gradient or during fast temperature changes.

3.2. Stability of Reference Electrodes

Two aspects of reference electrode stability are considered here: stability of the potential and stability of the electrode construction. Stability cannot be defined without consideration of the interactions between the reference electrode and its applicational environment. Since the potential of a single electrode cannot be measured, the experimentally determined, apparent potential stability of the tested reference electrode includes the instability that the other reference electrode used for comparison may have.

Naturally, the signal-processing stability of the measuring system must be better than the potential stability of the electrodes. However, discussion of the factors that influence the instability and signal distortion of the measuring system are beyond the scope of this chapter.

The potential of a reference electrode, by definition, is zero or a known value relative to a standard reference electrode system. The user, however, has to know the limits within which this statement is true. The limits are defined by (i) the reproducibility of the electrode potential, i.e., the statistically defined deviations when potentials of identically prepared reference electrodes are compared to each other; (ii) the stability of the potential, i.e., the range within which the potential varies under normal conditions of application; and (iii) electrode life, i.e., the length of time within which the required accuracy and precision of the measurement can be preserved. Reproducibility depends on the inherent chemical and electrochemical properties of the reference electrode system and also to a great extent on the method of electrode preparation. The stability of the potential is controlled by the same factors, but also is influenced by the applicational environment. These two factors (i and ii) are combined together in many reports and referred to as the precision of the reference electrode potential. Precision, therefore, indicates the limit within which a reference electrode potential can be reproduced or maintained under the specified conditions. The useful electrode life is determined by such factors as physical conditions of the measurement, electrode design, and durability of the materials of construction used in the reference electrode; therefore, it can be considered as a measure of the constructional stability of a reference electrode device. One has to keep in mind that numerical values obtained for these three greatly depend on the details of the chemistry of the system and the temperature of the application.

Knowledge of the factors that control potential reproducibility and stability for reference electrodes of molten salt systems lags behind that of the aqueous systems. The most precise potential measurements have been made with aqueous systems, such as the hydrogen, calomel, and silver/silver chloride reference electrodes, as well as the electrodes of the Weston standard cell. These electrodes, when prepared with careful attention to purity and procedure, are characterized by a better than ±20 μV reproducibility and stability (Ref. 69, p. 222). Aqueous silver/silver chloride electrodes showed stable potentials within ±5 μV for periods of several months.[127] Reproducibility data reported in Section 4 indicate a more modest precision for reference electrodes of molten salts. The best stability for reference electrodes is found in the Weston Cell, which actually consists of two aqueous reference electrodes of the second kind and used as a primary voltage standard. The stability of the emf in this cell is measured in tenths of a microvolt. This stability, however, can be achieved only by exercising extreme care in electrode preparation and by maintaining constant physical conditions in the cell and for the potential measurement.

The molten salt reference electrodes are commonly thought to have inherently poorer potential stability than their aqueous counterparts. This is not the case. Although some investigators have reported difficulty in reproducing potentials to the nearest 10 mV or even 50 mV, figures are more commonly given to the nearest 1 mV, supposedly as an indication of reproducibility and precision. The reported precision of a few millivolts,[128] better than 0.5 mV,[129,130] indicates that no inherent reason exists for poor potential stability of molten salt reference electrodes. It is true, however, that high precision is more difficult to achieve in molten salt media (because of temperature fluctuations, impurity effects, etc.) than in aqueous media.

Table VII summarizes the requirements for reproducibility and stability of reference electrodes used in the four main applications of electrochemical measurements: thermodynamic quantity determinations, kinetics studies, analytical procedures, and industrial or laboratory process control. Knowledge of these requirements is very important for the reference-electrode user. These requirements, along with the ease of fabrication and use, are employed to decide which of the possible reference electrode choices is optimal for a particular application. The optimal reference electrode system or construction is not necessarily the one capable of the most reproducible or stable potential measurement.

3.2.1. Stability of Potential

Stability of the potential over time is one of the most important requirements for a reference electrode device. Ideally, the reference electrode potential should be invariant; in practice, however, this can be achieved only within certain limits. Every electrode shows some potential variation (instability). Basically, two types of reasons for this instability can be distinguished: internal reasons such as electrochemical noise and aging of the reference electrode, and external reasons such as chemical and/or concentration changes of the active material and the electrolyte.

Although the causes of the potential variation of a single electrode can be examined on theoretical grounds, experimental determination is possible only by potential measurement between electrodes. To avoid the uncertainty that the instability of a different electrode system may introduce, the stability determination must be executed between identical electrodes. To obtain statistically meaningful reproducibility and stability data, potentials of several identically prepared electrodes are monitored for a period of time in a cell that ensures identical conditions for all tested electrodes.

TABLE VII. Requirements of Reference Electrodes in Various Applications[a]

Type of application	Reproducibility of the potential	Knowledge of the potential value	Elimination of $\Delta\phi_j$	Short-range stability	Long-range stability	Elimination of $\Delta\phi_{ir}$
Thermodynamic quantity measurement	E	E	E	E	NR	NR
Kinetic studies	D, NR	D, NR	NR	E	NR	E
Analytical procedures	D	D	NR	E	D	D
Process control	D	D	D	E	E	D

[a] E, essential; D, depending on the purpose of application; NR, not required. Short-range stability for hours; long-range stability for days to months.

Electrochemical Noise. The electrochemical noise inherent in electrodes causes a random, low-amplitude fluctuation in potential. The average of the varying potential over time is constant, however. In a broad sense, the electrochemical noise includes all kinds of potential variations that do not result in a shift of the average value.

The electrochemical noise is caused by random, local disturbances of the electrical double layer (such as rearrangements of surface atoms in the crystal structure) and local, randomly distributed differences of the anodic and cathodic components of the exchange current. All these disturbances are confined within the limits of the global dynamic electrochemical equilibrium. Also thermal noise may be a contributive factor in electrodes of high impedance. The magnitude of the electrochemical noise is usually in the range of 10^{-6}–10^{-3} V, depending on the chemical composition, the capacity, and the impurity level of the reference electrode. The frequency of the various noise components varies generally in the 10^{-1}–10^{9} Hz range. Several superimposed frequencies cause irregular amplitude fluctuations in the noise level.

Factors other than the electrochemical noise (because of its low amplitude) control the precision of the potential measurement. In ultraprecise investigations, however, the electrochemical noise is significant because it limits the precision of the potential measurement. In recent years, electrochemical-noise studies gained considerable attention,[131,132] when it was discovered that the noise level and pattern can provide useful information on various electrochemical phenomena, such as metal corrosion and passivation.[133,134]

Different averaging techniques show different patterns of potential fluctuations. For example, potentials measured regularly at long time intervals (e.g., 4–5 times a day) as a mean value of short (5–10 min) observation periods usually show a random variation (Fig. 6).[130] These slow potential variations are caused most likely by local fluctuations of the mass transport fluxes and physical conditions (e.g., temperature).

Electrode Aging. In contrast to electrochemical noise and slow fluctuations, electrode aging results in a gradual shift of the electrode potential. The electrochemical noise usually does not seriously limit application of reference electrodes, provided that the limits of the fluctuation are known. This is not true for electrode aging, which manifests itself by three, usually clearly distinguishable periods of a reference electrode: (i) potential stabilization (an especially pronounced period of the membrane electrodes), (ii) a period of relatively stable potential, and (iii) electrode degradation accompanied by a continuous potential shift and poor electrochemical characteristics, which finally renders the reference electrode useless. The course of the electrode aging is not yet well known. It is influenced by phase transformations, recrystallization of the active materials interrelated with changes in

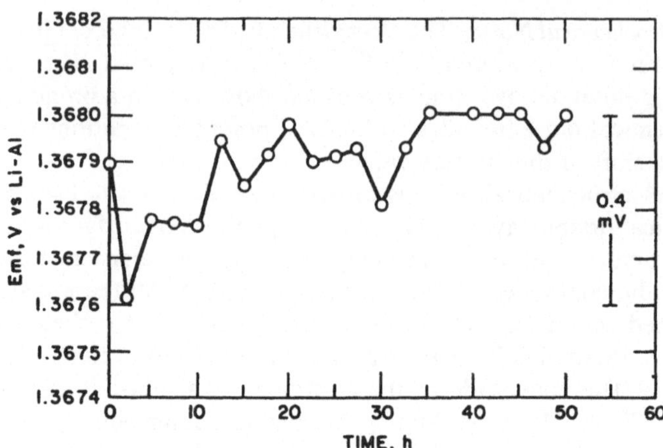

FIGURE 6. Potential stability of an Ni/Ni_3S_2 reference electrode monitored relative to a Li–Al electrode.[130] Circles indicate an average value of 5-min monitoring periods. Li_2S-saturated LiCl–KCl eutectic, 450°C. (By permission of the publisher.)

the kinetics and exchange current density, and in membranes by ion exchange and mechanical effects.

Chemical Changes. Chemical changes in the reference electrode system cause a potential shift. A good diffusion barrier is essential for preserving the original composition of the reference electrode. A diffusion barrier, however, cannot impede electroactive impurities being generated inside by corrosion processes as a result of poor construction material selection. Chemical changes in the reference electrode can also be caused by vaporization of certain components of the reference electrolyte, by decomposition, etc. The chemical changes can be retarded for a sufficiently long time by using a large volume of the electrolyte and large electrochemical capacity of the electrode. The capacity range of reference electrodes is usually 1–1000 mA h, depending on the available space and other design requirements. Microreference electrodes with less than a few milliampere-hours capacity are very sensitive to chemical changes.

In most cases, molten salt electrochemical work and reference electrodes require inert gas atmospheres or gas mixtures of controlled composition inside the electrochemical cell to preserve potential stability. Atmosphere-sensitive, open-cell experiments are carried out in a controlled inert-atmosphere (He, Ar, or N_2) glove box. Technical details of maintaining inert gas atmospheres of low impurity level are described in chapters of a previous volume[3] and in this volume.

The potential shift caused by aging and/or chemical changes can be detected by a master reference electrode. The master reference electrode is usually, but not necessarily, identical to the one used in the experiment and is used only occasionally for checking the working reference electrode potential. At other times, the master reference electrode is kept under conditions that prevent its contamination or deterioration. If a concentration change in the reference electrode, not an impurity effect, is the reason for the potential shift, then the reference electrode may be reconditioned *in situ* by electrochemical formation of the electroactive species. For example, loss of AgCl from an Ag/Ag^+ reference electrode can be replenished by anodic treatment of the silver electrode. Similarly, an Li–Al reference electrode can be reconditioned cathodically during long-term application (Section 4.1.4).

With widespread availability of modern voltmeters having 0.1–1 μV resolution, new potentiometric studies will probably result in a better understanding of the factors controlling potential stability and will provide guidance for better reference electrode preparation.

To document experimental results properly, one has to define the following properties of the reference electrode:

- Constituents and compositions of the electronic-conducting phase, the electroactive materials of the electrode and the reference electrolyte, and of the housing, diffusion barrier or membrane, if any.
- Results of the reversibility test.
- Reproducibility of the potential measurements, usually by standard deviation of the measured values.
- Short-range and long-range stability of potential.
- Electrode potential relative to a commonly used reference electrode of the same electrolyte system.
- Temperature and potential relationship measured versus a commonly used reference electrode indicating the tested temperature range.
- Electrode lead construction from the electrochemical double layer to the instrument terminal so that the thermo-emf contribution can be determined.
- Aging effects.
- Details of the construction that may influence the above-listed properties of the reference electrode.

3.2.2. *Constructional Stability*

The "constructional stability" is defined as the resistance of the construction materials against the effects that tend to destroy the physical integrity of the reference electrode device. It is ensured by proper design and selection of the construction materials that form the reference electrode. The construction materials of a reference electrode are either electrical insulators used as housing, diffusion barrier, Luggin capillary, or electrical conductors (metals or graphite) that are electrochemically inert in the required application and used as electrical leads or electrode holders. Corrosion and mechanical damage of these essential components may render a reference electrode useless. Corrosion of both the metallic and insulator material components—enhanced by the usually high chemical reactivity of the electrode materials or molten salt electrolytes—is more extensive at high temperature. For these reasons, reference electrodes with molten salt electrolyte require special construction materials.

Physical effects can result in mechanical damage. Nonuniform, sudden temperature changes, especially thermal cycling between room temperature and working temperature, cause mechanical stress. Quartz is especially sensitive to stress; other materials such as alumina cannot endure large temperature gradients and/or thermal shock. Resistance to thermal shock is the term that describes the thermal-mechanical properties of high-temperature construction materials. The resistance to thermal shock is a derived quality, depending principally upon the coefficient of thermal expansion, thermal conductivity, mechanical strength, elastic modulus, and plasticity. A low coefficient of thermal expansion and high thermal conductivity are advantageous properties of a ceramic material when the thermal-mechanical properties are considered. Pertinent values for some important materials are tabulated in Ref. 135 (p. 64).

Contraction of a solidifying electrolyte and/or its subsequent expansion during remelting are detrimental to glass, quartz, or ceramic enclosures. Generally, slow cooling of the reference electrode to room temperature is necessary to prevent cracking of the reference electrode housing. Usually, the only way to remelt the electrolyte, without breaking a glass or quartz container, is the application of a fast and uniform temperature rise to the container wall by using a large flame or a preheated furnace. This procedure produces a uniform liquid salt layer between the container and the solid salt before the bulk of the salt reaches too high a temperature and expands to a dangerous extent. Later expansion of the solid salt core is not a problem because the liquid shell can accommodate to the volume change.

Chemical stability of various refractory materials such as oxides, graphite and carbides, nitrides, and sulfides is discussed in Ref. 136.

The wide potential range (Table IV) within which a molten electrolyte is stable (the wide stability window) is one of the important features of the molten salt electrochemistry. Practical utilization of the wide potential stability window, and construction of a reference electrode that is applicable in the full range of the potential stability window, however, may be limited in certain applications by the lack of adequate construction materials.

The components of the test and reference electrolyte and the dissolved constituents of the active materials of the test and reference electrode obviously determine whether a construction material will be usable in a reference electrode. For example, high-density, high-purity alumina is available in the convenient forms of tubes, crucibles, etc.; therefore, it is excellent for reference electrode or electrochemical cell fabrication. Yet its application is limited in certain cases. Alumina cannot be used at potentials that are close to that of the lithium electrode. At potentials less positive than approximately 0.25 V versus Li, lithium alloys cause severe corrosion of alumina which may completely disintegrate. Since metallic Li dissolves in lithium-salt-containing molten electrolytes, this corrosion takes place even without direct contact between alumina and the bulk of the lithium metal or alloy. Similarly, boron nitride is excellent at higher potentials but may suffer from lithium attack in the 0–0.20 V versus Li potential range.[137] High-density BeO, MgO, or Y_2O_3, on the other hand, are stable even at the lithium electrode potential. The term "high-density" refers to almost pore-free, gas-tight ceramic materials fabricated with close to theoretical density. Ceramic materials are usually very stable even at the high positive potential region of the electrolyte stability window.

The terminal metal of the reference electrode should be inert in the electrochemical environment, i.e., its potential should be controlled by the active materials of the reference electrode system. Otherwise, the corrosion current, which develops between the terminal metal and the electronic conductor active material, inevitably creates mixed potential conditions that cause false potential readings and may change the chemistry of the reference electrode. Inertness of a metal in the solvent electrolyte can be tested by cyclic voltammetry or a technique similar to the micropolarization test (Section 3.1, Fig. 2a). In both techniques an acceptable metal shows high polarizability ($\eta/i > 10^4\ \Omega\,\text{cm}^2$) in the reference electrolyte, at least around the potential of the reference electrode. The impurity level of the electrolyte is critical in this experiment because impurities may maintain an electrode process which could be misinterpreted as the dissolution of the inert metal. Applicability of a metal is limited by anodic dissolution; the corresponding critical potential can be estimated on the basis of the standard emf series.

Plastics and glass, which are universally used in ambient temperature work, have limited applicability at high temperature because of chemical incompatibility and thermal degradation or softening. Polytetrafluoroethylene and related materials can be used only in low melting electrolytes below 250–280°C. Vitreous materials such as Pyrex, Supremax, and fused silica can be used up to 500–1300°C (depending on the properties of the material) in the absence of reactive electrolytes or active materials such as sulfides, carbonates, fluorides, or lithium. Boron nitride, graphite, and ceramic materials such as high-density beryllia, magnesia, alumina, and yttria are thermally stable at very high temperatures ($T > 1500$°C), but require careful evaluation of the chemical environment. Listed in the references are sources of some selected construction materials (mainly refractories and carbon products).[135,138] These lists—although they are limited in scope and include U.S. sources only—may help experimenters in finding useful brochures containing materials description and vendors for the purchase of materials.

Presenting an exhaustive guide on materials selection is not possible because of the wide variety of combinations of the solvent–salt systems, active materials, and operating temperature ranges. Nevertheless, there are excellent compilations and monographs on the subject,[1,3,62,136,139] in addition to the reference electrode descriptions in Section 4. References 135 and 140 discuss the mechanical and physical properties of the high-temperature refractory materials. A very useful review on the properties and laboratory application of selected refractory materials is available in which melting points, coefficients of thermal expansion, thermal conductivities, electrical resistivities, as well as compatibility with molten salts are listed along with trade names and manufacturers (Ref. 135, pp. 64–67). Materials for application in LiCl–KCl electrolytes that contain high lithium and/or sulfur-activity materials have been reviewed.[141] Information on container materials for molten salts, fused oxides, and liquid metals with indication of temperature range, duration, and atmosphere requirements are tabulated.[142] Observations on resistance to corrosion attack of various containment materials are presented in the literature.[114,143–145] Corrosion and containment properties of molten salts were reported in an annotated bibliography with ~500 references.[146]

With increasing temperatures, glasses and ceramic materials show increasing ionic conductivity, which is another limitation besides chemical and mechanical instability. Resistivities of various glasses are in the range of 10^{12}–10^{17} Ω cm at room temperature, but rapidly decrease with increasing temperature; the logarithm of the resistivity is a linear function of $1/T$. The resistivity of Pyrex glass is about 10^6 Ω cm at 350°C and decreases to 10^4 for a 100 K temperature increase. Figure 7 shows the electrical resistivities of the presently available best high-density, high-purity insulating materials as a function of temperature.[147] The figure is only for general orientation because minor variations in composition, impurity content, heat treatment, stoichiometry, etc. can have a significant effect on the high-temperature electrical resistivity.[148] Also, the molten-salt environment and oxygen can induce a dramatic drop in resistivity, as demonstrated by the use of high-density alumina in a membrane-electrode construction.[149] Figure 7 explains why electrochemical experimentation becomes extremely difficult above 1000–1300°C and requires special methods. If the resistivity of an insulating material drops below approximately 10^6–10^5 Ω cm, the insulating or conducting properties must be reevaluated from the standpoint of applicability, especially in reference electrode construction.

In addition surface conductance may be a problem with some insulators as a result of gradually developing electrically conducting corrosion layers.

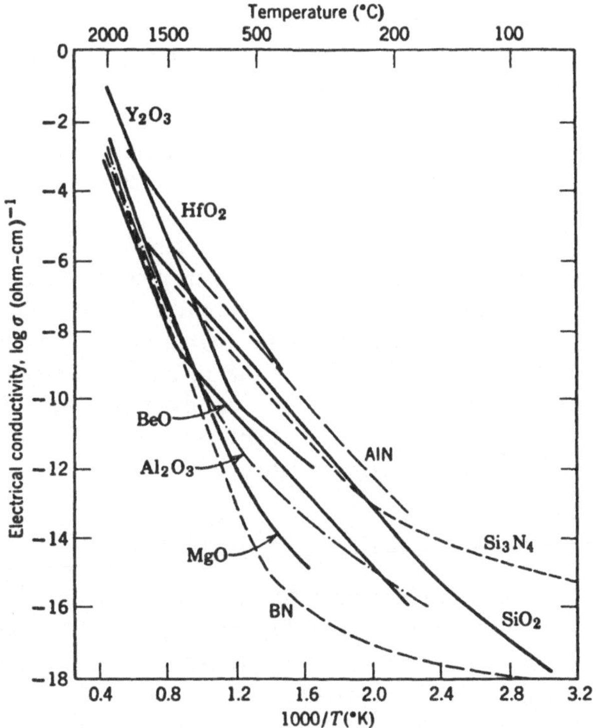

FIGURE 7. Electrical conductivities of the best insulating materials.[147] (By permission of the publisher.)

The usual cause is a strong reducing environment, e.g., lithium attack on Al_2O_3. Surface conductance and high-temperature bulk conductance of the otherwise insulating materials, as well as creeping electrolyte layers on "dry" walls of the container or reference electrode housing produce electrical connections, "short-circuits," and parasitic currents between the reference electrode and other electrodes or the ground. The parasitic current causes an IR-voltage drop and composition changes in the reference electrode by electrolysis, resulting in a progressive potential shift. An effective way to eliminate the problem of the creeping electrolyte is to extend the wall of the critical insulating components to the cold zone. An example for a cell design that eliminates these problems is shown in Fig. 23 (Section 4.1.5).

3.3. Interaction between Reference Electrode and Instrumentation

The measured voltage signal, i.e., the potential difference between the reference electrode and test electrode, carries the useful information about the electrochemical system being studied. Consequently, the instrument "readings" (real readings on a meter or stored data in a computer memory) must reflect the actual potential difference and its variations. Modern voltage signal measurement techniques or potentiostatic control involving sophisticated instrumentation can satisfy almost any potential measuring requirement but may yield erroneous results unless proper care is exercised during the measurement.

The interaction between a reference electrode and the connected instrumentation is viewed in this section with respect to the four basic requirements of potential measurement and control: (i) negligible measuring current load on the reference electrode, (ii) proper reference electrode impedance, (iii) proper positioning of the reference electrode, and (iv) elimination of electrical noise. In equilibrium and steady-state electrochemical techniques, only (i) and (iv) are important, but in transient techniques all requirements must be satisfied.

3.3.1. *Requirements of Potential Measurement and Control*

A measurement must not alter the conditions of the signal source; therefore, the electrode potential must be measured with a negligible current load on the affected electrodes. That is, the current that is required by the voltage-measuring device in order to generate an output signal, must not cause higher deviation from the true potential value than the required accuracy or precision of the measurement. Generally, 1 mV deviation is acceptable, since other effects (such as even very carefully "eliminated" liquid-junction potentials or instability of the potential or measuring instrument) cause similar uncertainty. However, higher accuracy is achievable if needed.

Tolerance to the current load depends on the chemistry and electrochemistry and size of the reference electrode. As a rule of thumb, large electrodes (1–10 cm^2) may be loaded with several microamperes, maybe even with hundreds of microamperes, while small electrodes (10^{-2} cm^2) may tolerate only loads less than a few nanoamperes. The usual reversibility check of the reference electrode, the micropolarization test (Section 3.1), reveals the exact numerical relationship between the current load and polarization (Fig. 2a). The current load on the reference electrode during potential measurement is dictated by the input impedance of the voltage-measuring instrument. Fortunately, the last decades have seen a great improvement in accuracy, zero point and gain stability, as well as increased input impedance of voltage-measuring instruments. There are voltage followers available, including operational amplifiers, electrometer amplifiers that have field effect transistors as inputs and input impedances of 10^{12}–10^{14} Ω, and digital multimeters (DMM) or high-precision oscilloscopes with usually $10^7\ \Omega$ or higher input impedances. Finding the proper instrument is not too difficult—if price is not a concern.

Chart recorders usually have lower than 10^6-Ω input resistance. Therefore, direct measurement of potential with them requires caution. A voltage follower inserted between the cell and the recorder, nevertheless, can satisfy the most exacting requirements.

There are additional reasons for applying measuring currents as low as possible: (i) in long-term experiments, even a moderate current load may exhaust the capacity of a small reference electrode; (ii) in a diffusion barrier of high resistance, the measuring current causes significant IR-voltage drop, which appears as an error in the potential measurement in addition to the electrode polarization; and (iii) membrane electrodes are very sensitive to current loads. Therefore, to avoid excessive, unstable asymmetry potentials and IR-voltage drops, membrane electrode potentials must be measured with an instrument that has at least 10^{12}-Ω input impedance.

A time constant is an important characteristic of both the measuring instrument and the reference electrode circuit. The time constant is expressed by the RC product, where R—for the reference electrode circuit—includes the resistance of the reference and the test electrode, the electrolyte, the diffusion barrier, and C includes the double layer capacitances of the electrodes and also the stray capacitances to the ground or shield. Functionally, a reference electrode may have either a passive or an active role in the measuring circuit. In the former case, the reference electrode simply serves as a reference for measurement of the electrode potential difference. In the latter case, however, it becomes a functional auxiliary component of a control system (such as a potentiostat) by providing the control circuit with an input signal. In both cases, the longer of the two time constants (i.e., of the reference electrode circuit and of the measuring instrument) determines the fastest possible signal processing that is free of distortion. In the control application, the impedance of the reference electrode or, more precisely, the impedance of the input circuit must be carefully tailored to the electrical parameters of the potentiostat or any other control device to avoid signal distortion or potential oscillation and to fully utilize the capability of a fast potentiostat.[(150–152)] This aspect of the experiment design becomes increasingly important because of the wide-spreading application of ultrafast transient techniques in kinetics studies. The revolutionary improvement in electronic instrumentation seen in the last decade enables experimentation with rapidly changing signals. Potential control with time constants and slew rates of less than a microsecond and several V/μs range is now commonplace. In these circuits, reference electrodes of low impedance are required. Potentiostat manuals usually provide guidelines for optimum control-circuit impedance selection. One must keep in mind, however, that low reference electrode impedance is accompanied by a less effective diffusion barrier. To satisfy these conflicting requirements and to save simultaneously the precise reference potential function, a dual reference-electrode system has been designed.[153,154] In this arrangement, a reference electrode that has a satisfactory diffusion barrier to ensure the correct potential-measuring function is coupled via a resistor–capacitor combination to a low-impedance reference electrode. The latter takes over the control function during fast transients or in high-frequency application. The specific criteria for reference electrodes and electrochemical cells to be used in fast-potentiokinetic techniques[155] and relationships between cell/reference-electrode design and frequency limitations[156] have been discussed.

3.3.2. Electrical Noise as Error Source in Potential Measurement and Control

The term "electrical noise," in a broad sense, includes all forms of electrical signals in the measuring circuit that distort or, in extreme cases, may completely obscure the actual signal and cause erroneous readings. Electrical noise may be picked up from the environment of the measuring setup or generated in the measuring instrument and the attached circuit. Fortunately, the effect of the noise on the measurement is greatly reducible by careful experimental design.

The electrical noise must be distinguished from the electrochemical noise; the former appears in the measured signal as a result of improper measuring practice, while the latter—as was discussed in Section 3.2.1—is a property of the electrochemical system.

Another form of electrical disturbance in measuring circuits is thermal noise, which is generated in resistive circuit components by the random thermal motion of charged particles. Because of the low voltage, the effect of the electrochemical and the thermal noise is usually not significant in electrochemical measurements. However, electrochemical noise analysis necessitates use of special reference electrodes and instrumentation.[134,157] Also, extreme care is required for elimination of electrical and thermal noise from the measurement.

Effectiveness of the electrical-noise reduction greatly influences the reliability of the potential measurement and control. The electrical-noise reduction, however, is an especially difficult problem in high-temperature measurement techniques.

Sources of the electrical noise include the following:

i. Ground loop.
ii. Capacitive and inductive pickup from power sources.

iii. Conducted ac-line transients.
iv. Thermo-electric drift.
v. Varying contact resistance.

The importance of these electrical problems is quite obvious when one considers that reference electrodes, which are usually high-impedance devices, are especially prone to noise pickup.

A ground loop occurs when multiple connections are formed intentionally or accidentally between any components of the instrument setup and ground. The ground is a point in the measuring system to which inactive components of the instrumentation (e.g., the instrument chassis, enclosures, and racks; the cable shields) or, in some cases, a certain component of the measuring circuit are connected with virtually zero resistance by heavy cables. It is important that each component of an instrumental setup should be connected only at a single point to a ground lead, and that each ground lead should reach to the same ground point, called the system reference ground. In the majority of instruments, the amplifiers and other circuit elements of the electronics are not connected permanently to the instrument enclosure or chassis. The user has an option to connect the active part of the instrument to the chassis or to let it float electrically within the grounded protective shell of the enclosure and shields (called a floating circuit). Unfortunately, for operational reasons, the majority of the available potentiostats seem to be exceptions to this option.

High-quality instruments designed for low-level signal measurement have an electrically isolated, floating shield (amplifier guard shield) within the enclosure which is isolated electrically from both the amplifier circuit and the enclosure. The floating shield is shorted to the guard terminal, a connecting point to the cable shields. Only the enclosure is grounded.

The system reference ground or the chassis of the component instruments of the setup are electrically connected to the "earth." The "earth" can be a heavy cable buried into the soil, a water line or—as is usually the case—the third conductor lead in the main current outlet. As the safety rule requires by law, an instrument system that is powered by the main current must be earth grounded to the conductor provided for this purpose in the outlet. The system reference ground must also be connected to this conducting body. Again, the same rule of the proper measuring practice applies—namely, only one connecting lead should be installed between the system reference ground and the earth ground, and the latter should be the only earth-grounded point in the entire instrumental setup. This principle can be implemented by using a common main current distributor strip for powering all instruments in the measuring system. Battery operated instruments are gaining popularity and availability, since they can avoid many of the noise problems.

There are situations that hide multiple grounding conditions. For example, an often-overlooked problem is that instrument enclosures or racks are in electrical contact with improperly earth-grounded support structures, such as metal frames or wet reinforced-concrete walls. Because different "earth" points may be at different potentials, multiple earth-ground contacts may introduce ac and dc error signals and active earth-ground loops. Ground loops and earth-ground loops are active when they maintain a current flow as a result of potential differences at various points of the loop.

These loops result in an erroneous signal being added to the actual electrode-potential signal. Figure 8 shows active ground loops and other electrical noise sources of an improper circuit design. The erroneous signals from the ground and earth-ground loops are unpredictable and usually vary in time. They are generally ac-voltage signals up to a few volts with a significant dc component of frequently more than 100 mV.

High-temperature work may involve circumstances in which the single-ground principle cannot be implemented. The source of the problem is the test cell itself, which could intentionally or accidentally contact a grounded structural element, such as the metal body of a furnace well or glove box.

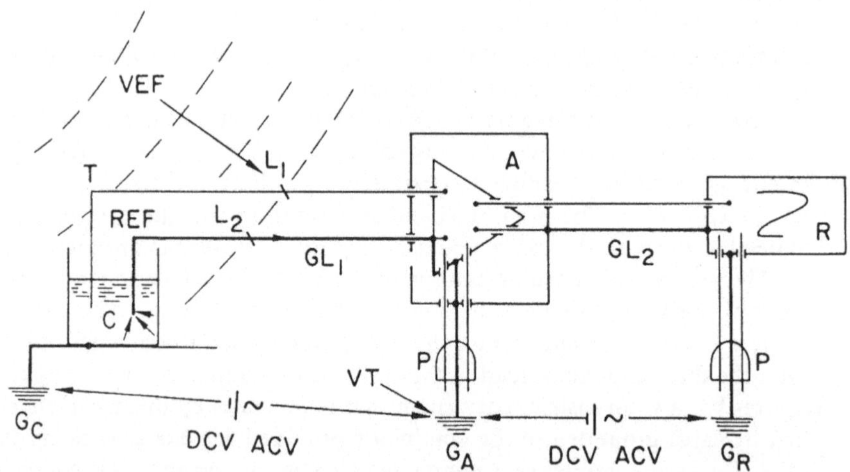

FIGURE 8. Severe noise problems in a measuring circuit. C, graphite or metal cell container in electrical contact to ground through support; T, test electrode; REF, reference electrode being discharged by ground-loop current; A, amplifier (illustrates signal input and output cables, as well as line cord for power input); R, recorder; G_C, G_A, and G_R, grounds at different potentials (multiple grounding); GL_1 and GL_2, two ground loops indicated by heavy lines (G_C-REF-A-G_A) and (G_A-A-R-G_R). As a result of the presence of ACV and ADV sources, ac and dc IR-voltage-drop contributions between REF–A and A–R distort voltage measurement. VT, voltage transients from AC main. Unshielded signal leads placed far from each other (L_1 and L_2) respond to variation of the electromagnetic field (VEF) by capacitive/inductive pickup. P, three-prong safety plug and line cord.

The main reasons for this situation occurring are as follows: (i) the significant electrical conductivity of insulating materials at high temperatures, (ii) "creeping" electrolyte layers on the surface of the insulating materials used for cell construction or isolation of the cell from a metal support, and (iii) an electrically conductive test-cell container (metal or graphite) that is supported by earth-grounded structures.

These occurrences cause three kinds of problems. First, the multiple ground connections introduce ground loops and, consequently, errors in the potential measurement. Second, if current is applied to the test electrode—as in kinetic studies or coulometric measurements—it may find additional pathways through the ground loop. As a result, a significant part of the measuring current may bypass the ammeter or current integrator or the test electrode, depending on the circuit arrangement. This condition affects not only the potential measurement but also the current-intensity and coulomb measurements. Furthermore, the intensity of the leakage currents usually varies during the experiment, and this prevents a correction from being made. Third, leakage current that finds a pathway through a reference electrode rapidly alters the original ratio of the potential-determining species and renders the experiment useless. The leakage currents may discharge the reference electrode or permanently damage it.

Potentiostats that utilize a grounded test/working-electrode terminal frequently introduce problems with ground or earth-ground loops. Unfortunately, a majority of the potentiostats presently on the market suffer from this problem. Only a few models are exceptions, e.g., the Stonehart potentiostat, which has chassis-independent, floating input terminals and also provisions for fine-tuning the reference electrode circuit impedance.[150] A potentiostat with a grounded test electrode cannot drive a cell properly if any cell component other than the test electrode is in electrical connection with the ground or earth ground.

Transients from ac power lines (i.e., large voltage fluctuations of the main-current source) are frequently conducted into the electronic measuring system, causing temporary erroneous operation. Switching power sources, on-off operation of heating elements, and relay solenoids are the common sources of these transients. These problems can be minimized by using separate outlets for the noise source equipment and for the measuring instrumentation, and by installing power-line filters, conditioners, or isolation transformers at the outlets that supply current to the measuring instrumentation.

Another common noise problem is the very elusive capacitive/inductive pickup from nearby power sources. This type of noise, in contrast to the above discussed noise, does not require electrical contact between the instrumentation and the noise source. The capacitive/inductive pickup includes noise signals reaching the measuring circuit through the air from power lines, motors, and transformers, as well as from fluorescent lights, electric arcs, and electromechanical relays or solenoids. All these sources are very common in the sophisticated environment of a high-temperature experimental setup. A preventive measure against the capacitive/inductive noise pickup is careful shielding of the sensitive circuit elements. Shielding is used primarily on connecting cables between the cell and the components of the instrumentation. In a very noisy environment, shielding of the cell may become equally important by placing the cell in a grounded box made of metal sheet or screen. A grounded metal glove box also may serve as a shield if properly isolated from the cell. In addition to these noise sources, which are very common in every measuring environment, the heater elements and the on-off regulation of the heating current may cause additional noise problems in the high-temperature work. For example, problems arose due to ac voltages induced in high-resistance (0.1–1 MΩ) membrane electrodes from furnace heater elements.[158] Rectification took place at the electrodes, and the resulting dc potentials completely vitiated attempts to measure electrode potentials. These effects were reduced to below 1 mV by winding the furnace noninductively and interposing an earthed metal screen between the furnace windings and the cell.[158]

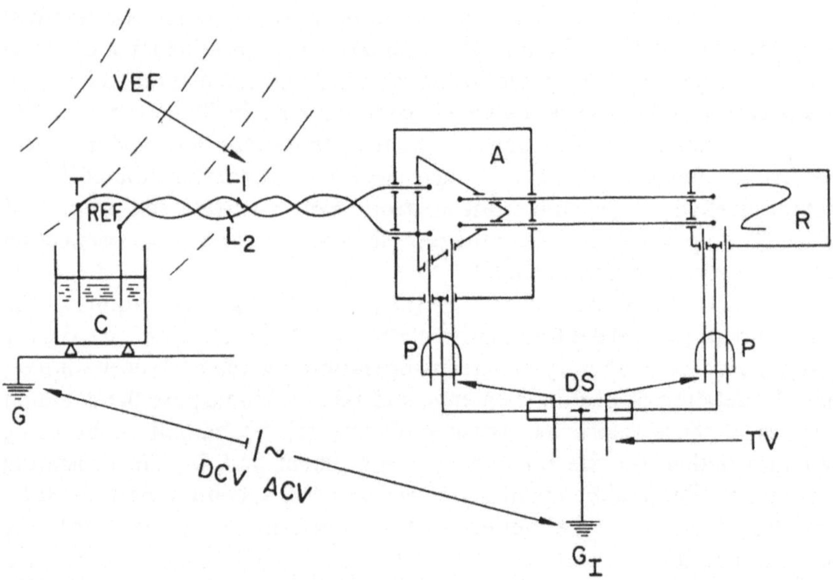

FIGURE 9. Instrumental setup with noise reduction. G, grounded cell support; cell is isolated from support; electrical isolation breaks ground loop. Twisted cable reduces capacitive/inductive pickup. DS, distributor strip for earth grounding; components of instrumentation are connected to the same ground, G_I; voltage transients (TV) from AC main, however, are in effect.

Figures 9 and 10 show ways to reduce or eliminate the noise effects discussed above. Since there are many variations of the amplifier grounding/shielding configurations, all versions of the proper noise elimination cannot be discussed here. An excellent brochure[(159)] describes the various amplifier–input configurations, shielding techniques, and the proper connections between signal sources and amplifiers.

The thermo-electric potential drift, a thermo-emf error signal, is an often overlooked error source in high-temperature electrochemistry. Since large temperature gradients and temperature differences exist between the high-temperature cell and the instrumentation (Section 3.1.4), a careful evaluation of the temperature fields around dissimilar-metal connections and their possible effects upon the potential measurement cannot be neglected. Good practice requires that connections between dissimilar metals be in the thermostated zone.

Variable contact resistances in the high-temperature zone between electrode components or electrodes and current leads are another elusive noise source. This problem can exist in any electrical circuit and at any temperature; but the high temperature, the usual thermal cycles of the test cell, and the unavoidable mechanical vibrations only exacerbate the problem in molten salt systems. To eliminate this noise source, instead of simple

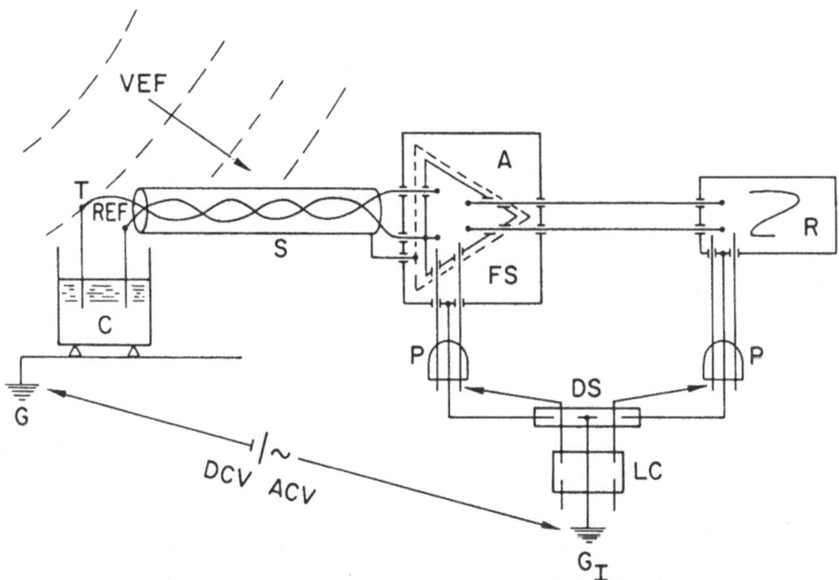

FIGURE 10. Instrumentation setup with noise elimination. Cell is isolated from support, no ground loop. S, shielded signal leads; FS, floating shield; LC, line conditioner and filter; DS, distributor strip.

mechanical contacts, exclusively welded or brazed junctions are recommended in the high-temperature zone between the current-leading metal components.

Another form of the electrical noise is the cross talk between channels of a multichannel instrument such as analog recorders and digital data loggers and other data acquisition systems equipped with multiplexers.

Minimizing the internal noise and the noise sensitivity of an instrument is the obligation of the designer and manufacturer. Effective noise reduction during use of the instrument, however, is the task of the user. By carefully analyzing the possible effects of the various noise sources on a particular measurement and following some simple rules, one can achieve a satisfactory signal-to-noise ratio. The general rules of noise elimination, which apply in any electrical measurement work, are as follows: (i) Use a stable system reference ground and earth ground; (ii) eliminate ground loops and earth-grounded loops; (iii) ground each signal-cable shield at only one point; (iv) consider whether the amplifier circuits of the instruments used in the setup are isolated from or connected to the chassis and the ground; (v) pay very close attention to the connections for the system reference ground and earth ground, and the guard shield; (vi) use twisted cables and proper shielding to reduce the inductive or capacitive pickup; (vii) use the shortest possible cable connections; (viii) place the measuring circuit as far as possible from the external noise sources; (ix) measure the noise contribution; (x) experiment to find the optimal solution for noise reduction. Detailed discussions on various noise sources and noise-elimination techniques are available in Refs. 159 and 160.

The noise elimination, however, is only one part of the proper use of the reference electrodes. Equally important are the proper cell design, measurement technique, and instrument usage. Since discussion of these important questions is beyond the scope of this chapter, the authors can only direct the attention of the reader to reviews of electrochemical instrumentation,[151,161-164] cell design,[1-3] and techniques.[165,166]

4. Description of Reference Electrodes

The previous sections have discussed the relevant theories and the underlying principles in the construction and application of reference electrodes in molten salts. In this section, descriptions of various reference electrodes in molten salt systems are presented. For each electrode, special emphasis is placed on electrode characteristics, construction details, and operating procedures. The information that is presented illustrates the importance of the theories and principles when they are applied in practice. In many instances, the information reveals the difficulties in implementing the theories and principles; however, it also shows how those difficulties can be overcome. Most of the data presented here are mainly from reported literature, and, in many cases, no independent confirmation is available. Therefore, it is important that the experimenter carry out tests on his own fabricated electrodes if clarification or confirmation of any information is necessary.

In this section, reference electrodes are described for eight common molten-salt systems: chlorides (also bromides and iodides), chloroaluminates, fluorides, cryolites (fluoroaluminates), nitrates, carbonates, sulfates, and hydroxides. Although chloroaluminates belong to the molten chloride system, their chemical properties and their working temperature range are quite different from those of alkali chlorides. Thus, reference electrodes in molten chloroaluminates are different and need to be treated separately. The same is true for molten cryolites. In mixed molten salt systems such as chloride–fluoride mixtures, reference electrodes described for either chlorides or fluorides can be used; which one is applicable depends on the composition of the mixture. In general, if one component of the mixture is minor, then reference electrodes described for the major component are applicable. However, there are no definite rules, and experimentation is required to determine the applicability of any particular electrode in a mixed molten salt mixture.

4.1. Chlorides, Bromides, Iodides

The chlorides represent the most extensively used and investigated molten salt electrolyte systems. They have been used in every field of the molten salt electrochemistry such as thermodynamic studies, kinetics investigations, industrial processes, and high-performance batteries. The chlorides, bromides, and iodides are good solvents for other salts and provide stable working media between the melting point and the temperature of noticeable vaporization. They evaporate without decomposition. Generally, the Li, Na, K, and Mg salts and their mixtures are used as solvent electrolyte media. Among them, the alkali chlorides are the most important. The chloride-salt mixtures allow selection of different working-temperature ranges (Table IV). Bromides and iodides offer lower melting points than those of the corresponding chlorides and special anion effects for study. However, they do have some disadvantages, including higher density and cost, as well as a narrower potential-stability window. The chloride, bromide, and iodide melts are electrochemically inert between the potentials of the metal deposition and the halogen evolution.

Although the LiCl–KCl mixtures have probably been used most frequently, the NaCl–KCl melts have the advantage of relatively easy preparation because of the nonhygroscopic nature of the salts involved.

Lithium and magnesium salts are very hygroscopic and retain water even in the molten state and at high temperature. Water impurity in the melt is the source of many complications in molten salt electrochemical work. Consequently, to get meaningful results, one must purify the salts and the melts before the experiment. The chemical and electrochemical consequences of water-related impurities, the various procedures of purification, and also the electrochemical methods used to characterize the purity of the melt have been discussed in a detailed review.[108] A typical procedure of chloride-salt purification is treatment with dry HCl followed by dry Cl_2 prior to preelectrolysis at 2.7 V between inert electrodes; finally, filtration is recommended. Similar procedures, with suitable modifications to the gas purges and electrolysis voltages, may be adopted for bromide and iodide melts.

Table VIII lists the most frequently used, well-established reference electrode systems for chloride melts and their standard potentials. As seen, the reference electrode selections embrace almost the entire available potential stability window of the chloride solvent electrolyte systems—from the strongly reducing Li–Al to the powerfully oxidizing chlorine electrode. The reference electrode systems discussed in Sections 4.1.1–4.1.7 will be described for chloride media. However, all these reference electrodes, except the chlorine electrode, are applicable also in bromide- or iodide-containing electrolytes, in most cases without modification. The Li–Al and the Ni/Ni_3S_2 reference electrodes tolerate some fluoride content of the reference electrolyte, as will be shown in the respective sections.

4.1.1. The Chlorine Reference Electrode

A chlorine reference electrode utilizes the equilibrium between chlorine molecules and chloride ions brought about on the electrochemically inert but catalytically active graphite or glassy carbon. The electrode reaction is

$$Cl_2 + 2e^- \rightleftharpoons 2Cl^- \tag{19}$$

TABLE VIII. Most Frequently Used Reference Electrode Systems in Chloride Melts[a]

Reference electrode system	Conditions	Temperature range used (°C)
Cl^-/Cl_2	Pure chloride $P_{Cl_2} = 1$ atm	25–900
Ni/Ni_3S_2	Electrolyte saturated with Li_2S	300–500
Ag/Ag^+	$N_{AgCl} = 10^{-4}$–10^{-1}	25–1100
Pt/Pt^{2+}	0.03–0.08 *M* $PtCl_2$	350–500
$Li\text{-}Al/Li^+$	20–42 atm% Li in Al	300–550

[a] Electrode entries in the order of their potentials. Conditions ensuring well-defined, reproducible potential.

The chlorine electrode should be regarded as the fundamental, thermodynamically well-defined reference system for electrode potential scales in chloride melts. In the melts of simple chlorides, which do not take part in Lewis acid–base equilibria and complex formation, the chloride activity is considered to be unity. Establishing the standard conditions for the electrode requires only that the gas pressure be maintained at unity. In practice, however, the chlorine reference electrode is seldom utilized except in thermodynamic studies, because of operational difficulties with the gas electrode.

The great importance of the chlorine reference electrode has been the foundation of a thermodynamically well-defined potential scale and emf series in various electrolyte systems(15) for metal electrodes, including important reference electrodes designed for easier application, such as the Ag/Ag^+, Pb/Pb^{2+}, and Pt/Pt^{2+} electrodes. These metal reference electrodes will be discussed in the later sections.

Although the presence of the trichloride ion in melts, as a result of the equilibrium

$$Cl_2 + Cl^- \rightleftharpoons Cl_3^- \tag{20}$$

was identified spectroscopically to be about 10^{-3}–10^{-4} m concentration,(167) the thermodynamically defined condition is still valid as long as the melt composition and the chlorine pressure are defined precisely.

For gas electrodes in general (and for the chlorine electrode in particular), a constant flow of the gas must oe maintained under precise pressure control. The incoming gas must be (i) thermally equilibrated before it reaches the electrode, and (ii) presaturated with the electrolyte to minimize the possibility of a volatile component of the melt being carried off with the exit gas. Gas electrodes require an inert electronic conductor with sufficient catalytic activity. The catalytic activity is particularly important because gas electrodes, even at high temperature, tend to show irreversibility. All these requirements must be satisfied in an acceptable chlorine reference electrode.

Application of high-purity graphite or glassy carbon as the inert electronic conductor has been common in all chlorine reference electrodes, apart from some rare use of Pt for chlorine electrode in short experiments.(158,168,169) The differences among the various designs and techniques reported have been related to (i) the methods of pretreatment of graphite, (ii) the manner of obtaining the necessary three-phase contact among the melt, graphite, and chlorine gas, and (iii) the electrochemical procedures for establishing equilibrium and for testing reversibility.

Pretreatment of graphite in a Cl_2 atmosphere at high temperature was found to be very important.(170,171) A unique method to enhance the catalytic activity of the graphite electrode has been developed.(16) The end of the graphite rod was coated with a concentrated sucrose solution and then carbonized in a chlorine atmosphere at 800°C for 24 h. The result was a thin, porous, and highly adsorbent carbon deposit.

Saturation of the graphite with chlorine gas in the assembled reference electrode and establishment of the three-phase contact have been accomplished by prolonged passing of chlorine gas over the graphite from an external source and a short anodic chlorine evolution to precondition the electrode prior to the potential measurement.[172-177] Much attention has been given to the method of delivering the chlorine to the three-phase contact zone. In essence, two ways have proven satisfactory. One method involves passing the chlorine down through the center of a hollow graphite tube immersed in the melt, allowing the gas to bubble out at the bottom hole (Fig. 11)[177-180] or through the pores of a graphite plug when a closed-end tube is used.[171,181] The other method involves flushing the gas over the surface of the graphite, as illustrated in Figs. 12[176] and 13.[170]

The purity of the electrolyte is important for the reliability of the chlorine electrode; water and traces of metal oxide are especially deleterious.

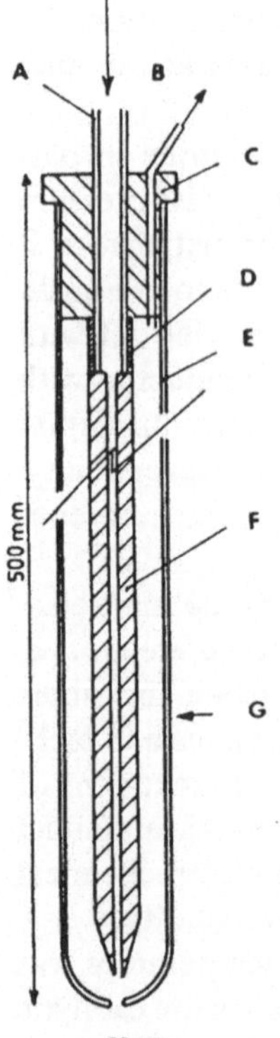

FIGURE 11. Chlorine reference electrode with graphite tube.[178] A, stainless-steel Cl_2 inlet tube; B, Cl_2 inlet and outlet; C, Teflon plug; D, graphite cement; E, fused silica reference-electrode housing; F, graphite; G, electrolyte level. (By permission of the publisher.)

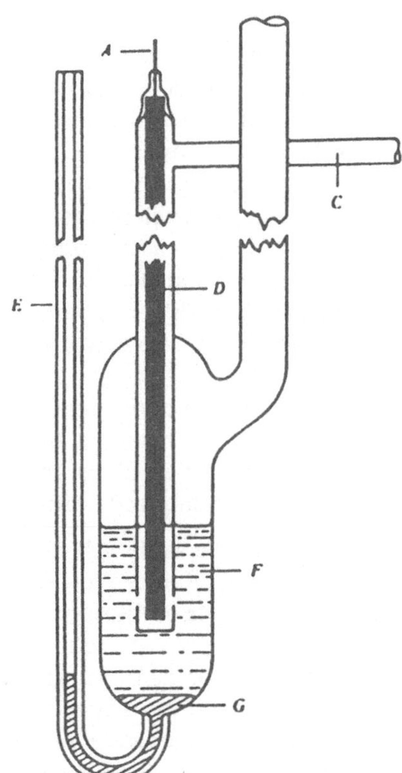

FIGURE 12. Formation cell with chlorine reference electrode.[176] A, platinum wire; C, chlorine gas inlet; D, graphite rod 6–10 mm in diameter; E, side tube for making electrical contact to the metal electrode; F, electrolyte; G, molten metal electrode. (By permission of the publisher.)

The potential of a satisfactorily prepared chlorine electrode is reproducible within 0.1 mV.[182] Lengthy treatment of graphite with chlorine probably has another important role in addition to saturation, namely, the elimination of water and oxide/hydroxide impurities from the melt and the graphite. The mechanism of the purification of chloride melts by means of chlorine treatment has become obvious from recent works.[106,183,184]

The schematic diagram of a simple cell used to study molten-metal and halogen electrode pairs is shown in Fig. 12.[174–176] The electrodes were reversible within ±0.5 mV over a wide range of temperature and chlorine pressure. This simple design, however, limits the applicability of the cell because of the reaction between the metal and chlorine.

In order to decrease the rate of direct reaction between the metal and chlorine, which decreases the emf, and also to isolate the reference melt from the electrolyte under study, various separations have been incorporated into the cell between the two electrodes. The most popular electrode separators are fritted glass or quartz, but other methods and materials have been utilized:

- The chlorine electrode is separated by asbestos fiber.[16]
- MgO powder at the bottom of a U-shaped glass cell is used to separate Mg and chlorine electrodes.[168]
- Two half-cells are used, each equipped with a side arm for accepting an electrolyte bridge as electrolyte junction.[185]
- The chlorine-electrode compartment is separated from the bulk of the melt by a test tube, which has a 4- to 8-mm-long capillary connection at the end.[180]
- A test-tube-like compartment is used, equipped with a small hole on the side for providing junction between the electrolytes of the two electrodes.[172,173]
- A glass membrane has also been used in $PbCl_2$ and AgCl formation cells[158,186]; however, the emf of these cells has been found to be consistently lower by about 20–30 mV compared with those of the corresponding cells without a membrane.

The most versatile cell and chlorine reference electrode designs for thermodynamic studies were reported by Laitinen and Pankey.[170] The chlorine electrode is illustrated in Fig. 13. A platinum wire (A) sealed through Pyrex glass at the top of the inner sheath provides contact and support from outside to the graphite electrode (B). The latter is a porous rod of pure graphite, 1/8 in. diameter and 6 in. long (Grade U-1, Ultra Purity Spectroscopic Graphite Electrode, United Carbon Products Inc., Bay City, Michigan). A detailed description of the pretreatment of graphite is given in the original article. The reference electrode is fixed in the cell so that the graphite rod is only partially submerged into the melt, the level of which is indicated by the dotted line (C). Chlorine admitted to the glass sheath (D) at the delivery tube (E) flows over the graphite at a rate of 6 cm^3/min, bubbles up through the melt at the bottom of the sheath, and exits through a side tube (F). A silicone-rubber stopper (G) holds the inner sheath in place concentric with the outer jacket. At the bottom of the electrode assembly, which is 24 in. long, a Pyrex frit provides the electrolyte junction. Chlorine obtained from a commercial cylinder is dried by passage over anhydrous $Mg(ClO_4)_2$. Before the potential measurement, the graphite rod is anodized at 1 A for 3 min versus another graphite rod (e.g., J in the cell assembly, Fig. 14). Figure 14 shows a classical example of a versatile cell similar to the one used by Laitinen and Pankey for thermodynamic studies. The figure caption explains the functions of the various cell components. This type of cell was used to relate the chlorine electrode to a Pt/Pt^{2+} reference electrode (Section 4.1.3), which, in turn, served as a convenient tool in the same cell to establish standard emf series for LiCl–KCl eutectic.[28]

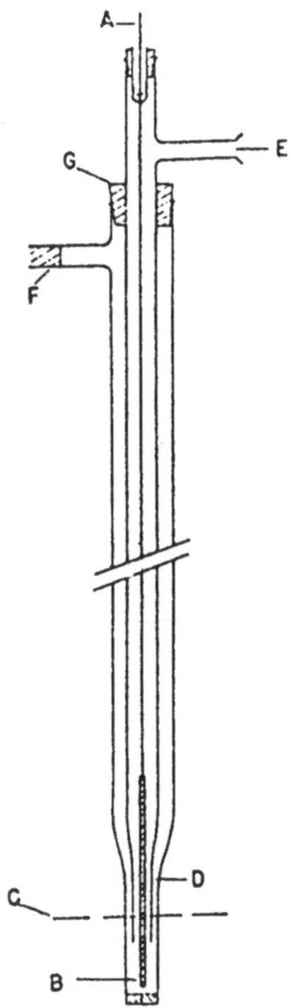

FIGURE 13. Universal halogen reference electrode for thermodynamic studies.[170] A, platinum wire; B, graphite electrode; C, electrolyte level; D, reference electrode housing with Pyrex frit; E, chlorine gas inlet; F, gas outlet; G, silicone rubber stopper. (By permission of the publisher.)

4.1.2. The Ag/Ag^+ Reference Electrode

The Ag/Ag^+ electrode is a very important reference system for molten chlorides. Silver does not react with molten chlorides and has very low solubility in its own molten chloride (0.06 mol% at 700°C in AgCl[187]). Low polarization resistance, about 50 mΩ cm^2, has been demonstrated for the Ag/Ag^+ system. Reversibility, reproducibility, and stability of the electrode potential have been reported in many solvent–salt media. On these grounds Plambeck recommends it as a universal reference electrode system for molten salts.[188] Because of the high solubility of the silver halides in molten chlorides, the Ag/Ag^+ electrode is a reference electrode of the first kind in chloride melts. The Ag/Ag^+ reference electrode consists of a 99.99% or higher purity silver wire dipped into the reference electrolyte. This latter

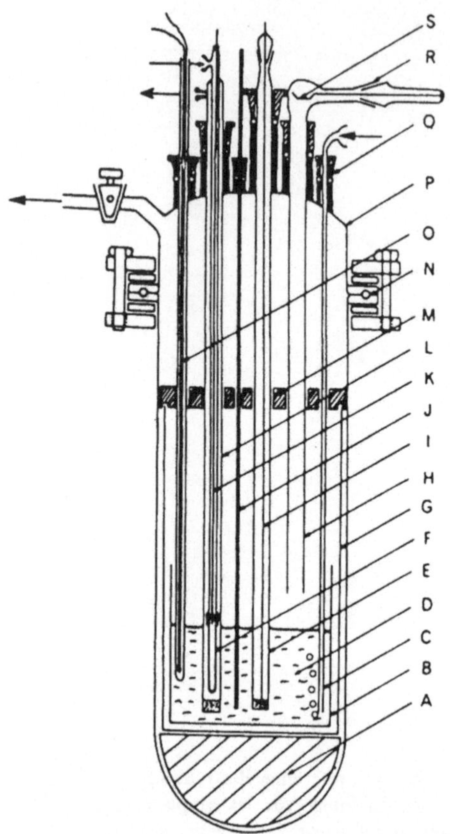

FIGURE 14. A typical electrochemical cell with gas reference electrode and controlled atmosphere.[3] A, firebrick support; B, crucible; C, gas bubbling; D, electrolyte; E, reference electrode housing with fritted disk; F, gas reference electrode; G, splash guard; H, feeding tube; I, reference electrode; J, counter electrode; K, inner tube for gas inlet; L, reference electrode housing; M, electrode guide; N, O-ring; O, thermocouple in sheathing; P, cell header; Q, one of the joints; R, ground joint; S, rotating spoon for additional solute introduction. (By permission of the publisher.)

is a solution of a precisely known AgCl concentration. The reference electrolyte is produced either coulometrically *in situ* within the reference electrode or gravimetrically by dissolving AgCl. The silver electrode shows Nernstian behavior in a relatively wide concentration range.[16,189] Generally, a concentration of 10^{-4}–10^{-1} N AgCl is applied. At lower concentrations, the potential is not defined because the potential of the electrode is affected by the impurities in the melt. At higher concentrations, the potential may be stable, but deviates from a true Nernstian response because of the change of the activity coefficient.

To maintain a constant concentration of the silver salt during the experiment, various designs have been devised. For example, Pyrex diaphragms have been used in LiCl–KCl eutectic.[170] The higher working temperatures (600–900°C) of an equimolar NaCl–KCl mixture require fused silica apparatus. With this melt, a firmly packed asbestos plug can be used as a diffusion barrier in a silver reference half-cell shown in Fig. 15.[190] Electrical contact between the reference electrolyte and the electrolyte under study can also be made through an asbestos fiber that is sealed into the end of the side-arm tube of the reference electrode.[27] Preparation of this diffusion barrier is a difficult procedure because of the high softening temperature of silica, but the barrier gives excellent results. It has been found that about 1.5 cm length of the asbestos fiber ensured constant concentration for several weeks at 700°C, as indicated by a less than 1-mV potential drift. The following diagram symbolizes the construction of this reference electrode in a cell:

$$\mathrm{Ag}\left|\begin{matrix}\mathrm{AgCl}(N = 6.103\times 10^{-2})\\ \mathrm{NaCl,\ KCl\ (equimolar)}\end{matrix}\right\vdots\ \text{asbestos fiber}\ \left\vdots\begin{matrix}\mathrm{MCl}\\ \mathrm{NaCl,\ KCl\ (equimolar)}\end{matrix}\right|\mathrm{M}$$

Cell 13

where M is the metal of the test electrode. The Nernst equation was obeyed

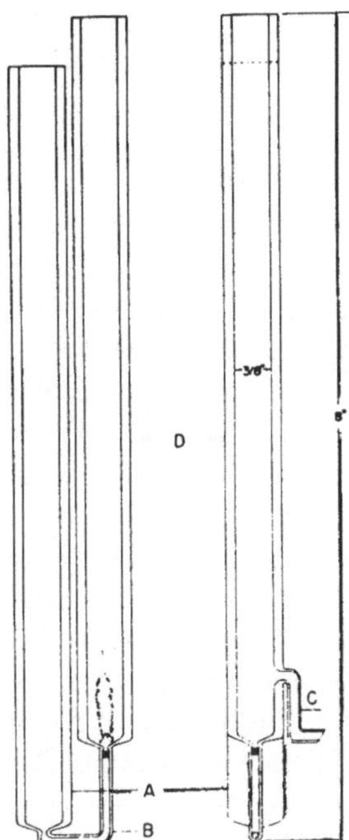

FIGURE 15. Double reference electrode system for Ag/Ag$^+$ electrode.[190] A and D, two compartments for electrodes; B, interconnecting tube with asbestos plug; C, connecting tube to test electrolyte. (By permission of the publisher.)

at low concentrations of AgCl ($N = 1.1 \times 10^{-3}$ to 6.35×10^{-2} mole fraction).[16] At similar low concentrations of MCl, the liquid-junction potential was found to be negligible.

An Ag/Ag^+ reference electrode with porous diaphragm junction has two disadvantages. First, the silver concentration in the reference electrolyte may decrease through diffusion. To avoid the error that the uncertain concentration of AgCl may cause, subsequent analysis of the silver content was reportedly necessary. Second, silver ions appearing in the test electrolyte may interfere with the potential-determining process of the test electrode.

To eliminate the two problems associated with the porous diaphragm, various membrane silver-electrode designs have been developed. A Pyrex membrane electrode has been developed for LiCl–KCl eutectic electrolyte.[42] The diagram of the cell (Cell 10) is shown in Section 2.3.2. The Pyrex membrane bulb blown at the end of any commercially available Pyrex tube had about 2–5 kΩ resistance in the temperature range of 350–550°C (bulb diameter was approximately 10 mm). The electrode potential was found to have a linear relationship with the logarithm of the concentration of AgCl up to the saturation limit (~1.3 molal at 510°C), but above 0.4 m AgCl concentration it required correction for the increasingly significant membrane potential (5–15 mV). This electrode must have been kept continuously above the melting point of the electrolyte, otherwise the membrane would have fractured on the solidification of the melt. Apart from this inconvenience, the electrode worked for a very long time provided that oxygen and moisture were excluded from inside the tube. A cell for studying silver reference electrodes and Pyrex membrane behavior in LiCl–KCl eutectic is shown in Fig. 16.[45]

A Jena Supremax glass membrane was used in equimolar NaCl–KCl electrolyte.[158] The electrode consisted of a closed-end Supremax tube (5 mm diam, 1 mm wall thickness, 0.5–1 MΩ resistance at the operating temperature) filled with AgCl solution, into which a silver wire was immersed. At least 6.9×10^{-2} mole fraction of AgCl was needed to achieve a satisfactory potential reproducibility (±8 mV). At lower concentrations of AgCl, the electrode potential became highly irreproducible and showed 100–200 mV variation in the potential. The potential was sensitive to air or moisture contamination and required careful gas atmosphere control. Also, keeping uniform temperature along the silver wire was important to avoid the thermocell effect, which altered the shape of the electrode by transferring silver from the warmer zone to the lower-temperature region. The author was able to eliminate both adverse effects by extending the silver wire with a platinum wire and keeping the silver portion totally submerged in the reference electrolyte. A membrane junction potential of about 20 mV was observed between solutions of 0.271 and 8.38×10^{-2} mole fractions of AgCl at 700°C. The electrode was calibrated against a chlorine electrode, and the measured standard potential corrected for the membrane potential agreed

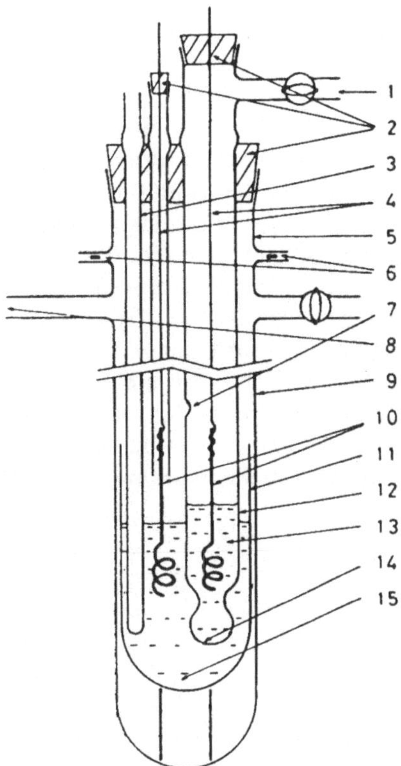

FIGURE 16. Cell with membrane silver electrode.[45] 1, argon inlet; 2, rubber stoppers; 3, Pyrex thermocouple well; 4, 0.8-mm-diam Ni wires; 5, Pyrex tube, 45 mm o.d.; 6, Teflon O-ring; 7, hole; 8, argon outlet; 9, quartz cell holder, 45 mm o.d.; 10, 2-mm Ag wire; 11, electrolyte container, Pyrex, 38 mm o.d.; 12, Pyrex tube, 17-20 mm o.d.; 13, reference electrolyte; 14, Pyrex glass membrane; 15, test electrolyte. (By permission of the publisher.)

within 1 mV with the value that was calculated from thermochemical data.[17] The potential was stable for more than 30 h. The thick membrane survived freeze-thaw cycles of the electrolyte. Supremax[191] and fused silica[192] membranes have been used in other studies of NaCl-KCl melts.

Measurements on various membranes in LiCl-KCl give insight to the magnitude of the membrane potential under various experimental conditions.[48] The authors attributed the differences (ΔE) found between the emfs of two types of cells (Cell 14 and Cell 15) to a membrane potential. In the first type, a membrane was used:

$$\underset{N_{MCl}}{M|MCl_z|eut}\,\vdots\,\text{membrane}\,\vdots\,\underset{N_{AgCl}}{eut, AgCl|Ag}\text{---}\underset{N_{AgCl}}{Ag|AgCl_z|eut}|Cl_2\ (\text{graphite})$$

Cell 14

The second type:

$$\underset{N_{MCl}}{M|MCl_z|eut}|Cl_2\ (\text{graphite})$$

Cell 15

In these cell diagrams "eut" means LiCl–KCl eutectic and the membrane is either of Pyrex, Vycor, or quartz. The left side of the double cell (Cell 14) is shown in Fig. 17.[48] The metal electrode (M) was produced *in situ* by cathodic deposition from the solution of MCl_z. A temporary graphite electrode was used as anode (not shown in Fig. 17) in this procedure and later withdrawn from the melt.

Figure 18 contains a summary of the ΔE measurements for a Pyrex membrane. We see that the nature of MCl_z has an important role; if M is Ag, ΔE is small except at higher than 0.6 mole fractions of AgCl in the reference electrolyte. The authors concluded that the salts that have a strong tendency to form complex ions with the Cl^- ions of the solvent salt exhibit high ΔE. For those melts having high ΔE values, ΔE decreases as the concentration of MCl_z decreases. This effect, along with the influence of the membrane material on ΔE, is shown in Table IX. Vycor membrane permits a higher working temperature than Pyrex or Supremax does.[48,193]

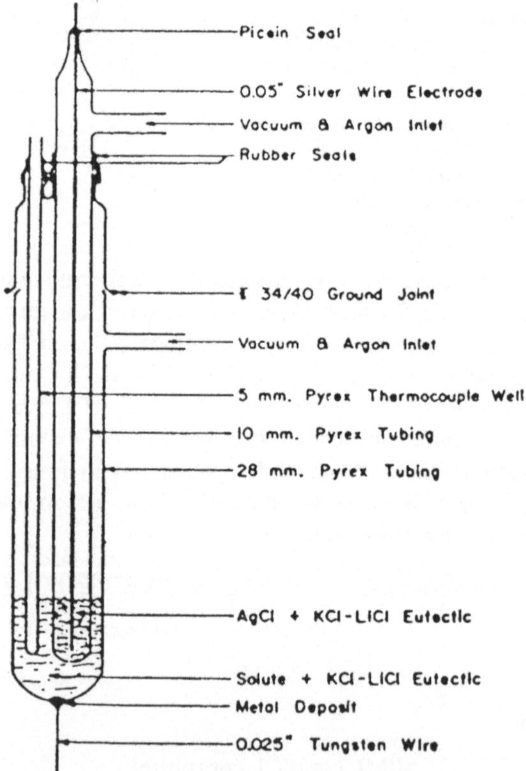

FIGURE 17. Daniell cell with membrane silver reference electrode for membrane potential study.[48] (By permission of the publisher.)

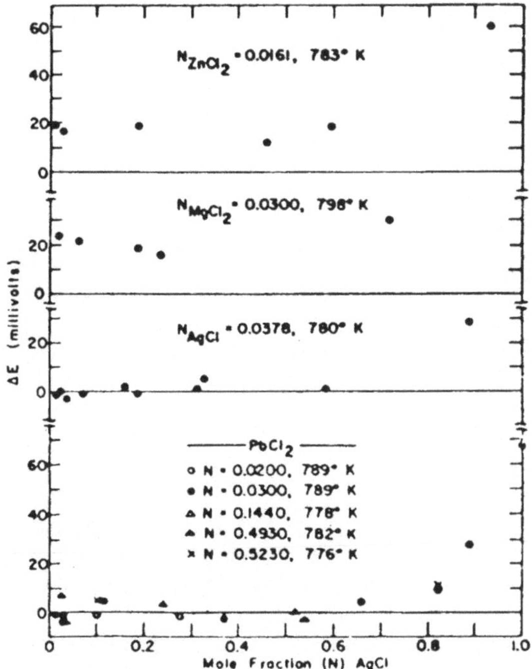

FIGURE 18. Pyrex membrane potential under various conditions.[48] (By permission of the publisher.)

TABLE IX. Effect of Membrane Materials on ΔE^a

MCl_z	Mol fraction N_{MCl_z}	Mol fraction N_{AgCl}	Temperature (°C)	ΔE (mV) Pyrex	ΔE (mV) Vycor	ΔE (mV) Quartz
AgCl	0.016	0.245	507	2	12	—
			514	2	11	—
			543	1	8	—
$PbCl_2$	0.030	0.136	522	4	9	10
$ZnCl_2$	0.016	0.026	505	16	42	42
	0.084	0.072	505	37	56	55
			526	35	51	52
			542	33	52	54

[a] Explanation for ΔE given in text. Data are taken from Yang and Hudson.[48]

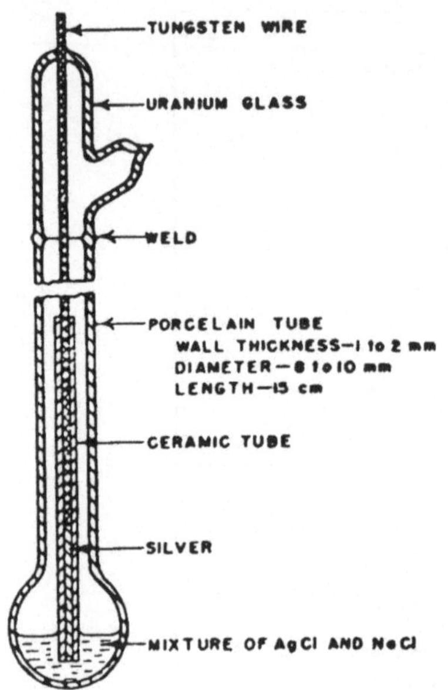

FIGURE 19. High-temperature silver reference electrode with porcelain membrane.[194] (By permission of the publisher.)

A porcelain membrane has also been used for constructing the Ag/Ag^+ reference electrode (Fig. 19).[194] The Na^+-ion conducting porcelain (Na_2O-SiO_2-Al_2O_3, 2.5-73.1-24.4 wt%) was made "in-house." The reference electrolyte was equimolar NaCl-KCl with 4.76 mol% AgCl content. The electrode required less than 30 min to reach a steady potential. A detailed description of the membrane preparation and electrode operation is given in the cited reference. The substitution of porcelain for glass extends the applicability of the Ag/Ag^+ reference electrode to 900–950°C.

4.1.3. The Pt/Pt^{2+} Reference Electrode

The platinum reference electrode was developed and extensively used by Laitinen and his co-workers for thermodynamic studies of metal[28] and halogen (Cl_2, Br_2, and I_2) electrodes[170] in LiCl-KCl melts. The cell and electrode arrangement are shown in Fig. 14, where I is the platinum electrode. For each experiment, a fresh platinum reference electrode was prepared by anodization of a 2-cm^2 platinum foil versus an electrode such as F or J at 20-mA constant current to generate Pt^{2+} ions coulometrically. The exact concentration of the platinum salt in the reference electrolyte (usually about 0.03 *M*) was either calculated from the coulometric data or analyzed subsequently. The platinum foil was boiled in $HClO_4$ and dried before insertion into the molten electrolyte. Potentials of the platinum electrodes were reproducible to 1 or 2 mV at sufficient Pt^{2+} ion concentration (0.03–0.08 *M*).

Later studies have indicated that certain instability problems of platinum electrodes relate to passivation and sensitivity to impurities.[195] The anodically formed passive layer has been identified as $K_2[PtCl_6]$.[196] Oxide, hydroxide, and silicate ions (the latter may come from the glass container) passivate platinum in LiCl–KCl eutectic either by the formation of a protective chloroplatinate layer or by the formation of a PtO layer at higher than 10^{-4} oxide-ion concentration.[197] In platinum-ion-free electrolyte, however, platinum may behave as a quasi-reference electrode (Section 2.4) or as an oxygen electrode in the presence of oxides and oxygen.

Platinum reference electrodes were also used successfully in other chloride melts. A detailed experimental evaluation of the performance of the chlorine, silver, and platinum reference electrodes in $MgCl_2$–KCl (32.5–67.5 mol%) electrolyte is given in Ref. 30. The use of the platinum reference electrode, however, appears to be restricted to $T < 500°C$ because of the volatility of $PtCl_2$.

4.1.4. *The Li–Al Reference Electrode*

The Li–Al alloy reference electrode has been developed for investigating electrode processes in experimental and full-scale engineering Li–Al/FeS cells[65,124] and for other molten-electrolyte battery development and testing.[198] In the course of this work, the Li–Al reference electrode has been refined to a precisely working device in the very demanding environment and operational conditions of these high-temperature batteries.

Of the alloy phases that constitute the Li–Al phase diagram[199] (Fig. 20), the α-Al and the β-LiAl phases are the most important for reference electrode construction. There is a wide composition range of the alloy, between approximately 8 and 48 at.% Li content, in which these two phases coexist and the alloy shows constant, stable electrode potential in Li^+-ion containing electrolytes. The alloy in this composition range is designated Li–Al.

The potential of the Li–Al electrode has been related to a lithium electrode in several studies. Figure 21 shows the potential of a Li–Al electrode versus lithium in LiCl-containing electrolyte as function of temperature. This diagram includes the results of several investigators, and indicates the rather close agreement of their results and also some discrepancies.[200–204] The straight solid line gives the potential of the Li–Al electrode (designated E in the figure) in millivolts on the lithium potential scale as follows:

$$\varepsilon_{\text{Li-Al}} = 297 - 0.022(T - 450), \qquad T \text{ in °C} \tag{21}$$

This relationship has been found valid in all investigated Li^+-ion-containing halogenides irrespective of the concentration of the Li salt and the tem-

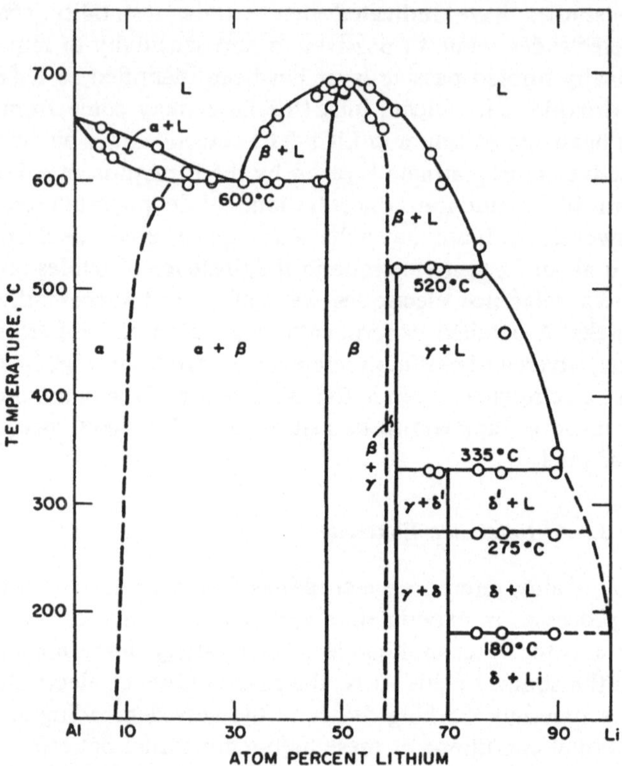

FIGURE 20. Phase diagram of the lithium-aluminum system.[199]

perature. As indicated in the diagram, the different electrolytes used in the studies covered different temperature ranges according to the melting points of the salt mixtures.

The emf of the cell

$$\text{Li}|\text{Li}^{+}\text{-ion containing electrolyte}|\text{Li–Al}$$

Cell 16

measures the Li activity in the alloy ($a_{\text{Li alloy}}$) relative to that of the pure lithium (a_{Li}), but is independent of the Li-salt activity of the melt:

$$E = -\frac{RT}{F}\ln a_{(\text{Li alloy})} \tag{22}$$

However, unlike the chlorine reference electrode, the lithium electrode cannot serve as a thermodynamic standard reference system. The lithium electrode is not in thermodynamic equilibrium because of the solubility of

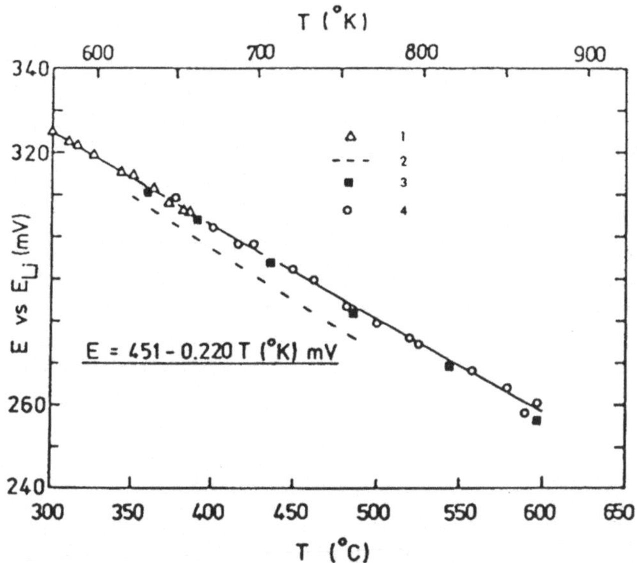

FIGURE 21. Li-Al potential versus lithium as measured by several authors; 1, Ref. 203; 2, Ref. 204; 3, Ref. 202; 4, Ref. 201. (By permission of the publisher.)

the metal in the electrolyte and the displacement reaction with other cations [Eq. (18), Section 2.5]. These effects explain the observed emf instability in Cell 16 and the differences between the reported Li–Al electrode potential values. The displacement reaction between Li and K [Eq. (18)] produces a high concentration of potassium metal in the melt and results in an appreciable partial pressure of potassium at these temperatures[205] (0.95 mbar at 450°C in equilibrium with lithium). Consequently, the displacement reaction can proceed to a considerable extent and create unstable conditions on the electrode surface; it also tends to decrease the KCl content of the electrolyte in the long run. In an all-Li^+-cation electrolyte, the displacement reaction is not relevant, but the lithium solubility and the resulting significant electronic conductivity of the electrolyte[206] cause problems.

The effects of the displacement reaction and lithium dissolution in the electrolyte, as well as of the electronic conductivity, become less important when the activity of lithium is decreased by alloying with aluminum. In the γ-phase region, only ±2 mV instability has been observed.[207] At even lower lithium activity, such as that of the α-Al + β-LiAl two-phase alloy (a_{Li} = 0.007 at 427°C), the potential of the alloy electrode is very stable. The Li–Al electrode may be considered in thermodynamic equilibrium with the electrolyte.

The Li–Al reference electrode utilizes an alloy that has approximately 20–42 at.% lithium content. A wide range of the alloy composition within the indicated limits satisfies the standard condition of this electrode. Theoretically, a broader range (8–48 at.%) is permissible, but for practical reasons the range is shortened at both ends. The LiCl–KCl electrolyte composition is optional as long as it is precisely defined and the lithium-salt content is not too low.

The standard state of the electrode can be produced easily and reproducibly by electrochemical means.[208] Figure 22a shows the potential variation of an aluminum wire charged with lithium intermittently with a

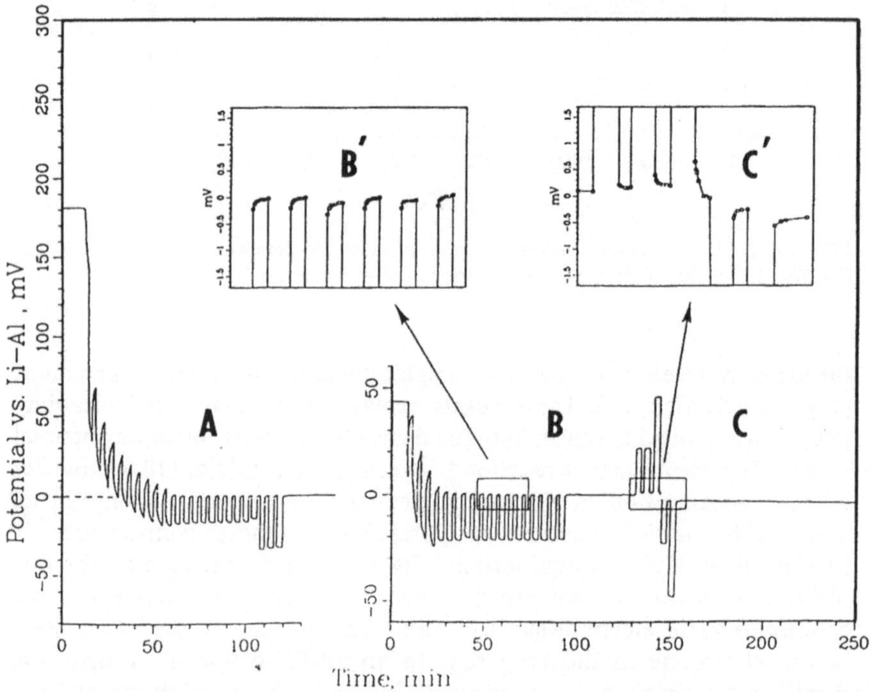

FIGURE 22. Intermittent coulometric titration curve of Li–Al formation as recorded by an analog chart recorder. Conditions: LiCl–KCl eutectic, 450°C, surface area of Al wire 0.15 cm², 100 and 200 μA charging current, 3-min on and 2-min off periods. Section A: cathodic charging of Li into an Al wire. Electrode potential is high at the beginning and gradually shifted to the negative direction; constant open circuit potentials indicate the presence of an alloy of standard composition; doubled current at the end of Section A does not effect the open circuit potential. Section B: Reconditioning of a slightly exhausted, off-composition electrode. The shape of the curve is identical to that of Section A. Computer plots of digitally recorded potential versus time data sets in insets B′ and C′ indicate the high precision of the potential measurement and stability and reproducibility of the potential. Section C: Proof that the standard condition is being observed.

constant current of approximately 0.5–5 mA/cm^2 current density. Duration of the current-interruption period is constant within the course of this intermittent coulometric titration and is usually between 1 and 3 min. After current interruption, the potential relaxes toward a steady value, which is determined by the alloy composition. The imaginary envelope curve that connects the open-circuit potentials recorded just prior to the next charging period shows a characteristic shape. The long plateau of the envelope curve (shown in the figure by a broken line which is an extrapolation of the actually measured plateau values to the potential axis of the diagram) indicates that the standard alloy composition is present. This phenomenon has great practical importance; the standard state of the reference electrode is apparent and does not require knowledge of the concentration of the electroactive species, as is the case for the platinum or silver reference electrode. The standard state condition and, consequently, the standard potential can be reproduced within at least ±1 mV precision under most experimental circumstances going from either the undercharged or the overcharged condition of the Li–Al alloy. Furthermore, an aged reference electrode, which has been gradually exhausted by the measuring current or other effects and shows higher potential than the standard, can be reconditioned *in situ* in the test cell by this technique (Fig. 22b). Any other electrode of the test cell that has a stable potential during the procedure and is not influenced by the titrating current can serve as a counter electrode for the conditioning of the reference electrode. The reversibility test is a part of the technique (Fig. 22c). Because of the reliability of this technique, the Li–Al reference electrode can be used as a "primary" standard for calibration of other reference electrodes in lithium–halogenide solvent–electrolyte systems. The high potential excursions observed during the application of the 100- and 200-μA cathodic or anodic currents, as seen in Fig. 22c, are caused by the high resistance of the diffusion barrier that was used in the construction of the tested Li–Al reference electrode. Because of the fast electrode kinetics,[209] the polarization resistance of the Li–Al electrode is low.

Construction of the Li–Al reference electrode is similar to the Ni/Ni_3S_2 electrode described in the next section.

4.1.5. The Ni/Ni_3S_2 Reference Electrode

The Ni/Ni_3S_2 reference electrode is based on the equilibrium

$$3Ni + 2S^{2-} \rightleftharpoons Ni_3S_2 + 4e^- \tag{23}$$

The reversible properties of the electrode were first attributed to the Ni/NiS couple.[210] Later, however, x-ray diffraction examinations revealed[130] that Ni_3S_2 forms on the surface of the nickel electrode in Li_2S-containing LiCl-

KCl electrolytes. Ni_3S_2 is the first stable phase formed during anodic sulfidization of nickel and the Ni/Ni_3S_2 electrode is an excellent reference system.

The electrochemical properties of the Ni/Ni_3S_2 electrode were investigated in a Li_2S formation cell,[208] which utilized the overall cell reaction

$$4LiAl + Ni_3S_2 \rightleftharpoons 2Li_2S + 3Ni + 4Al \tag{24}$$

The cell was constructed as shown in the following cell diagram:

$$\underset{T_1\ T_2}{Cu|Ni|}(Li\text{-}Al)|LiCl, KCl, Li_2S_{(s)}|Ni_3S_2\underset{T_2\ T_1}{|Ni|Cu}$$

Cell 17

where T_1 is the temperature of instrumentation, and T_2 is that of the cell. According to the primary standard concept (p. 189), the potential of the Li-Al electrode is defined as zero at any temperature in order that this alloy serve as a thermodynamic reference state in any chemically compatible electrolyte. Equation (25) describes the emf of Cell 17 and the potential of the Li_2S-saturated Ni/Ni_3S_2 reference electrode on the Li-Al potential scale in millivolts as a function of temperature:

$$E = 1369.7 + 0.097(T - 450), \qquad T \text{ in } ^\circ C \tag{25}$$

Thermal cycling between 360 and 500°C caused no hysteresis and indicated excellent reproducibility. In other than eutectic compositions of the LiCl-KCl mixture and in an LiF-LiCl-LiBr (22-31-47 mol%) electrolyte, the same emf values and temperature coefficients have been observed.[130]

The Nernstian response of the emf of the concentration cell

$$Ni|Ni_3S_2|LiCl, KCl, Li_2S_{(s)} \,\vdots\vdots\, LiCl, KCl, Li_2S(N)|Ni_3S_2|Ni$$

Cell 18

to the sulfide activity has been confirmed using the experimental cell shown in Fig. 23. Important features of this cell[211] include multiple reference electrode construction for comparison of electrode potentials (of electrodes 1, 4, 5, and 9 on the figure); means for coulometric manipulation of Li_2S concentration between the Ni/Ni_3S_2 cell-container electrode (1) and an auxiliary Li-Al electrode (9), which can be raised to remove it from the electrolyte; stirring of electrolyte for maintaining uniform temperature within the cell; and perfect electrical isolation from the high-temperature furnace well by the hanging cell support, which contacts only cold insulating components.

The Ni/Ni_3S_2 electrode acts as an electrode of the second kind in responding to the sulfide activity of the electrolyte.[212] The low solubility

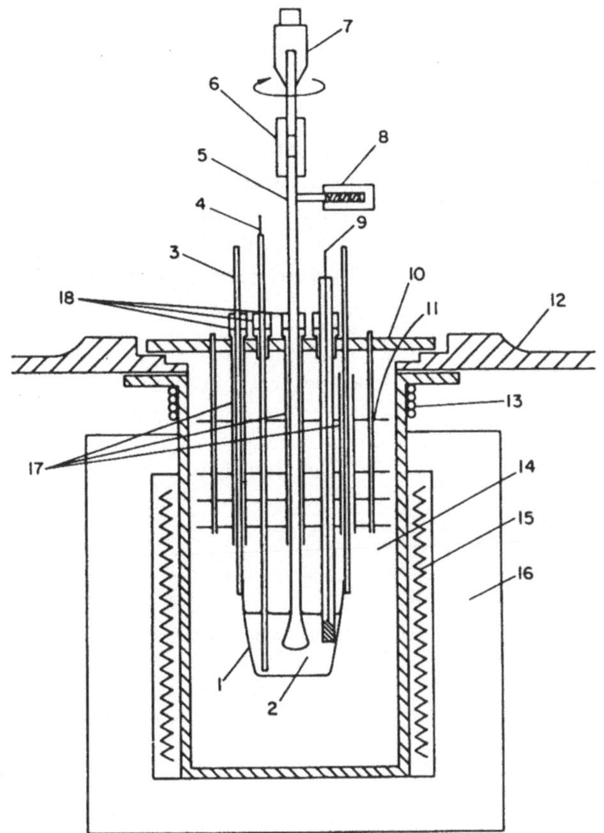

FIGURE 23. **Cell with multiple reference electrode system for Nernstian studies of Ni/Ni_3S_2 electrodes. 1, Ni crucible as main counter electrode; 2, electrolyte; 3, holder and current lead to Ni crucible; 4, Ni/Ni_3S_2 reference electrode with thermocouple (Fig. 24); 5, Ni/Ni_3S_2 indicator electrode as stirrer rod; 6, insulator connecting sleeve; 7, chuck of stirrer motor; 8, sliding electrical contact; 9, Li-Al counter electrode; 10, Plexy glass cell-holder plate; 11, holder of heat reflector plates (all stainless steel); 12, glove-box floor; 13, Cu tubing for water cooling; 14, high-purity He atmosphere; 15, electrical heating coils; 16, thermal insulation; 17, high-density alumina tubes for electrical insulation; 18, Swagelok compression holders.**

of the Li_2S in the tested electrolytes ensures constant sulfide activity when solid Li_2S is present. Solubilities of Li_2S in several molten electrolytes have been measured[(213)] and found to be in agreement with the calculated values[(214)]; solubility is in the range of 500–8000 ppm, depending on the temperature and composition of the electrolyte. Adjustment of the Li_2S saturation to a new temperature is reasonably fast. The very low Ni^{2+}-ion activity in Li_2S-saturated electrolytes ($\sim 10^{-12}$) emphasizes the merits of this reference electrode of the second kind.

A typical construction of the Li_2S-saturated Ni/Ni_3S_2 reference electrode[65] (Fig. 24) utilizes a high-density alumina tube (3.2-mm o.d., 1.6-mm i.d.) of the required length (long enough that the upper end is at room temperature) and a Ni tube with its lower end closed by electron-beam welding. This nickel tube serves as sheathing for a 0.25-mm o.d. chromel-alumel thermocouple and, at the same time, as the base metal for anodic formation of Ni_3S_2. The diffusion barrier is made by inserting a 10–15-mm-long solid piece of alumina rod into the alumina tube. End sections of both the rod and the tube are polished to make a near perfect fit. Friction holds

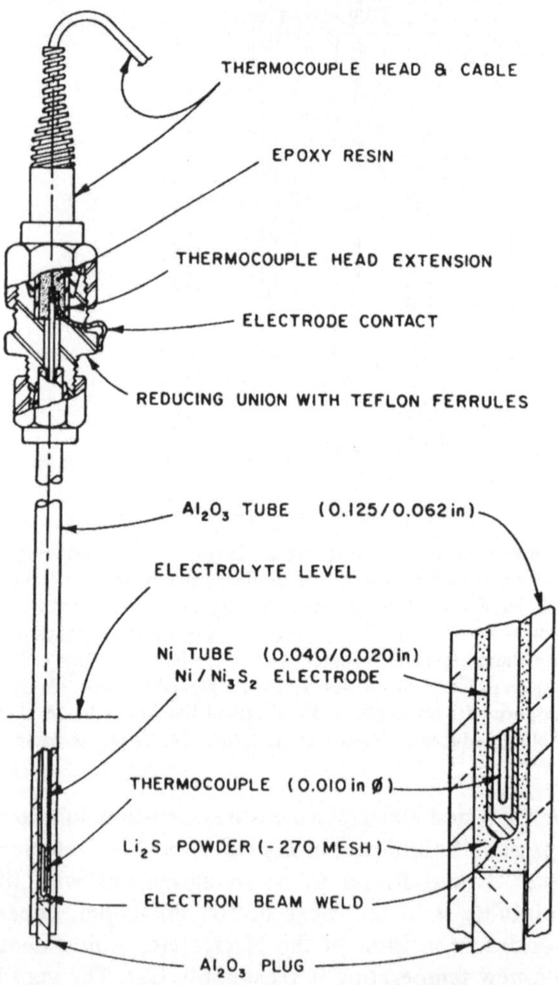

FIGURE 24. Design details of a Li_2S-saturated Ni/Ni_3S_2 reference electrode.[65] (By permission of the publisher.)

this plug in place, and the narrow gap between the two pieces forms a fine capillary that creates an excellent liquid junction with high resistance and thus a very low diffusion rate. The resistance of the diffusion barrier is adjustable between 0.2 and 5 kΩ. The electrodes are assembled "dry" in a high-purity helium-atmosphere glove box and are activated by allowing molten electrolyte to penetrate through the diffusion barrier. Because of the small masses of its active components and housing, the device responds rapidly to temperature changes. A typical response time for 90% of the total change (starting from electrode temperatures below the melting point of the electrolyte) is 12 s; the potential approaches the final mean value to within 1 mV in 50 s. These reference electrodes are used for high-temperature battery testing,[124,198] e.g., in an experimental cell shown in Fig. 4.

Reproducibility of the potential was studied with four identically prepared Ni/Ni_3S_2 electrodes immersed in Li_2S-saturated LiCl–KCl eutectic.[208] During a 30-day test period after electrode activation, no more than ±0.3 mV potential variation relative to the mean value was detected. The potential remained stable even after many months of use.

Construction of the Li–Al reference electrode discussed in the previous section is similar to the Ni/Ni_3S_2 reference electrode described above and shown in Fig. 24, except that either powdered Li–Al alloy or a piece of high-purity aluminum wire wound around the nickel current lead is substituted for Li_2S. In the latter version, the required alloy composition is produced by the described electrochemical method (Fig. 22). Both the chemical and the constructional stability of these electrodes are excellent. The construction of these electrodes is insensitive to freeze–thaw cycles.

4.1.6. *The Na/Na^+ Reference Electrodes*

The high solubility of sodium in a sodium halogenide electrolyte prevents reference electrode constructions with diffusion barriers. A membrane must be used to separate the metal and the electrolyte. Sodium membrane electrodes are used in cells of the type

$$\text{Na (or Na alloy)}|\text{Na}^+\text{-ion conducting membrane} \vdots \text{NaCl, MCl}_n|\text{M}$$

Cell 19

A similar arrangement can be used to measure sodium activity in alloys relative to pure sodium if the alloy in question is substituted for the melt and the metal (M) on the right-hand side of the above cell diagram.[215–217] The Na^+-ion-conducting membrane is either glass or β''-alumina. The membrane acts as a Na^+-ion conductor and the electrode reaction is $Na \rightleftharpoons Na^+ + e^-$ on the metal|membrane interface. The electrode potential is in agreement with the Nernst equation if the membrane conducts exclusively Na^+ ions, i.e., $t_{Na^+} = 1$.

A sodium electrode consists of a glass tube terminated at the low end with a thin-walled glass bulb that contains oxide-free sodium or sodium alloy. The electrical contact is made by a wire sealed into the glass at the upper end of the electrode. Different metal lead wires are preferred for different alloys: Pt for Na–Hg, Mo or W for Na–Sn and Na–Pb, Armco iron for Na and Na–Pb. Sodium borate glass is needed when pure metal is used because sodium reacts with a silicate glass

$$4Na + SiO_2 = 2Na_2O + Si \tag{26}$$

The drawbacks of the glass electrode filled with pure sodium have led to the development of sodium-alloy electrodes. The decreased activity of sodium in the alloy permits use of Pyrex or even quartz. Initially, the Na–Hg amalgam electrode was popular,[218–222] although the high vapor pressure of Hg limited its use to below 300°C. The Na–Sn alloy electrode[223,218] in Pyrex can be used up to 500°C. For even higher temperatures, a fused silica support tube and membrane can be used. To increase the electrical conductivity, the silica membrane may be conditioned by immersion into molten $NaNO_3$ at 500°C prior to construction of the electrode.[224] The Na–Sn alloy expands during solidification and usually cracks the membrane bulb. An improved version with Na–Pb alloy eliminates this problem and constitutes a sodium electrode with a very small emf temperature coefficient.

4.1.7. Miscellaneous Electrodes for Chloride Melts

The Pb/Pb^{2+} Reference Electrode. This electrode combined with a chlorine electrode has been used in thermodynamic studies in numerous investigations.[172,175,186,225–228] The Pb/Pb^{2+} reference electrode has a reversible, reproducible, and stable potential. One of its features is the moderate solubility of lead in molten $PbCl_2$, only 0.12 mol% at 800°C[187] and even less in $PbCl_2$–MCl_z mixtures, where M is an alkali metal.

Oxide Electrodes. The following oxide-electrode systems behave reversibly as electrodes of the second kind in LiCl–KCl eutectic: $Cu|Cu_2O|O^{2-}$, $Pt|PtO|O^{2-}$, $Pd|PdO|O^{2-}$, and $Bi|BiOCl|O^{2-}$.[64] Because of the relatively high solubility of these metal oxides, the applicability of the electrodes to the measurement of oxide activity is limited to solutions containing O^{2-} ion in concentrations comparable with or greater than those contributed by dissolution of the metal oxide being studied.

The potential response for oxide activity of a calcia-stabilized zirconia membrane electrode has been found to be in accordance with the Nernst equation in NaCl-KCl electrolyte at 720–800°C.[229] The electrode works well in O^{2-}-ion rich, basic electrolyte, but shows deviation from the Nernst equation in acidic electrolyte at low oxide activity, supposedly because of the increasingly significant role of the chloride-ion conductance of the membrane. Because of the fast kinetics, this electrode can be used in coulometric titrations as a sink or source of O^{2-} ions for the determination of the solubility of metal oxides, e.g., of NiO and Cu_2O in equimolar NaCl-KCl melt between 700 and 1000°C.[230]

The Oxygen Electrode. This electrode system has been studied repeatedly in LiCl-KCl melt with little success. Unworkability of this electrode has been reported,[231] but under certain conditions satisfactory results have been obtained on Au and Pt.[232] At the higher temperatures applicable in NaCl-KCl electrolyte (700°C), the platinum electrode is reversible with respect to changes in oxide content and partial pressure of oxygen.[233]

4.1.8. Reference Electrodes for Bromide and Iodide Melts

Bromine and iodine electrodes respond to the bromide or iodide activity of the respective melts. Similarly to the chlorine electrode, they need an inert electronic conductor, which is graphite of sufficient catalytic activity carefully saturated with the bromine or iodine. For studying the thermodynamic properties of the corresponding salts, bromine and iodine reference electrodes have been used.[170,172,176,234-237] For studying the thermodynamic properties of pure bromides or iodides, formation cells were used.[173,176,238,239] Since bromine and iodine are gaseous at the temperature of the investigation, the same cell and electrode designs (Figs. 12–14) and similar operating procedures as for a chlorine electrode are applicable for all halogen electrodes.[170]

The relationship between chlorine and bromine reference electrodes was investigated in a cell that utilized a glass membrane junction:

$$\text{(graphite)}Br_2|MBr \,\vdots\, \underset{\text{membrane}}{M^+\text{-ion-conducting}} \,\vdots\, MCl|Cl_2\text{(graphite)}$$

Cell 20

where M is either of Li, Na, K, or Rb.[240] Measurements of the potential difference between chlorine and bromine and between chlorine and iodine electrodes in the melts of corresponding alkali metal halides have been reported.[241] Also, membrane electrodes have been used in bromide and iodide melts.[220-222,242]

Besides the halogen electrodes, the Ag/Ag^+,[243-247] Li-Al,[65,130] and Ni/Ni_3S_2[65,130] reference electrodes can also be used in bromide- and iodide-containing melts.

4.2. Chloroaluminates

The chloroaluminates belong to the broad family of salts that are characterized by a complex anion with aluminum atom center: $[AlX_4]^-$ or $[Al_2X_7]^-$, where X is Cl, Br, or I atom. This section deals mainly with the so-called high-temperature chloroaluminate melts,[248] which have melting points that are above 70°C depending on the melt composition. Salts with organic cations, such as *N*-(*n*-butyl)pyridinium or various imidazolinium ions, constitute the room-temperature melts.[102,103,248,249] The properties and preparation of these melts are reviewed in the given references. It should be emphasized that when $AlCl_3$ is used to prepare chloroaluminate melts, it must be carefully purified by sublimation prior to use. Failure to purify the $AlCl_3$ may result in the introduction of oxide contaminants, which could substantially alter the chemistry and electrochemistry of the melt and the potential-determining process of the reference electrode.[250]

The acid–base equilibrium in the chloroaluminate melts is a characteristic feature of this electrolyte system. In sharp contrast to the previously described halide-salt solvent-electrolyte systems, the electrochemical and other properties of the chloroaluminate solvent systems are very sensitive to the mole ratio of the components, i.e., to the mole ratio of the cation-forming salt (basic component), and the aluminum salt (acidic component). The $AlCl_3$–NaCl system will be discussed here as an example. The acid–base chemistry of the $AlCl_3$–NaCl melt can be represented by the following equilibria:

$$[AlCl_4]^- + AlCl_3 \rightleftharpoons [Al_2Cl_7]^- \qquad (K = 2.4 \times 10^4) \tag{27}$$

$$2[AlCl_4]^- \rightleftharpoons [Al_2Cl_7]^- + Cl^- \qquad (K = 1.06 \times 10^{-7}) \tag{28}$$

where the equilibrium constants (K) are given on a mole fraction scale at 175°C,[251] $[Al_2Cl_7]^-$ ion is a Lewis acid, and the Cl^- ion is a base. The second equilibrium is an autosolvolysis reaction, which is analogous to the autoprotolysis of water. Consequently, the $pCl = -\log[Cl^-]$ of a neutral melt ($Na[AlCl_4]$) is ~3.5 at 175°C. This melt can be made more acidic by coulometric generation of aluminum ions from an aluminum anode in a compartment of the experimental cell or by gravimetric addition of $AlCl_3$; also, the melt can be made more basic by coulometric removal of aluminum or gravimetric addition of NaCl.[251] The autosolvolysis constant is larger at higher temperature. The greater the polarizing power of the inorganic or organic cation in various melts, the larger the constant at a given temperature; a wide range of the autosolvolysis constants [for Eq. (28)] has been demonstrated.[252] For example, this equilibrium constant for the *N*-(*n*-butyl)pyridinium chloroaluminate melt was measured by potentiometric titration to be 3.8×10^{-13} at 30°C. The ion-fraction distribution

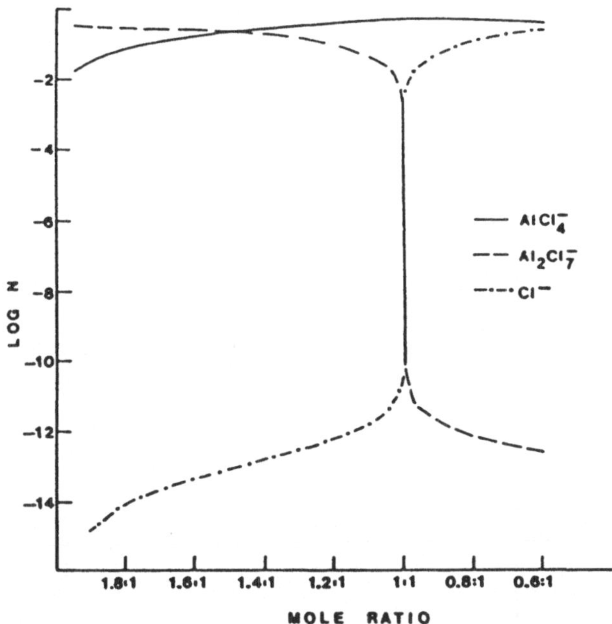

FIGURE 25. Anion-fraction distribution in an *N*-(*n*-butyl)pyridinium chloroaluminate melt at 30°C as function of $AlCl_3$ concentration.[253] (By permission of the publisher.)

as measured in this experiment is shown in Fig. 25.[253] This figure emphasizes the importance of the exact specification of the $AlCl_3/RCl$ ratio, where R is either an organic or an inorganic cation. Close to the 1:1 ratio, the melt is very sensitive to minor variations in composition or to impurities, this behavior is similar to the unbuffered aqueous solutions that have a pH about 7. Consequently, the neutral melt, which is almost impossible to prepare or maintain exactly at neutrality, is not a convenient electrolyte medium. For practical reasons, generally either the highly acidic melts (usually with a $AlCl_3$:RCl ratio of 2:1, Fig. 26b) or the NaCl-saturated basic melts (Fig. 26a) are commonly used in electrochemical studies to maintain a stable reference electrode potential.

4.2.1. *The Aluminum Reference Electrode*

The aluminum reference electrode, which is a natural choice in chloroaluminate melts and used almost exclusively, responds to the pCl of the melt. In this respect, it is an electrode of the second kind, since its potential is controlled through the half-cell reaction:

$$[AlCl_4]^- + 3e^- \rightleftharpoons Al + 4Cl^- \tag{29}$$

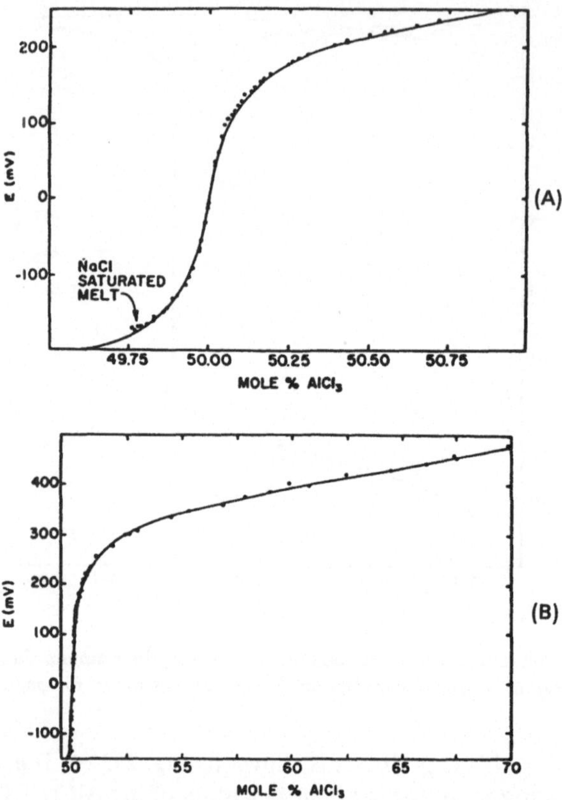

FIGURE 26. Potential of an Al electrode as function of melt composition versus an Al reference electrode prepared with a 1:1 $AlCl_3$:NaCl electrolyte. A, close to neutral range; B, acidic range.[251] (By permission of the publisher.)

according to the following equation:

$$\varepsilon = \varepsilon^{\circ} - \frac{RT}{3F}\ln\frac{(a_{Cl^-})^4}{a_{[AlCl_4]^-}} \tag{30}$$

The aluminum reference electrode has been investigated in an $AlCl_3$ concentration cell[251] and in combination with other metal electrodes.[254] The liquid-junction potential in this experiment across a Pyrex diaphragm was found to be less than 1 mV. Figure 26 shows the emf of this cell in two concentration ranges. Figure 26a indicates why an NaCl-saturated (or more precisely, a close-to-saturated) melt is preferred in the basic region. The use of an NaCl-saturated melt is technically unattractive owing to a slow NaCl dissolution/precipitation equilibrium, which may create nonequilibrium conditions during and after thermal excursions of the cell. On the acidic side (Fig. 26b), the emf change is relatively small around the usual 2:1 $AlCl_3$:NaCl composition of the melt; a large, about 10 wt%, change of $AlCl_3$ would cause a roughly 25-mV shift.

In NaCl-saturated basic melts, the potential of an aluminum electrode is fixed by the presence of the solid NaCl. An $AlCl_3$-concentration cell that utilizes such an electrode is defined as

$$E = E^0 - \frac{RT}{F} \ln \frac{a_{AlCl_3}}{a^3_{NaCl}} \tag{31}$$

where the activity term is written for the nonsaturated compartment; the activities in the saturated compartment are included in $E°$. For this type of cell, pCl is expressed as the negative logarithm of the NaCl activity under the conditions defined in Ref. 262.

In basic melts, the aluminum electrode potential was studied[(255)] with an Al–Cl_2 electrode pair in a "high-temperature" Na[$AlCl_4$] formation cell schematized as

$$Al|NaCl, Na[AlCl_4]|Cl_2(graphite)$$

Cell 21

utilizing the following cell reaction:

$$Al + NaCl + 1.5Cl_2 = Na[AlCl_4] \tag{32}$$

The relatively low concentration of the $AlCl_3$ in basic melts permits a high-temperature cell (400–700°C) study, because of the reduced vapor pressure of the $AlCl_3$. The experimental cell employed a molten aluminum reference electrode. Figure 27 shows the measured emf as function of the

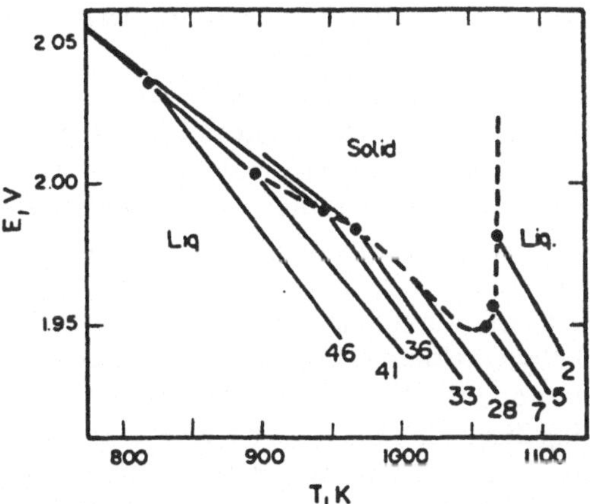

FIGURE 27. Emf of cell Al|NaCl, Na[$AlCl_4$]|Cl_2(vitreous carbon) as a function of temperature at various concentrations of $AlCl_3$ as indicated in mol% at the lower end of the emf versus *T* plots. Solid circles indicate liquidus temperature for the given composition; - - -, liquidus line.[(255)] (By permission of the publisher.)

electrolyte composition and temperature. Because of the high temperature, the higher solubility of NaCl permits a study in a wider pCl range in basic melts than is available at low temperature (compare Fig. 26a and Fig. 27).

For lower-temperature work (less than 400°C), an aluminum reference electrode is of simple construction. It consists of a high-purity (99.99% or higher) Al wire, which is immersed in the reference electrolyte. A Pyrex tube equipped with a glass frit or a fine capillary at the low end provides the electrode housing and junction for the reference electrolyte. As mentioned above, the reference electrolyte is usually either a melt with 2:1 mol ratio of the $AlCl_3$:base component (acidic) or a NaCl-saturated chloroaluminate melt (basic).[251,254,256–258] Figure 28 shows a special H-shape cell used to study electrode polarization with respect to an aluminum reference electrode (7) equipped with a capillary diffusion barrier.[259] The double-cell construction permits precise work without using an inert-atmosphere glove box. The ampule containing prepurified electrolyte is opened in the "purifying" cell (right-side cell). The last purification step of the electrolyte and the adjustment of the $AlCl_3$:base ratio are carried out in the purifying cell, and only the purified and tested electrolyte is transferred into the test cell (left-side cell), by rotating the H-cell in a proper angle. The interconnecting tube may have a glass frit (not shown) to filter the electrolyte.

The aluminum reference electrode in chloroaluminate melts has good reproducibility (usually within a few tenths of a millivolt) and excellent stability; it has been found satisfactory for all kinds of studies with inorganic chloroaluminates. If the $AlCl_3$ loss by evaporization is prevented, the potential remains stable within ±0.5 mV for about a half of a day. Less than ±2 mV variation of the potential has been observed for a given reference electrode over a period of two months.[254]

The aluminum reference electrode is applicable also in organic chloroaluminates.[253,257,260,261] The applicability of the aluminum reference electrode, however, may be limited by reactions between the organic cation and the Al electrode. Sensitivity of the organic cation to this reaction varies according to the molecular structure, but also pCl of the melt and temperature are important factors. For example, the basic 1-butylpyridinium melt is unstable at the aluminum electrode potential for this reason[253]; but, the basic methylethylimidazolium chloroaluminate melt ($AlCl_3$: RCl = 40 : 60) is stable to about −1.1 V versus aluminum.[261]

The foregoing discussion indicates the extreme importance of considering the acid–base characteristics when potentials are measured and defined in haloaluminate melts. As the given examples show, in order to specify a thermodynamically meaningful aluminum electrode potential scale in chloroaluminate melts, the pCl or the $AlCl_3$/RCl ratio in the melt must be precisely defined. Problems arise, however, when one would like to compare

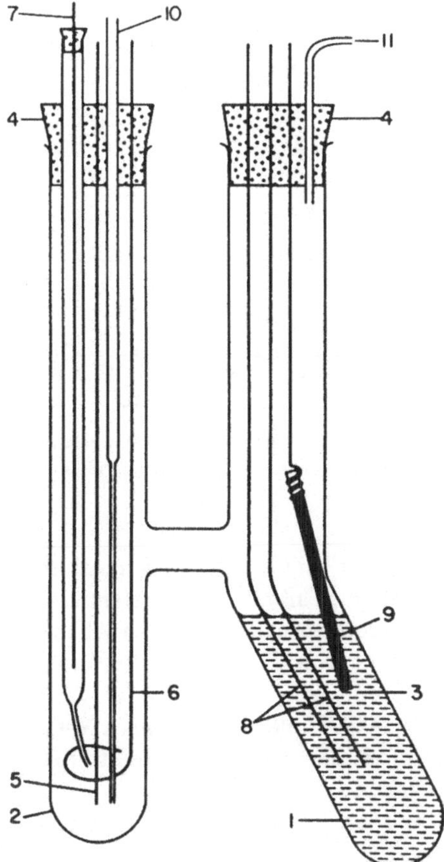

FIGURE 28. Double cell equipped with an Al reference electrode.[259] 1, cell for electrolyte purification; 2, cell for electrochemical studies; 3, electrolyte to be purified; 4, rubber plugs; 5, test electrode; 6, counter electrode; 7, Al reference electrode with capillary junction; 8, Al electrodes for melt purification; 9, graphite electrode; 10, inert gas inlet; 11, to vacuum.

potentials of melts having a different base component (RCl). The pCl scale is only relative in different melts, because the autosolvolysis constants greatly differ in different melts.[252]

4.2.2. Other Reference Electrodes for Chloroaluminates

Although the aluminum reference electrode has been used exclusively in chloroaluminate melts, one may consider other reference electrode systems. For example, because silver ions do not take part in the complex-forming equilibria, the Ag/Ag^+ electrode exhibits smaller variations of the

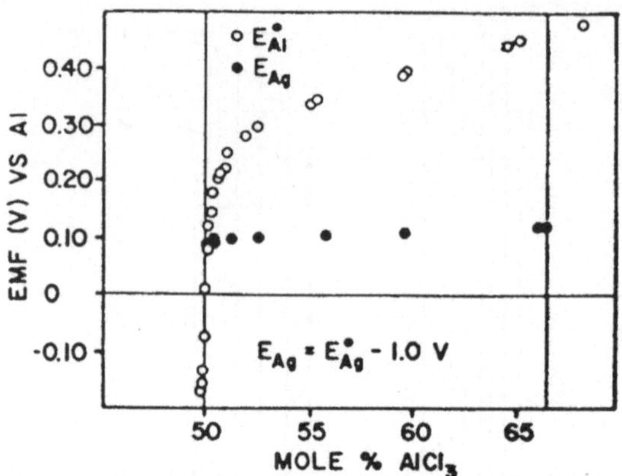

FIGURE 29. Standard potentials of aluminum and silver electrodes in neutral and acidic melts as function of $AlCl_3$:NaCl ratio at 175°C.[(254)] (By permission of the publisher.)

potential in the range of the acidic melts than the aluminum electrode does. Figure 29 shows this effect clearly.[(254)] The potentials of the aluminum and the silver electrodes were measured versus an Al reference electrode prepared with $Na[AlCl_4]$ (1:1 $AlCl_3$:NaCl) reference electrolyte in a cell shown in Fig. 30.

The potential of an $M|M^{z+}$ electrode versus a NaCl-saturated aluminum electrode in basic melts is described[(262)] by the following equation:

$$\varepsilon = \varepsilon^{\circ} - (RT/zF) \ln (a_{MCl_z}/a^{z}_{NaCl}) \tag{33}$$

where MCl_z is the chloride of the metal in low enough concentration to obey Henry's law, and NaCl is the basic component of the melt. Further discussion of the consequences of this equation, although very important for the correct use of reference electrodes in chloroaluminate melts and interpretations of the results, is beyond the scope of this chapter. These important features of the metal electrodes in basic chloroaluminate melts are discussed in the original article.[(262)]

A few studies have been reported of aluminum bromide and aluminum iodide containing melts.[(263–265)]

4.3. Fluorides

Interest in molten fluorides stems from their importance in nuclear technology and their use in the production of fluorine, electrodeposition of refractory metals, formation of corrosion-resistant diffusion coatings, and fluorination by electrochemical techniques. Most studies in alkali metal

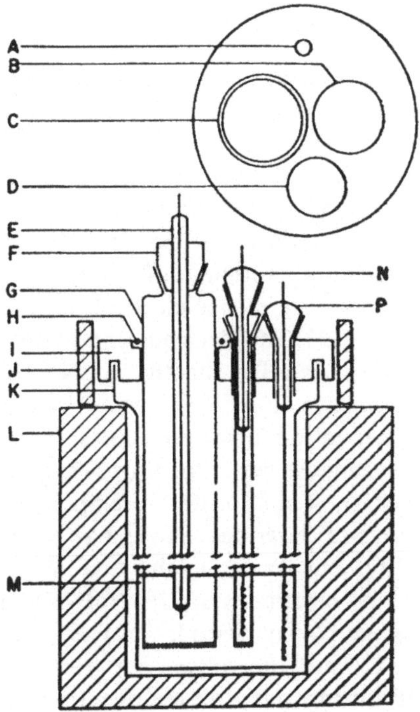

FIGURE 30. Experimental cell equipped with Al reference electrode to study electrodes in chloroaluminate melts.[254] A, thermocouple well; B, 19/22 ground joint for reference electrode; C, 25-mm hole for test electrode compartment; D, 14/20 ground joint for counter electrode; E, test electrode; F, Teflon adaptor; G, fritted test-electrode compartment; H, O-ring; I, Teflon cell top; J, ring furnace to prevent $AlCl_3$ precipitation; K, Pyrex glass cell; L, furnace; M, melt; N, reference electrode compartment and Al electrode; P, tungsten counter electrode. (By permission of the publisher.)

fluorides and other fluorides are rather recent and in connection with the development of molten salt reactors and electrodeposition of silicon and the refractory metals.

Various reference electrodes have been developed for use in molten fluorides. As in many other molten salt systems, the most serious problems encountered in the development of reference electrodes for fluorides deal with containment and materials of construction. Electrically insulating materials that are resistant to attack by molten fluorides are scarce. Boron nitride (BN), which is an insulator but permits impregnation by molten fluorides, has been frequently employed as a diaphragm in molten alkali fluorides. Among the reference electrodes developed for use in alkali metal fluorides, the Ni/Ni^{2+} electrode appears to be the most popular.

4.3.1. The Ni/Ni^{2+} Reference Electrode

The Ni/Ni^{2+} reference electrode is well studied in fluoride melts and has been successfully used in electrode potential measurements and kinetics studies.[247,266–273] The Ni/Ni^{2+} system is a good reference system since it shows Nernstian reversibility and fairly large exchange currents, and its standard potential is essentially in the middle of the potential span in molten fluorides. Two common materials used to provide ionic contact for the Ni/Ni^{2+} electrode are BN and lanthanum fluoride.

The design of a Ni/Ni^{2+} reference electrode with a BN diaphragm is shown schematically in Fig. 31. The electrode consists of an Ni/Ni^{2+} couple in a fluoride melt contained in a BN sheath. The Ni^{2+} concentration in the reference melt is of the order of 10^{-4}–10^{-2} mole fraction. The BN sheath generally is made thinner in the lower portion (as shown in the figure) to achieve wetting in a reasonable period of time. This reference electrode is recommended for use in molten fluorides at temperatures up to 800°C.

The Ni/Ni^{2+} electrode described above has long-term stability and excellent reproducibility. For example, the emf of a concentration cell comprised of two Ni/Ni^{2+} electrodes in LiF–NaF–KF at 502°C is constant to within ±3 mV over an 11-day period.[267] The electrode also has low polarizability. Figure 32 shows the voltage between two identical Ni/Ni^{2+} electrodes in LiF–NaF during and after anodic and cathodic galvanostatic polarization. It can be seen from the figure that the voltage returns to zero in a few minutes. The reversibility of the electrode has been shown to be

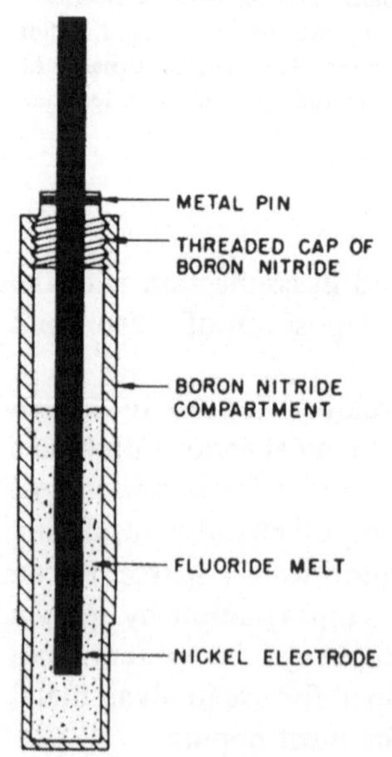

FIGURE 31. Ni/Ni^{2+} reference electrode with boron nitride diaphragm (Ref. 267). (Reprinted by permission of the publisher.)

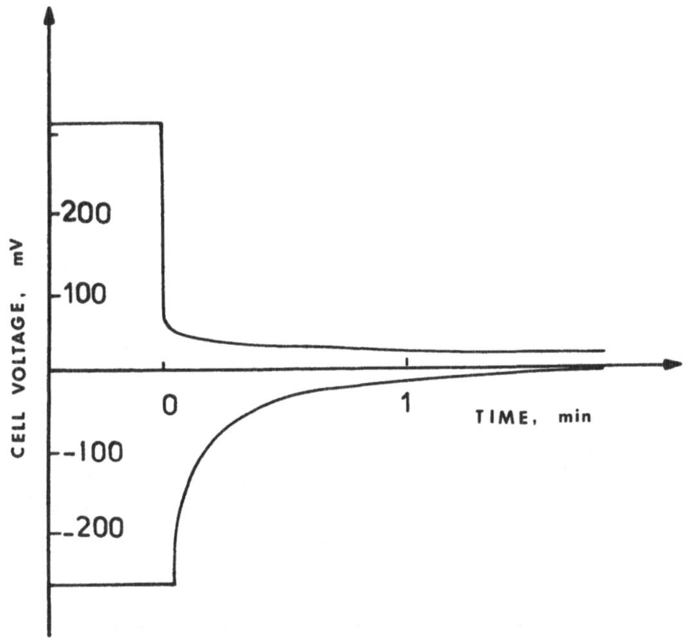

FIGURE 32. Potential between two identical Ni/Ni^{2+} electrodes in LiF–NaF melt at 750°C during and after anodic and cathodic galvanostatic polarization (current density = 30 mA/cm^2, time = 1 min) (Ref. 272). (Reprinted by permission of the publisher.)

adequate, as indicated by the applicability of the Nernst equation to the Ni/Ni^{2+} system (Fig. 33). The useful life of the Ni/Ni^{2+} reference electrode in fluorides may range from many hours to many days, depending on the operating temperature and the melt employed. In general, shorter life is expected at higher operating temperatures.

Commercially available BN (grade HP, made by Carborundum Company) is commonly used for construction of electrode sheaths. This grade of BN is somewhat porous and contains a few percent of boric oxide. Boron nitride, normally an insulator in fluoride melts (e.g., 2.3×10^{10} Ω cm at 450°C), is slowly impregnated by the melt, thus providing ionic contact. Different melts require different times to wet the BN, and the wetting rate depends on the free oxide concentration of the melt and the operating temperature. For example, the LiF–NaF melt wets the BN at 750°C in 6 h;[272] the time required for wetting in LiF–NaF–KF and LiF–BeF_2 melts at 500°C is 24 to 48 h.[267]

It has been shown that a junction potential exists across the BN diaphragm of the Ni/Ni^{2+} reference electrode. This junction potential is either negligible (e.g., in LiF–NaF at 750°C) or quite small (of the order of 10 mV in LiF–NaF–KF eutectic at 600°C); however, in some cases, values

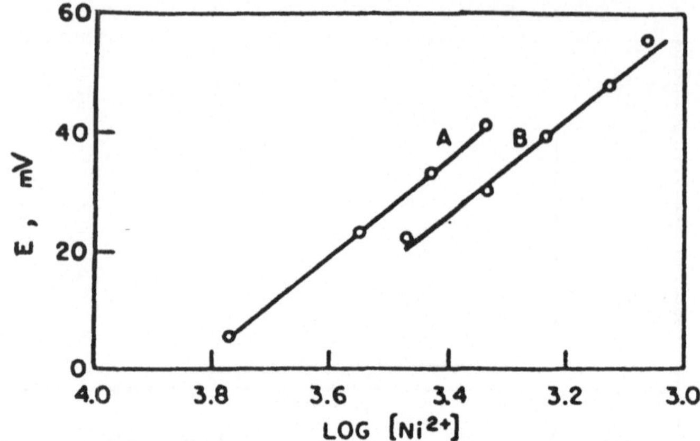

FIGURE 33. Nernstian plot for the Ni/Ni^{2+} equilibrium in molten fluorides (Ref. 267). A, LiF-NaF-KF at 517°C, slope of the line = 0.080, theoretical slope = 0.078. B, $LiF–Be_2F–ZrF_4$ at 507°C, slope of the line = 0.081, theoretical slope = 0.077. (Reprinted by permission of the publisher.)

up to 100 mV have been observed. The potential, however, is generally stable, reproducible, and constant at constant temperature. Thus, for thermodynamic studies, correction for junction potential is necessary.

Prolonged contact with fluoride melts will eventually deteriorate the BN. This is probably due to the reaction between NiF_2 in the reference melt and BN

$$\tfrac{3}{2}NiF_2 + BN = BF_3 + \tfrac{3}{2}Ni + \tfrac{1}{2}N_2 \tag{34}$$

Diffusion of solute species in the fluoride melts into the reference electrode compartment also occurs. The extent of diffusion appears to be related to the oxide content of the melts as well as the previous history of the BN (e.g., amount of exposure to atmospheric moisture during machining). Therefore, it is recommended that BN be kept in a desiccator except for the time required for machining.

An Ni/Ni^{2+} reference electrode for high temperatures (900–1100°C) has also been developed.[274–276] Several designs were proposed, and one preferred and recommended design is shown in Fig. 34. The reference compartment is made of two BN envelopes. The inner envelope is formed by sintering (under pressure at 700°C) a mixture of equal portions of fine (grade HPF, Carborundum Company) and coarse (grade HPC, Carborundum Company) BN powder with about 2.4% boron oxide. Commercial BN is used for the outer envelope, which provides mechanical support to the inner one. A plug of BN "doped" with NaF forms the junction between the reference electrode and the melt. The behavior of the reference electrode

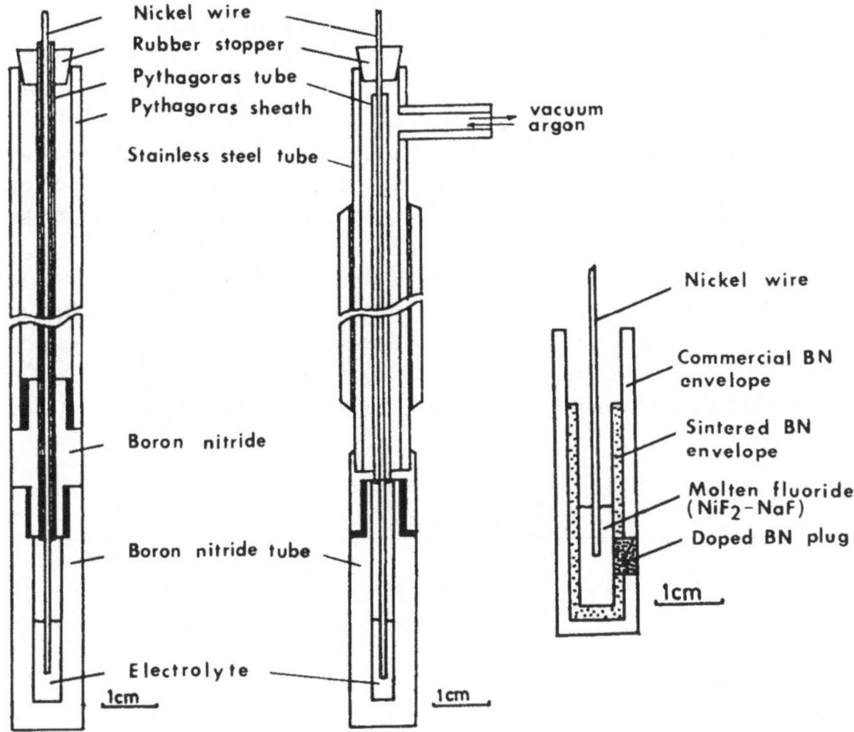

FIGURE 34. Schematic diagram of Ni/Ni^{2+} reference electrodes for use at high temperatures (900–1100°C) (Ref. 275). (Reprinted by permission of the publisher.)

differs slightly, depending on the concentration of NaF dopant of the sintered BN plug. At 40 wt% NaF, the stabilization period is fast (about 15 min); however, the lifetime of the electrode is short (about 2 h). The reference electrode with a plug containing only 5 wt% NaF stabilizes slowly (12 h) but can be used in molten fluorides for longer periods (24 h). The resistance of this electrode is of the order of 50 Ω. The stability, reproducibility, reversibility, and polarizability of this type of reference electrode were tested and found to be satisfactory.

One modification of the Ni/Ni^{2+} reference electrode utilizes single-crystal LaF_3 as the membrane. Lanthanum fluoride crystal is a fluoride-ion-conducting electrolyte and has small solubility in molten fluorides. The role of the LaF_3 crystal is thus primarily to provide an electrical connection via fluoride-ion conduction between the reference electrode melt and the melt under study. Figure 35 illustrates the construction of an Ni/Ni^{2+} electrode with an LaF_3 membrane.[277,278] The electrode consists of a LaF_3 cup made by drilling a single crystal. The cup contains $LiF-BeF_2$ melt saturated with

NiF_2. Contact to the melt in the cup is achieved with a nickel wire. The crystal is inserted into a nickel tube, but it is insulated from the tube by BN. (The BN in turn is shielded from the LaF_3 crystal by means of a thin nickel cup, as shown in Fig. 35.) The end of the nickel tube is welded to a nickel frit. The purpose of this frit is to protect the LaF_3 crystal from undue etching by the molten fluoride. Once the melt inside the frit and at the frit/crystal interface becomes saturated with LaF_3, further attack on the crystal is minimized.

Typical resistance of the electrode is about 500 kΩ over the temperature range 500–600°C. In general, the reference electrode makes electrical contact within 5–10 min after being immersed in fluoride melts. This type of electrode has been used successfully both for electroanalytical measurements and for obtaining thermodynamic data in LiF–BeF_2,[277] LiF–BeF_2–ZrF_4,[279] and

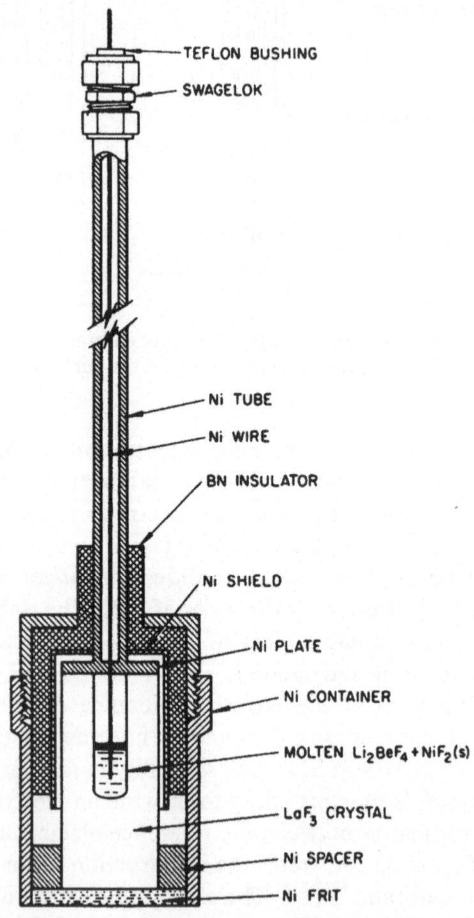

FIGURE 35. LaF_3-membrane Ni/Ni^{2+} reference electrode (Ref. 278). (Reprinted by permission of the publisher.)

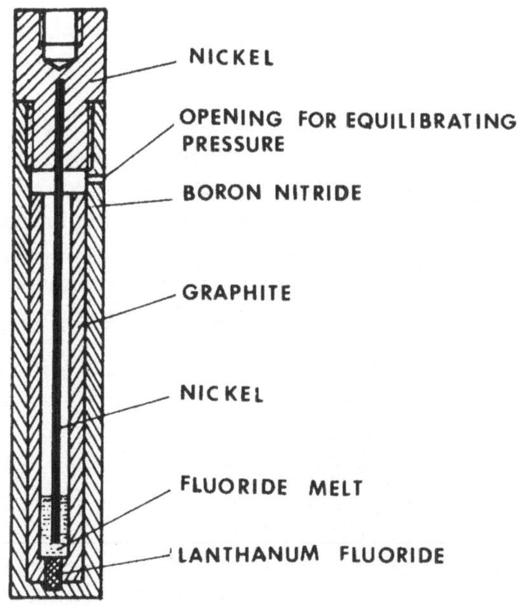

FIGURE 36. Ni/Ni^{2+} reference electrode with LaF_3 membrane (Ref. 281). (Reprinted by permission of the publisher.)

LiF–NaF–KF.[279,280] The LaF_3 crystals remain in excellent condition after contacting $LiF–BeF_2$ melt approximately 50 days at 500°C.[277] Rapid temperature cycling should be avoided to prevent cracking of the LaF_3 crystal. In several cases, asymmetry potential at the crystal interface is not negligible. The reasons for a significant asymmetry potential are not fully understood. The asymmetry potential, however, is stable and thus can be readily corrected for. The potential of this type of reference electrode exhibits a tendency to decrease with time in a reducing environment.

Another design of an Ni/Ni^{2+} reference electrode with an LaF_3 membrane is shown in Fig. 36.[281] This electrode consists of a nickel wire immersed in a molar solution of NiF_2 in a fluoride melt contained in a graphite tube. The graphite is isolated from the bulk melt by a BN sheath. An ionic junction is ensured by an LaF_3 crystal, which is inserted at the end of the graphite tube and in contact with the BN, as shown in the figure. This electrode has been shown to function as a reference electrode in LiF–KF and LiF–NaF–KF eutectic for at least 6 h at 500°C.[282]

4.3.2. The Ag/Ag^+ Reference Electrode

There are two types of electrodes for molten fluorides based on the Ag/Ag^+ system. In the first type, the reference melt is a solution of AgCl in a chloride melt and contained in a BN compartment.[274,275,283] The potential of the electrode is determined by the reaction

$$Ag^{+} + e^{-} = Ag \tag{35}$$

and the reversibility of the silver system in chloride is well known. This reference electrode can be employed at temperatures up to the melting point of silver (960°C). The behavior of the Ag/Ag^{+} electrode is similar to that of the Ni/Ni^{2+} reference electrode with a BN diaphragm.

The second type Ag/Ag^{+} reference electrode is shown schematically in Fig. 37.[284] A silver wire of 1 mm diameter is immersed in molten AgCl in a quartz tube. The tube, which is closed at the lower end by a plug made

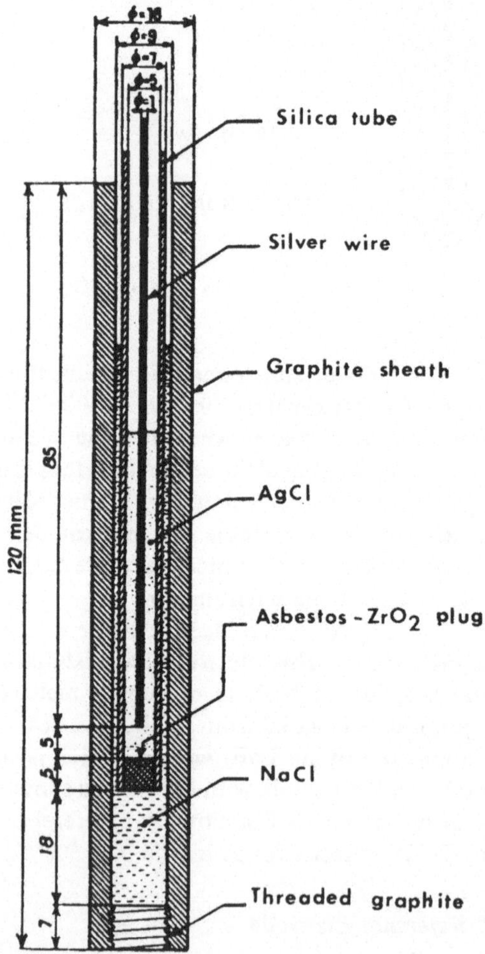

FIGURE 37. Ag/Ag^{+} reference electrode for molten fluorides (Ref. 284). (Reprinted by permission of the publisher.)

from a cement containing zirconia and asbestos, is placed inside a graphite tube filled with molten NaCl. A graphite plug is screwed into the bottom of the graphite tube. The levels of AgCl and NaCl are adjusted to prevent the passage of silver into the melt under study. The reproducibility and reversibility of this reference electrode have been verified; the graphite tube has no influence on the potential of the electrode. In a 1-h test in $KF-ZrF_4$ melts at 850°C, diffusion of KF or ZrF_4 through the graphite tube into the reference compartment was found to be negligible.[284]

The liquid-junction potential of the Ag/Ag^+ reference electrode in fluoride melts is not known.

4.3.3. The Ni/NiO Reference Electrode

This electrode consists of a nickel wire immersed in molten $LiF-BeF_2$ that is saturated with NiO and BeO.[285] The potential at the nickel wire is established by the reaction

$$NiO + Be^{2+} + 2e^- = BeO + Ni \qquad (36)$$

with

$$\varepsilon^0 = 1.9075 - 0.056 \times 10^{-3} T \qquad (37)$$

where ε^0 is in volts, versus the Be/Be^{2+} electrode, and $T = 780\text{–}980$ K.

The concentration of Ni^{2+} is held constant because of the low solubility product of NiO and the presence of common ion O^{2-}, which in turn is held constant by the presence of saturating BeO.

The electrode assembly is shown in Fig. 38. The compartment for the reference melt is constructed either of sintered BeO tube (10 mm diam) closed at one end by a plug of the same sintered BeO or of a silica tube closed at one end by a silica frit of 5-10 μm porosity. A nickel tube with holes drilled through for melt circulation is used to provide structural support for both silica and BeO compartments. The nickel electrode is fabricated from a nickel rod with a small rectangle of sintered nickel welded on the lower end. The sintered nickel is immersed in an aqueous slurry of nickel hydroxide, then fired at 800°C in air. This step provides a large surface area for the electrode reaction, and the NiO formed in this manner is in intimate contact with the electrode. Nickel electrodes, thus prepared, normally reach a steady potential in 1-2 h in contrast to the 4-8 h required by bare nickel wires. In the case of BeO tube, the walls are too dense for melt penetration; however, the seal between the tube and the plug has microcracks which permit ionic contact with the melt under study.

The reliability of the Ni/NiO electrode was tested in $LiF-BeF_2$ melt by potential measurements versus the Be/Be^{2+} electrode. Measured cell emfs over a temperature range of 500-700°C showed random fluctuations

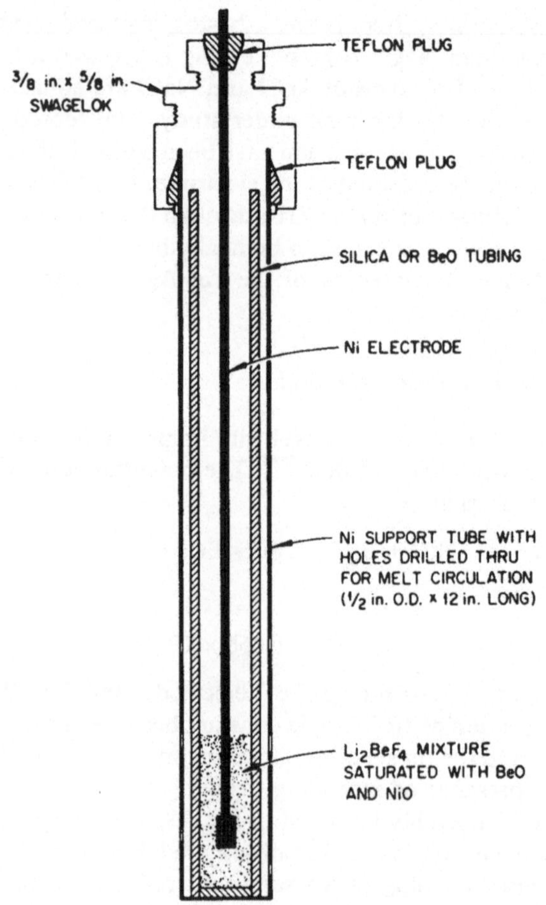

FIGURE 38. Ni/NiO reference electrode assembly (Ref. 285). (Reprinted by permission of the publisher.)

of ±0.5–2 mV for electrodes with BeO tubes and ±5–10 mV with SiO_2 tubes. Silica tubes show some attack by the melt. In general, the Ni/NiO electrode is usable in any molten fluoride which is not too basic, provided that the molten fluoride contains a cation of an oxide with low solubility.

4.3.4. Other Reference Electrodes

The Fe/Fe^{2+} Reference Electrode. The Fe/Fe^{2+} reference electrode, which utilizes single-crystal LaF_3 as a membrane, has the same electrode configuration described earlier for the Ni/Ni^{2+} electrode. The Fe/Fe^{2+} system is known to be reversible in molten fluorides from studies in LiF-

NaF–KF, $LiF–BeF_2$, $LiF–BeF_2–ZrF_4$, and $NaBF_4$.[279,286,287] Emf measurements on the system also show the applicability of the Nernst equation. This electrode is useful in melts such as $NaBF_4$, where the use of the Ni/Ni^{2+} system is not possible owing to the low solubility of NiF_2.

The H_2, HF Reference Electrode. This electrode uses a gas mixture of H_2 and HF that is bubbled at either an open-end palladium or a nickel or BN tube through which a platinum gauze or wire has been inserted.[288,289] The potential of the electrode established by the reversible reaction

$$H_2 + 2F^- + 2e^- = 2HF \tag{38}$$

is quite stable (±1 mV) in molten fluorides and is reproducible. The H_2, HF electrode can be used only in fluoride melts in which solution constituents undergo no oxidation by HF or reduction by H_2. A gas mixture of H_2 and HF of known and constant composition is required for the precision of this electrode.

The Be/Be^{2+} Reference Electrode. The Be/Be^{2+} electrode works well in any fluoride that contains beryllium ions.[290–292] The Be/Be^{2+} system was demonstrated to be reversible.[289] The Be/Be^{2+} reference electrode consists of a beryllium rod dipped into a solution of BeF_2 in a fluoride melt. The electrode compartment is made of a nickel tube closed at the lower end by nickel frit. Silica has also been used for the electrode compartment and frit in $LiF–BeF_2$ mixtures at BeF_2 concentrations greater than 0.33 mole fraction. Potential fluctuations of the electrode are less than ±0.1 mV at a temperature range of 500–900°C. The Be/Be^{2+} electrode is strongly reducing and can be easily contaminated by impurities that are reduced from the melt and affect the potential.

Quasi-Reference Electrodes. Several metals (such as Pt, Ir, W, Au, Ni, Al-alkali metal alloy) can be immersed in fluorides and used as quasi-reference electrodes.[293–302] Another practical quasi-reference electrode in niobium-containing fluoride melts is the redox electrode Cu(or Mo)/ Nb^{5+}, Nb^{6+} ($Nb^{5+}/Nb^{6+} = 1$).[273,303] The potential of these quasi-reference electrodes is not well defined; however, it generally reaches a steady-state value after a few minutes of immersion. The potential is generally stable (within about ±1 to ±20 mV) over periods of several days.

4.4. Cryolites

The production of aluminum metal by the Hall–Héroult process is the most important industrial application not only of molten salts but also of electrochemistry in general. In this process, aluminum metal is produced by the electrolysis of alumina dissolved in molten cryolite (Na_3AlF_6) at about 970°C. The fundamentals of the process have not changed since its discovery in 1886. However, over the past 100 years, significant progress has been made in understanding the chemistry and technology of the process.

Molten cryolites contain varying proportions of F^-, AlF_6^{3-}, AlF_4^-, and $Al_3F_{14}^{5-}$ as the anionic species and Na^+ as the cationic species. The most important acid-base equilibrium in molten cryolite is $2F^- + AlF_4^- \rightleftharpoons AlF_6^{3-}$. The dissolution of alumina in molten cryolites forms oxyfluoride complexes. Possible ionic complex species include the tetrahedral anion $AlOF_3^{2-}$ and the octahedrally coordinated complex $AlOF_5^{4-}$, although oxygen-bridged species cannot be excluded.

Several reference electrodes have been developed for use in cryolite and alumina-cryolite melts. The most critical factor in the development of reference electrodes in molten cryolites is the choice of construction material that is an insulator at 900–1050°C and is resistant to the highly corrosive effects of the molten salt. There is no electrode that is generally applicable as a reference electrode in cryolite melts. The C/CO_2 and the SnO_2 electrodes seem to be the best choice under oxidizing conditions; liquid aluminum electrodes seem best under reducing conditions.

4.4.1. The Oxygen Reference Electrode

In alumina-containing cryolite melts, a platinum wire flushed with oxygen and immersed in the melt behaves as a reversible oxygen electrode. The reversibility of such an electrode is the subject of several studies[304–314] that measured the emf at 957–1090°C of the following galvanic cell:

$$Al_{(1)}|Al_2O_3, Na_3AlF_6|O_2(Pt)$$

Cell 22

with the cell reaction

$$2Al + \tfrac{3}{2}O_2 = Al_2O_3 \tag{39}$$

Since the aluminum electrode is known to behave reversibly (Section 4.4.4), agreement between the experimental data and the calculated theoretical emf for the above reaction (based on thermodynamic data) indicates the reversibility of the oxygen electrode. Most of the earlier experimental results from emf measurements generally show potential values somewhat lower than those calculated from thermodynamic data (Table X). The reason for the discrepancy between observed and thermodynamically calculated emf values (or any deviation from the reversible behavior of the oxygen electrode) is probably due to depolarization effects caused by dissolved aluminum metal from the aluminum electrode. In alumina-cryolite melts in contact with aluminum, it is well known that there is some metal solubility; when the dissolved metal has access to the oxygen electrode, it influences the potential of the electrode. When the oxygen electrode is separated from

TABLE X. Comparison of Selected Experimental and Calculated Decomposition Potentials of Alumina in Cryolite

T (°C)	Al_2O_3 (wt%)	E_{exp} (V)	E_{calc} (V)	Ref.
1060	10	2.06	2.161	305
1015	Saturated	2.12	2.187	308
1020	10	2.11	2.184	309
957	Saturated	2.2	2.219	310
1000	Saturated	2.183	2.194	311

the aluminum electrode by a membrane (e.g., an alumina membrane which acts as a sodium-ion conductor at those temperatures), the electrode then behaves reversibly.[311] For instance, the variation of the emf as a function of p_{O_2} of the cell

$$(Pt)O_2|Al_2O_3, Na_3AlF_6 \vdots alumina \vdots Al_2O_3, Na_3AlF_6|O_2, Ar(Pt)$$

Cell 23

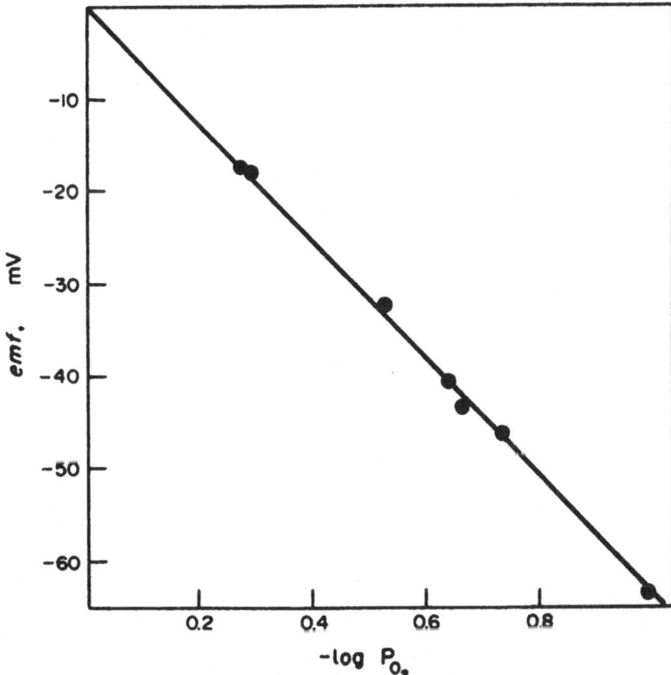

FIGURE 39. Emf as a function of the partial pressure of oxygen at 1000°C for the cell $(Pt)O_2|Al_2O_3$, Na_3AlF_6:alumina:Al_2O_3, $Na_3AlF_6|O_2$, Ar(Pt). The solid line is the theoretical emf (Ref. 311). (Reprinted by permission of the publisher.)

has been shown to be in agreement with the values from the equation

$$E = -\frac{RT}{zF} \ln p_{O_2} \tag{40}$$

(Fig. 39). The emf of the cell is stable within ±0.5 mV for several hours at constant temperature and pressure. The emf is also independent of the depth of immersion of the electrode. The reversibility of the emf is checked by shorting the cell for 10 s. The emf always returns to the original value within 1 min. A schematic diagram of a simple oxygen reference electrode with alumina membrane is shown in Fig. 40.

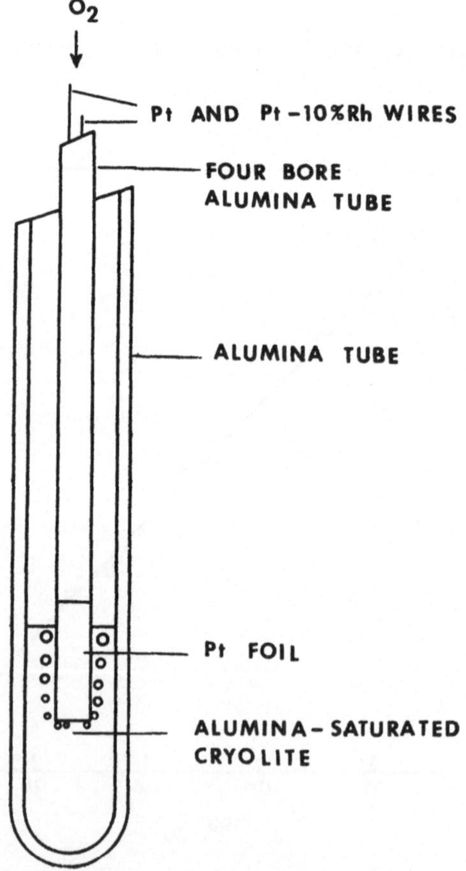

FIGURE 40. Oxygen reference electrode with alumina membrane for alumina-saturated cryolite melts (Ref. 311). (Reprinted by permission of the publisher.)

4.4.2. The Carbon/Carbon Oxide Reference Electrode

This type of electrode consists of a carbon (or graphite) tube dipped into alumina-cryolite melt and flushed with carbon dioxide, carbon monoxide, or a mixture of the two.

The C/CO_2 (or $C/CO_2, N_2$) reference electrode has been shown by several authors to be a reliable reference electrode.[308,314–321] The measured emf of the cell

$$Al_{(l)}|Al_2O_3, Na_3AlF_6|CO_2(C)$$

Cell 24

and its temperature dependence (Fig. 41) agree with the thermodynamic emf for reaction

$$Al_2O_3 + \tfrac{3}{2}C = 2Al + \tfrac{3}{2}CO_2 \tag{41}$$

The C/CO_2 electrode is normally stable within ± 10 mV. Continuous flushing with CO_2 is necessary in order to maintain a stable potential. Also, a certain flowrate of CO_2 is required. For example, for an electrode of 12 mm diam, the potential is independent of flowrate greater than 30 cm^3/min. The stability and reversibility of the electrode depend to some extent upon the carbon material.[317] Pyrolytic graphite and vitreous carbon usually maintain potentials above the stable value for long periods of time and show considerable hysteresis when a current is passed (Fig. 42); the baked carbon and graphite attain steady state rapidly and show less hysteresis (Fig. 42).

Before using a carbon electrode, one should heat it under vacuum, then keep it under argon at about 1100–1150°C for a few hours. Also, if a carbon electrode with a stable potential is removed from the melt and

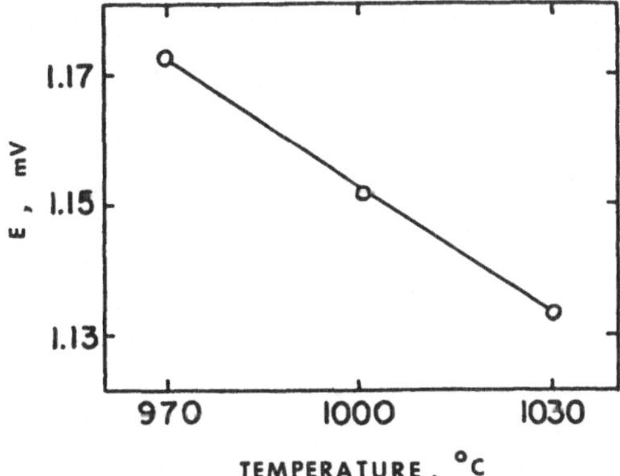

FIGURE 41. Temperature dependence of the emf of the cell $Al_{(1)}|Al_2O_3, Na_3AlF_6|CO_2(C)$. Slope of the line = 6.3×10^{-4} V/K; theoretical slope = 5.9×10^{-4} V/K (Ref. 315). (Reprinted by permission of the publisher.)

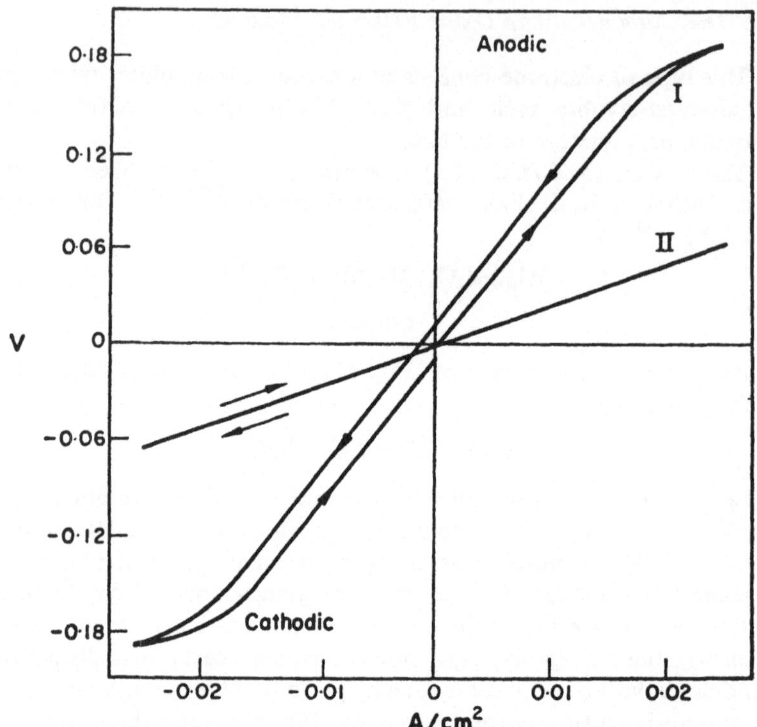

FIGURE 42. Current density-potential curves for C/CO_2 electrodes in alumina-cryolite melt at 1010°C (Ref. 317). I, pyrolytic graphite; II, baked carbon. (Reprinted by permission of the publisher.)

exposed to air, upon its reuse the electrode will give a high initial potential followed by a slow gradual decrease to a stable value. (This has been ascribed to the electrochemical activity of chemisorbed oxygen present at the electrode surface.)

For use in conjunction with a liquid aluminum electrode, the C/CO_2 electrode must be protected from the depolarizing effect of dissolved metal by a membrane (e.g., alumina, BN). On the other hand, if a sufficient flowrate of CO_2 is maintained (e.g., 40 cm^3/min for an electrode of 8 mm diam), the potential of the C/CO_2 electrode is not affected by the presence of dissolved metal. The dissolved metal is probably oxidized by CO_2 before reaching the C electrode. Although at 1000°C (the temperature at which the C/CO_2 reference electrode is commonly used) the equilibrium for the reaction

$$CO_2 + C \rightleftharpoons 2CO \tag{42}$$

is shifted strongly to the right, part of the electrode in contact with the alumina-cryolite melt is not appreciably attacked by CO_2. A certain wear is generally observed around the hole through which the gas passes, particularly with carbon. Graphite, being less reactive, is therefore more suitable as an electrode material.

Another carbon/carbon oxide gas reference electrode uses CO. A C/CO reference electrode that is based on the equilibrium

$$CO + 2e^- = C + O^{2-} \tag{43}$$

was tried.[317] Attack of the C electrode by CO was found to be little. One of the drawbacks of the C/CO electrode is that its potential appears to be less stable than that of the C/CO_2 electrode.

The final carbon/carbon oxide gas reference electrode uses a mixture of CO_2 and CO. There are two theories on the potential-determining reaction for this electrode. In the first, several authors postulate that the electrode acts as an inert, reversible oxygen electrode. The potential of the electrode is determined by the oxygen adsorbed on the carbon,[79,322–324] and therefore by the oxygen partial pressure (the oxygen partial pressure is determined by the equilibrium $CO + \frac{1}{2}O_2 \rightleftharpoons CO_2$). This theory implies that the carbon electrode does not play an active part in the potential-determining reaction. For example, the emf of the cell

$$Al_{(1)}|Al_2O_3, Na_3AlF_6|CO_2, CO(C)$$

Cell 25

is dependent on p_{O_2}, which in turn depends on p_{CO_2}/p_{CO} ratio according to the equilibrium

$$CO + \tfrac{1}{2}O_2 \rightleftharpoons CO_2 \tag{44}$$

In the second theory, the oxygen partial pressure is suggested to be determined by the reactions

$$C + O_2 = CO_2 \tag{45}$$

$$C + \tfrac{1}{2}O_2 = CO \tag{46}$$

i.e., reactions between carbon and oxygen are taken into account.[317] (The carbon-oxygen system is in equilibrium at only one composition, which is about 99.2% CO, 0.8% CO_2, and $5.4 \times 10^{-17}\%$ O_2 at 1000°C.) The overall reaction for the cell $Al_{(1)}|Al_2O_3, Na_3AlF_6|CO_2, CO(C)$ can be written

$$Al_2O_3 + \frac{3}{1+x}C = 2Al + \frac{3x}{1+x}CO_2 + \frac{3(1-x)}{1+x}CO \tag{47}$$

where $p_{CO_2} = x$ and $p_{CO} = 1 - x$. The cell emf at 1000°C is given by

$$E = 1.03 + \frac{0.26x}{1+x} + \frac{0.126}{1+x}[x \log x + (1-x)\log(1-x)] \tag{48}$$

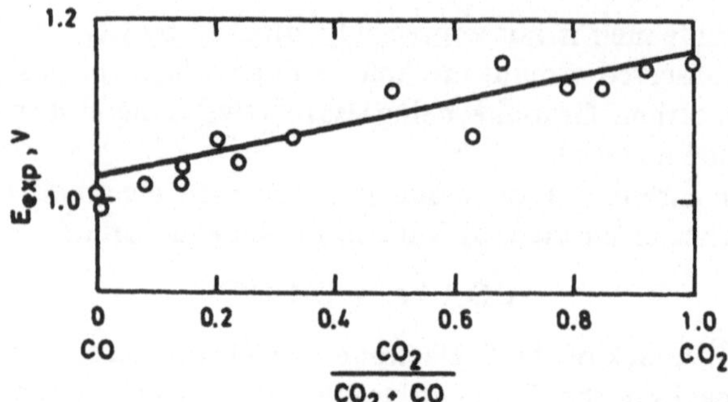

FIGURE 43. The dependence of the potential of the $C/CO_2, CO$ electrode (versus aluminum electrode) on the ratio of CO_2 to $CO_2 + CO$ in molten alumina-cryolite at 1000°C (Ref. 315). Solid line corresponds to Eq. (48). (Reprinted by permission of the publisher.)

In Fig. 43, this equation is represented by a curve which approximates to a straight line; also given are the experimental data reported in the literature.[317] It can be seen from the figure that there is a good agreement between the experimental points and the straight line. As an alternative to the treatment above, the $C/CO_2, CO$ electrode can be considered to be made up of C/CO_2 and C/CO half-cells in parallel, the half cells being located at different spots on the carbon surface. The potential of the mixed-gas electrode is intermediate between those for the pure gases, increasing with increasing CO_2 content.

4.4.3. The Carbon Reference Electrode

A carbon electrode that is not flushed with a gas and is immersed in alumina–cryolite melts has been used as reference electrode in a number of studies.[79,310,324–333] Stable potentials within a few millivolts have been claimed for this electrode. Because oxygen is present above the melt, the C electrode must be considered as a kind of $C/CO_2, CO$ electrode. At 1000°C, the cell $Al_{(l)}|Al_2O_3, Na_3AlF_6|C$ has a voltage of about 1.07 V. Figure 43 shows that the C electrode does indeed act as a $C/CO_2, CO$ electrode, since the 1.07-V value falls in the region of the line in the figure. As discussed earlier, the potential of the C electrode is thus dependent on the partial pressure of oxygen. The C electrode was also suggested to be a C/CO_2 electrode.[310,334] The reason given is that CO_2 dissolves in the alumina–cryolite melt to form CO_3^{2-}, and the partial pressure of CO_2 is determined by $CO_2 + O^{2-} = CO_3^{2-}$.

Although the C reference electrode is simple, it has a number of disadvantages. The most important one is that if the environment of the reference electrode is not kept constant, its potential may not be fixed. This drawback is highlighted by the result of a study[79] on the concentration cell using an MgO ionic conducting membrane

$$C|16\ \text{wt\%}\ Al_2O_3, Na_3AlF_6 \vdots MgO \vdots x\ \text{wt\%}\ Al_2O_3, Na_3AlF_6|C$$

Cell 26

The plot of the emf of the cell versus wt% Al_2O_3, which exhibits several maxima and minima, indicates the inadequate stability of the carbon electrode. However, if the electrode is enclosed to keep a stable atmosphere, it will keep its stability. This enclosure is also necessary to prevent the deleterious effect caused by any dissolved metal present in the melt.

4.4.4. *The Liquid Aluminum Reference Electrode*

The liquid aluminum reference electrode was first introduced by Drossbach[322] and has become the most widely used reference electrode in cryolite melts (e.g., Refs. 335–354). The electrode consists of pure molten aluminum held in an alumina, corundum, or BN container (alumina or corundum are used only in melts saturated with alumina). Electrical contact to the liquid aluminum is made by a Mo, Ta, or W wire sheathed in an alumina, corundum, or BN tube. (The free end of the wire, which dips into the molten aluminum, becomes corroded during operation because some metal dissolves in the aluminum and a solid intermetallic compound is formed at the surface. However, this corrosion does not affect the electrode potential appreciably.) The electrode either is open to the cryolite or has a small hole in the container to provide access to the melt. Boron nitride has also been used as a diaphragm for the aluminum electrode owing to the slight impregnation of the material by the cryolite. (The resistivity of BN at 1000°C is about 3×10^4 to $3 \times 10^7\ \Omega$ cm, but the resistivity decreases to a few thousand Ω cm after cryolite impregnation.) Several designs of the aluminum reference electrode are given schematically in Figs. 44, 45, 46.

The aluminum electrode has been shown to be reversible and has excellent stability and reproducibility. The emf of a cell made of two identical aluminum electrodes is usually zero, with fluctuations no larger than 2 mV. The electrode generally maintains a stable potential (±5 mV) for a considerable length of time (up to several days). An equilibration period from several minutes up to 30 min is necessary before a newly immersed aluminum electrode gives a steady potential.

The overall electrochemical reaction at the aluminum electrode is

$$AlF_4^- + 3e^- = Al + 4F^- \tag{49}$$

The reaction sequence may either occur in one electrochemical step

$$Al^{3+} + 3e^- = Al \tag{50}$$

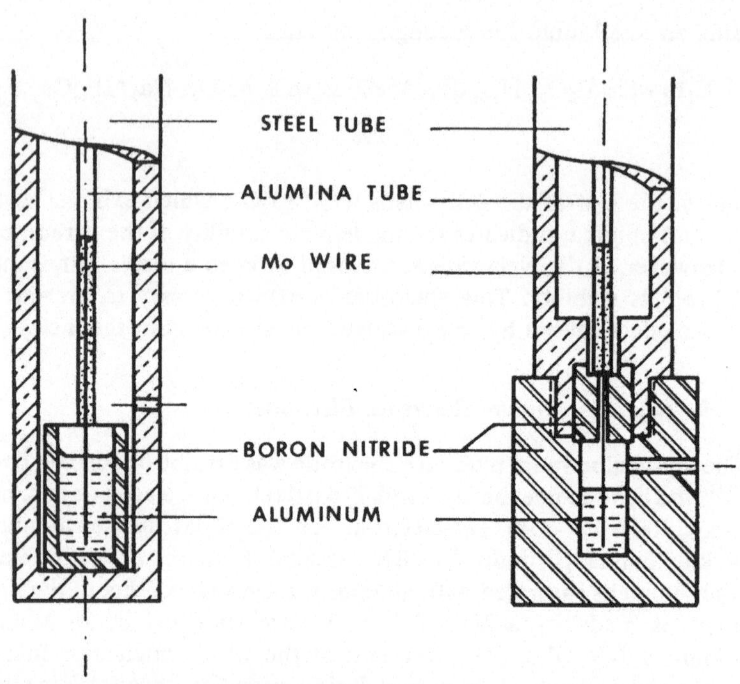

FIGURE 44. Liquid aluminum reference electrodes for use in alumina-cryolite melts (Ref. 344). (Reprinted by permission of the publisher.)

or via a mechanism involving subvalent species such as

$$Al^{3+} + 2e^- = Al^+ \tag{51}$$

followed by

$$Al^+ + e^- = Al \tag{52}$$

(any of the above reactions is preceded by a dissociation of aluminum complexes, $AlF_4^- = Al^{3+} + F^-$) or via a mechanism involving a sodium-containing intermediate (symbolized by Na*),

$$Na^+ + e^- = Na^* \tag{53}$$

and

$$3Na^* + AlF_4^- = Al + 3Na^+ + 4F^- \tag{54}$$

The reversible potential of the aluminum reference electrode in a cryolite melt is given by

$$\varepsilon = \varepsilon^0 + \frac{RT}{zF} \ln \frac{a_{NaAlF_4}}{a_{NaF}} \tag{55}$$

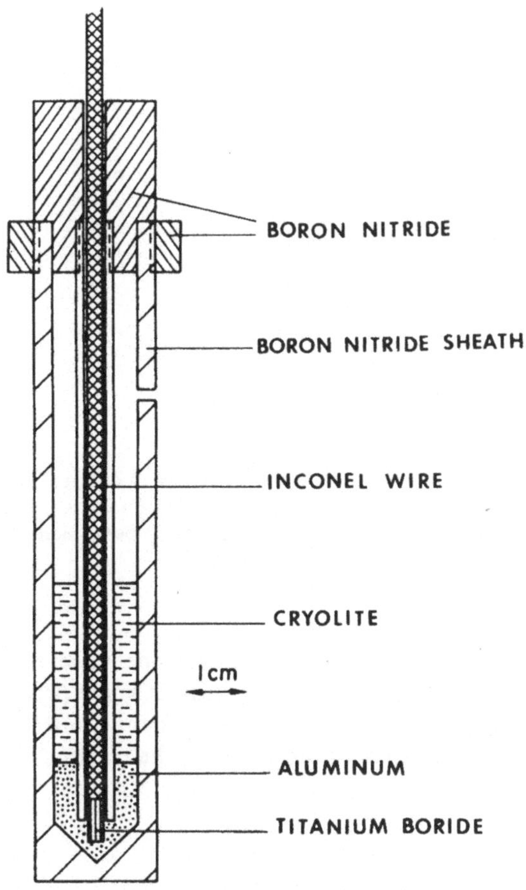

FIGURE 45. A design of liquid aluminum reference electrode with boron nitride diaphragm (Ref. 338). (Reprinted by permission of the publisher.)

The potential of the electrode depends on the NaF activity, which in turn depends on the so-called cryolite ratio (the ratio of NaF to AlF_3), since NaF and AlF_3 can be considered as fluoride-ion donor and acceptor, respectively. Thus, the potential of the electrode becomes increasingly negative when the cryolite ratio increases. As an example, the effect of cryolite ratio on the emf of the galvanic cell

$$Al_{(l)}|Al_2O_3, Na_3AlF_6\,\vdots\,\text{alumina}\,\vdots\,Al_2O_3, Na_3AlF_6, NaF|Al_{(l)}$$

Cell 27

is shown in Fig. 47.

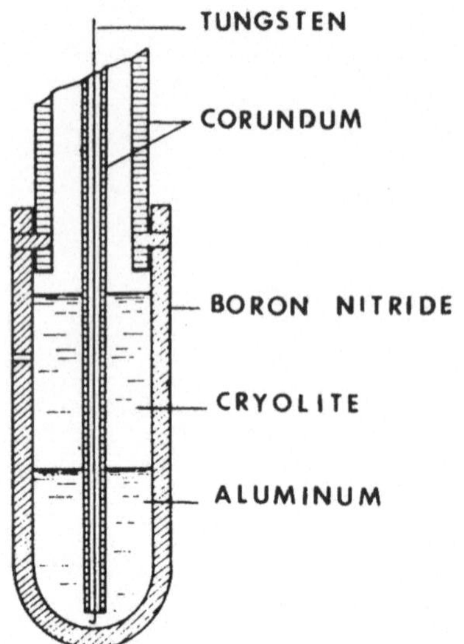

FIGURE 46. **A design of liquid aluminum reference electrode with BN diaphragm (Ref. 343). (Reprinted by permission of the publisher.)**

As mentioned earlier, the aluminum electrode is a reversible and reliable electrode. However, it cannot be used without some precautions. First of all, the electrode reacts with the cryolite melt to a certain degree. Several reaction schemes[355] have been proposed for the reaction of aluminum with cryolite melt:

$$2Al + AlF_3 = 3AlF \tag{56}$$

$$Al + 3NaF = AlF_3 + 3Na \tag{57}$$

$$Al + 6NaF = AlF_3 + 3Na_2F \tag{58}$$

Regardless of the reaction mechanism, the net effect of the reaction is an increased activity of AlF_3 and a reduced activity of NaF in the reference electrode compartment, which causes a negative shift in the potential of the aluminum electrode. Therefore, for thermodynamic studies, a correction is necessary to take this effect into account. For other applications, it is important to ensure that the electrode is at equilibrium, i.e., the cryolite melt is saturated with dissolved aluminum.

The use of an alumina membrane requires that the molten cryolite be saturated with alumina. On the other hand, alumina cannot be used for melts with a high cryolite ratio (>4). The crystal modification of alumina (α to β) in NaF-rich melts causes the breakdown of the material. Tests with BN,[356] which is commonly used in melts unsaturated with alumina,

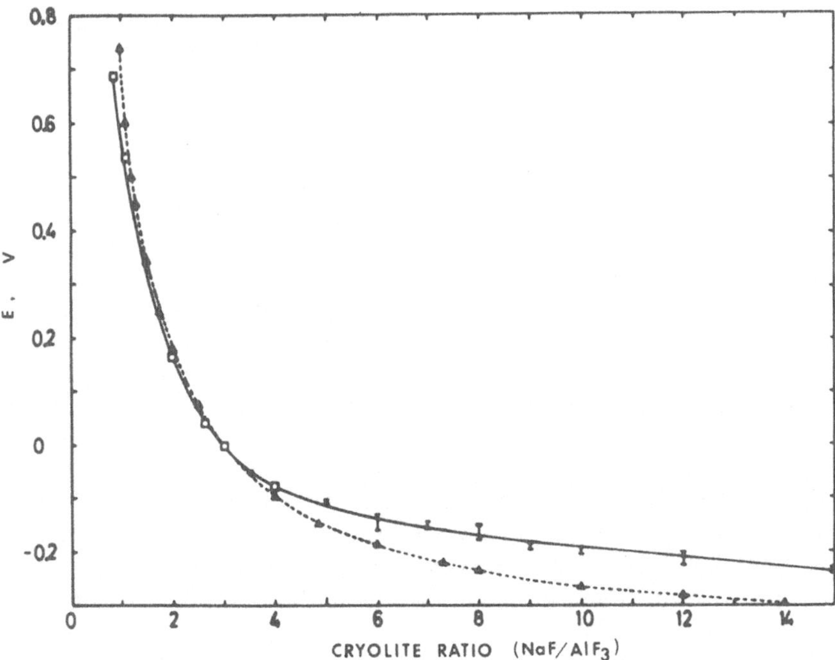

FIGURE 47. Emf between two aluminum electrodes as a function of the cryolite ratio in NaF–AlF_3 melts saturated with alumina at 1010°C (Ref. 341). Solid line: Refs. 339, 341. Broken line: Ref. 312. (Reprinted by permission of the publisher.)

have shown that it contaminates the melt owing to dissolution of the boric acid binding agent in the BN. Thus, for studies that do not tolerate contaminants, the aluminum electrode cannot be used either in oxide-free cryolite melts or in melts containing below 0.2 wt% Al_2O_3.

Electrical contact wires (Mo, Ta, or W) for the aluminum electrode become corroded during operation as some metal dissolves in the aluminum and a solid intermetallic compound is formed at the surface. However, this contamination has no noticeable effect on the potential of the aluminum electrode.

In addition to the common type of molten aluminum pool electrode described above, there is another important type of molten aluminum electrode. This type consists of a thin coating of molten aluminum on a relatively insoluble electrically conductive substrate such as graphite,[(357)] tungsten,[(125)] zirconium diboride,[(310)] or titanium diboride.[(358)] Coating of the substrate normally is done by electrodeposition, and the electrode needs periodic recoating to maintain the same potential. An aluminum reference electrode of this type, which has been used in industrial aluminum electrolysis cells, is in Fig. 5, Section 3.1.3.

4.4.5. The SnO_2 Reference Electrode

A sintered body of an electronically conductive oxide insoluble in alumina–cryolite melt behaves reversibly with respect to the oxyanions in the melt. An electrode based on this behavior has been used in the potentiometric determination of oxide activity in those media. The behavior of electrodes made of sintered oxides such as Fe_2O_3, Fe_3O_4, Cr_2O_3, and SnO_2 has been investigated.[359–364] Among these oxides, SnO_2 appears to be the most suitable as a reference electrode. The oxide is practically insoluble and not wetted by molten cryolite. The material also has excellent resistance to thermal shock and can be easily sintered. A schematic diagram of an SnO_2 electrode is given in Fig. 48. The electrode is about 5.5 cm long and 0.7 cm in diameter. Electrical contact to the electrode is made by a Pt wire which has an enlarged tip and is forced into a cavity drilled into the electrode. The cavity is then filled with SnO_2 powder. A mullite tube, sealed to the electrode with alumina cement, is used to protect the Pt contact wire from vapors of the molten bath.

The SnO_2 electrode is prepared by sintering SnO_2 powder containing about 2% ZnO (as sintering agent) at about 1200°C for 4 h. Four-hour heating to the sintering temperature and four-hour cooling afterwards is recommended (to prevent cracking of the electrode).

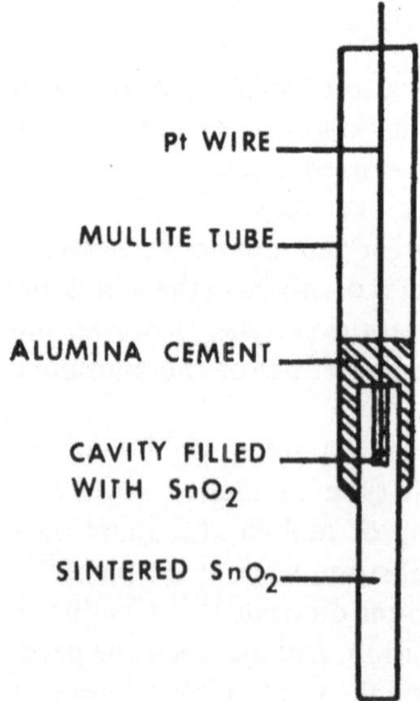

FIGURE 48. SnO_2 reference electrode (Ref. 361). (Reprinted by permission of the publisher.)

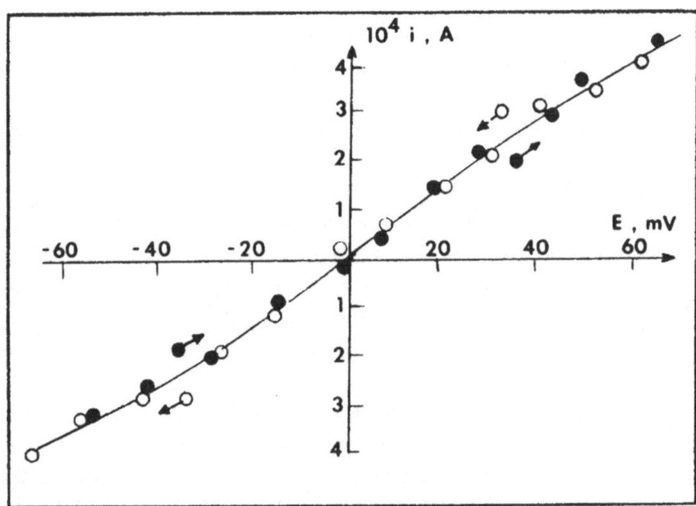

FIGURE 49. Current-potential curve between two SnO_2 electrodes in alumina-cryolite melt at 1025°C (Ref. 361). (Reprinted by permission of the publisher.)

The SnO_2 electrode is reversible (Fig. 49), reproducible, and stable within ±5 mV for many hours. When immersed in the alumina-cryolite melt, the electrode generally reaches its equilibrium potential in half to one hour. The electrode must be used under an inert atmosphere (e.g., nitrogen), since SnO_2 can be easily attacked by CO and CO_2. The partial pressure of CO or CO_2 in the atmosphere should not exceed 0.01 atm. Also, any dissolved aluminum metal present in the melt will attack the electrode.

4.4.6. Other Reference Electrodes

The Lead-Sodium Reference Electrode. Lead-sodium alloys have been used as reference electrodes for several equilibrium and activity measurements[312,365-367] in alumina-saturated cryolite. The electrode is made of a lead-sodium alloy and is contained in an alumina crucible. A tantalum wire shielded in an alumina sheath serves as the electrical contact. The electrode gives stable and reproducible potentials, especially when the surface is covered with a layer of solid alumina powder. In alumina-cryolite melts, this electrode would probably not act as a pure sodium electrode, but rather as an aluminum electrode. The equilibrium, given by Eq. (56), might be established at the electrode surface independently of the sodium content in the bulk of the lead-sodium alloy.

The Solid Metal Quasi-Reference Electrodes. Several solid metals or alloys have been used as quasi-reference electrodes in cryolite. In many cases, the reversible potential is not known.

Fe–Al Electrode.[368] The electrode is a rod of an Fe–Al alloy. Since aluminum is soluble in cryolite, the composition of the alloy tends to change during use. However, it appears that the potential of this alloy electrode is approximately constant at aluminum contents in the alloy above 50 wt%.

Pt–Al Electrode.[369] The alloy is prepared by cathodically diffusing aluminum from an alumina–cryolite melt to a Pt wire. The emf between two identical electrodes does not differ more than 20 mV over a period of several hours. The reversible potential is not known. However, this electrode may behave the same as an aluminum electrode.

Pt Electrode.[370–372] A platinum wire dipped in cryolite or alumina–cryolite melts has been found to give stable potential.

Ni Electrode.[373,374] The electrode is a nickel wire immersed directly in cryolite. The potential of the electrode is constant and reproducible within ±10 mV after many hours of immersion in the melt. The stable potential is probably due to formation of an insoluble NiO film on the wire. The potential of the nickel wire electrode is about 1.1 V versus the liquid aluminum electrode at 1030°C.

The Ag/Ag^+ Reference Electrode. This reference electrode is based on the Ag/Ag^+ couple in $NaF–AlF_3$ melt housed in a BN container.[350,375] The reversibility of the system has been confirmed. The potential difference of two electrodes prepared in an identical manner does not exceed 10 mV. The difference remains within ±5 mV over a period of 5 h. At a constant AgF concentration, the potential of the Ag/Ag^+ electrode depends on the composition of the melt.

4.5. Nitrates

Molten alkali nitrates represent a class of molten salts with relatively low working temperatures (about 150–550°C; nitrates decompose at high temperatures) and the desirable characteristics of low volatility, low viscosity, and high electrical conductivity. Because of these favorable physical properties, molten nitrates are used successfully in a number of industrial processes. They are utilized as heat transfer fluids in the petrochemical industry, heat storage media for solar energy collectors, and salt baths for metal heat treating. Nitrates are also of interest as reactive media for production of oxygen and for possible use in thermal batteries.

Molten nitrates have also been used as media for many fundamental investigations. Nitrate melts exhibit interesting oxide chemistry. Oxide ions have been shown not to be a stable species in molten nitrates. The predominant species in nitrate melts are superoxide (O_2^{2-}) and peroxide (O_2^-).[376–378] These species are interrelated by the electron transfer and disproportionation equilibria:

$$O_2 + e^- \rightleftharpoons O_2^- \tag{59}$$

$$O_2^- + e^- \rightleftharpoons O_2^{2-} \tag{60}$$

$$2O_2^- \rightleftharpoons O_2^{2-} + O_2 \tag{61}$$

Electrochemical systems that use nitrates employ the Ag/Ag^+ reference electrode almost exclusively. A description of the Ag/Ag^+ reference electrode and other minor electrodes is given below.

4.5.1. The Ag/Ag^+ Reference Electrode

First used by Flengas and Rideal,[25] the Ag/Ag^+ electrode has become the most popular in molten nitrates and has been used extensively.

The Ag/Ag^+ reference electrode is based on the Ag/Ag^+ equilibrium. The construction of the electrode is quite simple, consisting of a silver wire immersed in a solution of $AgNO_3$ in a nitrate melt. Because of the relatively low working temperatures of nitrates, Pyrex and Vycor are often used as container materials for the Ag/Ag^+ electrode. Ionic contact between the reference melt and the nitrate melt under study is made either via a liquid junction such as a pinhole through the glass container, glass frit, and asbestos fiber or via a glass membrane. Maintaining an inert atmosphere in the reference electrode is recommended, although exposure to air is known to have no significant effect on the potential stability of the reference electrode.

The silver ion in the reference melt can be introduced by either adding the $AgNO_3$ directly or anodizing the silver electrode, a process that is 100% current efficient. The common Ag^+ concentration for the Ag/Ag^+ reference electrode in nitrate melts is 0.07–0.10 molal (m). Other concentration scales have also been used in nitrates. The difference in standard potential on different concentration scales for 250°C is given below:

	ε^0 (V)
Ag/Ag^+ (0.07 *M*)	−0.15
Ag/Ag^+ (0.07 m)	−0.12
Ag/Ag^+ (1.0 *M*)	−0.03
Ag/Ag^+ (1.0 m)	0.00 (defined)

Very low and very high concentrations of Ag^+ are not suited for use in the Ag/Ag^+ reference electrode. In very dilute concentrations ($<10^{-4}$ mole fraction), a small amount of corrosion of silver will give rise to a relatively large change in silver-ion concentration,[379] thus affecting the potential of the electrode. At high Ag^+ concentrations (>0.1 mole fraction), the potential dependence on concentration does not follow the Nernst equation.

The applicability of the Nernst equation has been demonstrated for dilute solution of Ag^+ in nitrate melts in many studies.[26,379-393] For a concentration cell of the type

$$Ag|AgNO_3(N'), MNO_3 \,\vdots\vdots\, AgNO_3(N), MNO_3|Ag$$

Cell 28

where M is an alkali metal, the emf is given by

$$E = \frac{RT}{F} \ln \frac{N}{N'} \tag{62}$$

This equation has been found to be valid for dilute solution of $AgNO_3$ in molten nitrates. The result of the emf measurements of the cell

$$Ag|AgNO_3(N'), KNO_3 \,\vdots\vdots\, AgNO_3(N), KNO_3|Ag$$

Cell 29

is given here as an example.[381] Figure 50 shows the Nernst plot for this cell at 436°C. The agreement with the Nernst relationship is also demonstrated at different temperatures for this cell (in the temperature range of 344–423°C). Adoption of the standard state such that the activity coefficient is unity at infinite dilution leads to the conclusion that the activity coefficient

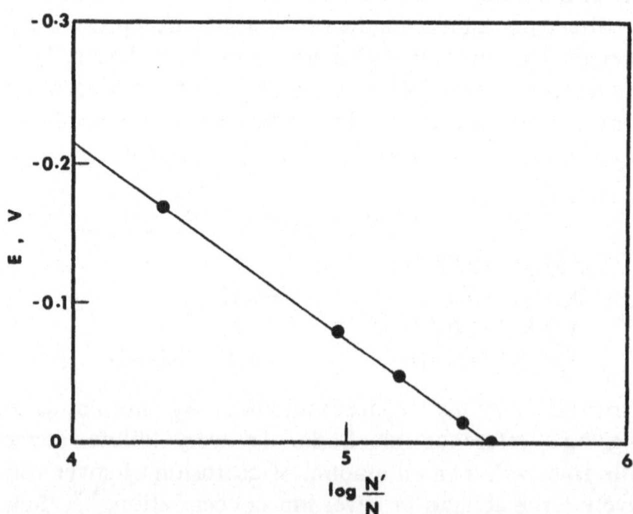

FIGURE 50. Nernstian plot for the cell $Ag|AgNO_3(N), KNO_3 \,\vdots\vdots\, AgNO_3(N'), KNO_3|Ag$ at 436°C (Ref. 381). (Reprinted by permission of the publisher.)

of $AgNO_3$ is unity over the concentration range of 2×10^{-4} and 5×10^{-3} mole fraction in the above cell.

The Ag/Ag^+ reference electrode shows excellent stability and reproducibility in nitrate melts. For example, the potential difference between freshly prepared and aged Ag/Ag^+ reference electrodes is within 0.5 mV or better for days. The potential of the Ag/Ag^+ electrode generally does not change more than 1% after days of immersion in nitrate melts.

As mentioned earlier, several ways have been employed to establish electrical contact between the reference melt inside the Ag/Ag^+ electrode and the nitrate melt under study. A pinhole through the tip of the glass container can provide the desired ionic contact, and the resistance through this hole is generally greater than 10 kΩ.(394) Although a simple technique, the use of a pinhole is not effective in preventing silver nitrate from diffusing into the investigated system. Asbestos fibres and glass frits, sealed to the glass container, are other common ways to provide contact while isolating the reference melt. Glass frits are often partially fused to reduce diffusion through the frits. Liquid-junction potentials across glass frits(31-34,382,384,395-398) and asbestos fibers(26,381,385,386) are shown to be negligible. Glass frits and asbestos fibres have been applied successfully as junctions for Ag/Ag^+ reference electrodes in a number of thermodynamic studies. However, these types of junction are difficult to maintain. Diffusion of silver into the nitrate system under study is slow but still possible. Recently, glasses, especially Pyrex, have found increasing use as membranes for the Ag/Ag^+ reference electrode.(42,399) Glass membrane electrodes are sturdy, and a thin membrane is sufficient to prevent diffusion of silver ions from the reference melt into the nitrate melt under investigation. A glass membrane provides electrical contact via its ionic conduction, usually via sodium ions (in lithium-ion-containing melt, mobile lithium ions which may diffuse into the glass may also conduct current). To ensure that sodium ion is the only current-carrying species, 2% of a sodium salt in the molten salt under study is sufficient.(43) A Pyrex glass membrane of 0.1-mm thickness is widely used.

Membrane potentials of Pyrex glass, Vycor, and fused silica have been measured and studied in molten nitrates.(52,53,56,395,400-403) Figures 51 and 52 show the emf at 340°C of the cell

$$Ag|AgNO_3 \vdots Pyrex \vdots AgNO_3, NaNO_3|Ag$$

Cell 30

and the membrane potential of Pyrex glass (the difference between the measured and calculated emf) as a function of silver nitrate mole fraction, respectively.(53) Figure 52 shows that the membrane potentials are of the order of a few millivolts over the entire $AgNO_3$ concentration range. At

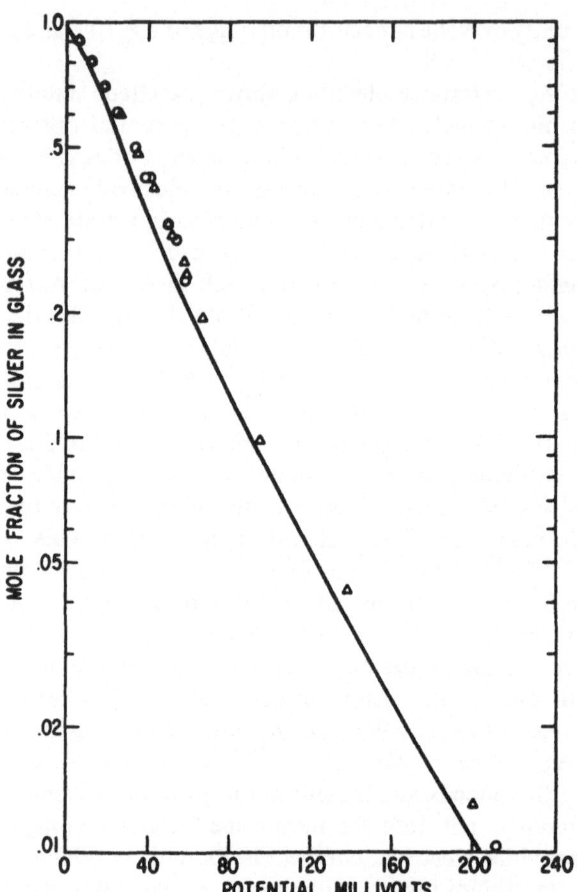

FIGURE 51. Emf at 340°C of the cell Ag|AgNO$_3$:Pyrex:AgNO$_3$, NaNO$_3$|Ag. Solid line corresponds to theoretical values (Ref. 53). (Reprinted by permission of the publisher.)

low silver concentrations, the membrane potential accounts for about a few percent of the cell emf. This result suggests that the emf of a cell with a Pyrex glass membrane in a molten nitrate using Ag/Ag^+ electrodes should follow Eq. (62) fairly well at low silver concentrations. This relationship explains the linear plot of cell emf versus the logarithm of silver nitrate activity observed at low $AgNO_3$ concentrations.[(399)]

A general and detailed discussion of the theories on membrane potential has been given in Section 2 on theoretical principles. As discussed in that section, membrane potential has been commonly interpreted on the basis of either the liquid-junction or ion-exchange models. In nitrate melts, the ion-exchange theory appears to give good agreement with experimental results of several studies[(53,56,400–402)] for Pyrex, Vycor, and fused silica

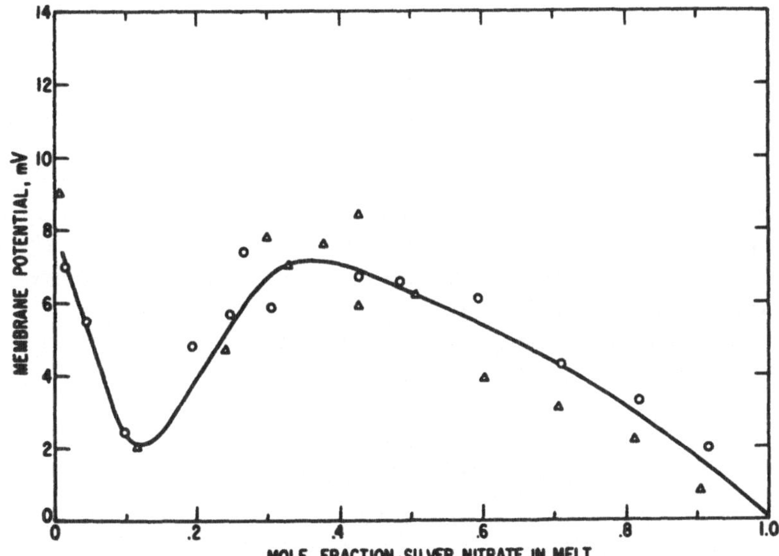

FIGURE 52. Membrane potential at 340°C of the cell Ag|$AgNO_3$⋮Pyrex⋮$AgNO_3$, $NaNO_3$|Ag (Ref. 53). (Reprinted by permission of the publisher.)

membranes. However, one study[403] on Pyrex membrane in $NaNO_3$ indicates the applicability of the liquid-junction model.

Glass membranes in molten nitrates and in molten salts in general may exhibit an asymmetry potential. This potential is generally attributed to an unequal distribution of sodium ions arising during glassblowing. If the glassblowing procedure is standardized, asymmetry potentials are reasonably reproducible. The asymmetry potential usually decreases with time. In thermodynamic studies using glass membranes, appropriate correction for asymmetry potential is necessary.

In order to obtain reproducible results with glass membrane electrodes, it is sometimes necessary to precondition the glass membranes by soaking them in molten nitrate for several days, occasionally passing a current of the order of microamperes alternately in opposite directions.[395]

4.5.2. The Silver/Silver Halide Reference Electrode

The silver/silver halide reference electrode has been used in thermodynamic investigations in molten nitrates, especially in the determination of activity coefficients of silver nitrate and alkali halides and in studies of complex formation between silver and other metals with halide ion.[67,404–409] This type of electrode is based on the low solubility of silver halide (chloride, bromide, iodide) in molten nitrates. The solubility of AgCl in equimolar

$NaNO_3$–KNO_3, for example, is 4.9×10^{-3} mol/kg solvent at 280°C.[410] In molten $NaNO_3$ at 357°C, the solubility is 1.29×10^{-3} mole fraction.[409] The silver/silver halide electrode consists of a silver wire covered with a thin layer of solid silver halide. The silver wire is immersed in a nitrate melt that is saturated with silver halide. Materials for containment and for ionic contact are the same as those described earlier for the Ag/Ag^+ reference electrode. Covering the silver electrode with solid silver halide is achieved by dipping it into molten silver halide. This can also be done *in situ* by immersing the wire in the nitrate melt saturated with the silver halide and allowing the precipitation of silver halide on the electrode surface.

The silver/silver halide electrode is reversible to halide ions in molten nitrates. The Nernst plot of the cell

$$\text{Ag}|\text{AgBr(s)}|\text{KBr(m')},\ \text{NaNO}_3,\ \text{KNO}_3 \,\vdots\vdots\, \text{KBr(m)},\ \text{NaNO}_3,\ \text{KNO}_3|\text{AgBr(s)}|\text{Ag}$$

Cell 31

is given in Fig. 53 as an example of the Nernstian reversibility of the electrode. The potential of the silver/silver halide electrode is generally stable within ±0.2 mV for days. The electrode also shows temperature reversibility.

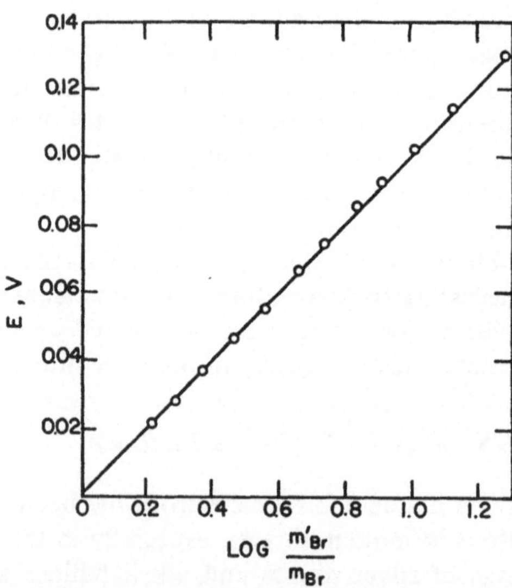

FIGURE 53. Test of the applicability of the Nernst equation for the cell $\text{Ag}|\text{AgBr}_{(s)}|\text{KBr(m')}, \text{KNO}_3, \text{NaNO}_3 \vdots\vdots \text{KBr(m)}, \text{KNO}_3, \text{NaNO}_3|\text{AgBr}_{(s)}|\text{Ag}$ at 261°C (Ref. 67). (Reprinted by permission of the publisher.)

4.5.3. Other Reference Electrodes

The Mercury Reference Electrode. One proposed version of the Hg reference electrode is a mercury pool covered with Hg_2SO_4 and anhydrous K_2SO_4. Ionic contact to the nitrate melt under study is made via a 1-mm-diam hole. A tungsten wire is used as electrical contact.(411) The liquid-junction potential of this type of electrode is not known. The other version of the Hg electrode is simply a mercury pool in contact with the nitrate melt via a hole(412) or via glass membrane.(413) The mercury electrode is only useful at low temperatures and was used only in some very early works in molten nitrates.

The Sodium Reference Electrodes. The electrode is a pool of molten sodium in a sodium-ion-conducting glass.(414) Applicability is limited to nitrate melts containing sodium ions. The sodium-ion conduction of glass of the sodium reference electrode has been studied using the cell

$$\text{Na} \vdots \text{glass} \vdots \text{Na alloy}$$

Cell 32

The emf has been shown to be stable and agrees with theoretical values. Besides the original work, there is no reported use of this electrode as a reference electrode in other nitrate studies.

Miscellaneous Reference Electrodes. Electrodes consisting of cobalt amalgam or $LiNO_3$-KNO_3 eutectic containing 0.8 mol% $CoCl_2$ have been proposed for use as reference electrodes in nitrate melts.(413,415–417) In these membrane electrodes Pyrex is used as the membrane, and Pt or W serve as electrical lead wire. Electrode reactions of these electrodes are unknown. An equilibration period is necessary to obtain stable potential. Two reversible gas electrodes, (Pt)NO_2, O_2 in nitrates(418–421) and (C)O_2 in nitrates saturated with peroxide,(422,423) can be used as reference electrodes. An electrode of a third kind, Pd/PdO, CdO/Cd^{2+}, has been used to monitor changes of cadmium-ion concentration in nitrate melts.(68)

4.6. Carbonates

Molten alkali metal carbonates have attracted considerable attention in the last 30 years as the electrolytes for the molten carbonate fuel cell.(424-426) Carbonate mixtures are the only molten electrolytes that are invariant with respect to the electrochemical oxidation of hydrogen and carbon monoxide, thus providing the basis for the fuel cell;

Anode reaction:

$$H_2 + CO_3^{2-} = H_2O + CO_2 + 2e^- \tag{63}$$

$$CO + CO_3^{2-} = 2CO_2 + 2e^- \tag{64}$$

Cathode reaction:

$$\tfrac{1}{2}O_2 + CO_2 + 2e^- = CO_3^{2-} \tag{65}$$

Other applications of molten carbonates are catalysts for coal gasification and destruction of toxic chemical wastes, oxygen regenerative systems, and sulfur removal from fossil fuel.

Among the very few reference electrodes for use in molten carbonates, the CO_2, O_2 and the CO_2, CO gas reference electrodes are most suitable, mainly because these electrodes are in direct equilibrium with the molten salt, and therefore no membrane is necessary. This is an important advantage since finding a suitable membrane in carbonate melts is a formidable problem. Also, in the presence of CO_2 at constant partial pressure, the thermal dissociation of carbonates and therefore their ionic composition can be maintained at equilibrium.

4.6.1. The CO_2, O_2 Reference Electrode

The CO_2, O_2 reference electrode is the most frequently used reference electrode in molten carbonate studies. The equilibrium potential of this electrode is determined by the overall reaction

$$\tfrac{1}{2}O_2 + CO_2 + 2e^- = CO_3^{2-} \tag{66}$$

In spite of the practical inconvenience commonly associated with construction and operation of gas electrodes, the CO_2, O_2 electrode has been in general use since the middle 1960s because of its excellent stability and reversibility. Compared with the CO_2, CO electrode (the other gas reference electrode for molten carbonates) the CO_2, O_2 electrode has an additional advantage: it does not involve CO. Thus, it is less toxic and fewer safety precautions have to be considered.

The most common design of the CO_2, O_2 electrode, illustrated in Fig. 54,[427] is a gold wire spiralled at the tip that dips into a carbonate melt contained in an alumina tube. The melt is in contact with a slow stream of oxygen and carbon dioxide. An inner alumina capillary tube situated a few centimeters above the melt is used to introduce the gas mixture into the reference electrode compartment. Ionic contact to the bulk melt is obtained through a fine pinhole in the alumina tube. Another design is shown in Fig. 3b. A number of variations and modifications on the above electrode design and operational details of the CO_2, O_2 electrode are discussed below.

Metal Substrate. Although Pt has been used as the metal wire for the CO_2, O_2 electrode in several studies,[428–440] gold is recommended as the substrate. Gold is not perfectly inert to molten carbonates in the presence of oxygen, but the potential taken by the gold electrode is somewhat less positive than that taken by platinum.[429] It has been shown that platinum

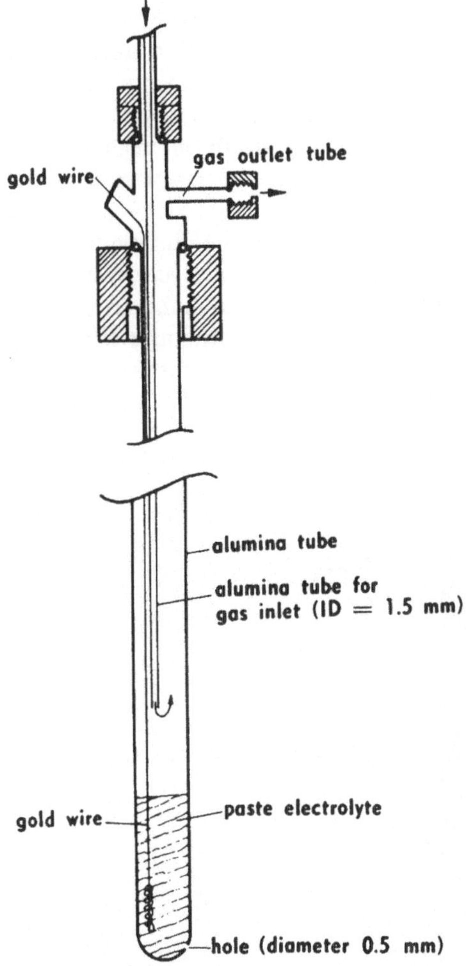

FIGURE 54. Schematic diagram of a CO_2, O_2 reference electrode design (Ref. 427). (Reprinted by permission of the publisher.)

is attacked by molten alkali carbonates in the presence of oxygen and forms complex oxides or oxysalts.[438,441,442] In Li^+-containing melts, formation of Li_2PtO_3 has been noted.[442] In general, equilibration periods for new gold electrodes are about one hour or less.

Reference Electrolyte. Elimination of traces of water from the reference electrolyte is important for the reliability of the CO_2, O_2 electrode. There are two common methods for removing small traces of water in carbonates. The first involves heating the carbonate salt slowly under dry CO_2, followed by bubbling dry CO_2 through the melt for at least 12 h.[443,444] In the second, two heating stages are used.[445] The carbonate is first slowly heated and melted under a CO_2 atmosphere and allowed to solidify. The second stage of heating is then conducted under vacuum over a period of 48 h to about 700°C; the vacuum is maintained at this temperature for a further 24 h. These drying procedures generally result in negligible residual current density (on the order of microamperes per square centimeter) with pure CO_2 above the melt.

In the CO_2, O_2 electrode design with a pinhole in the alumina sheath, a closely packed powder of immobilized electrolyte (a mixture of inert ceramic power, e.g., $LiAlO_2$, and a molten carbonate) can be used instead of a molten carbonate. The purpose is to minimize diffusive and convective mixing between the reference electrolyte and the electrolyte under study.

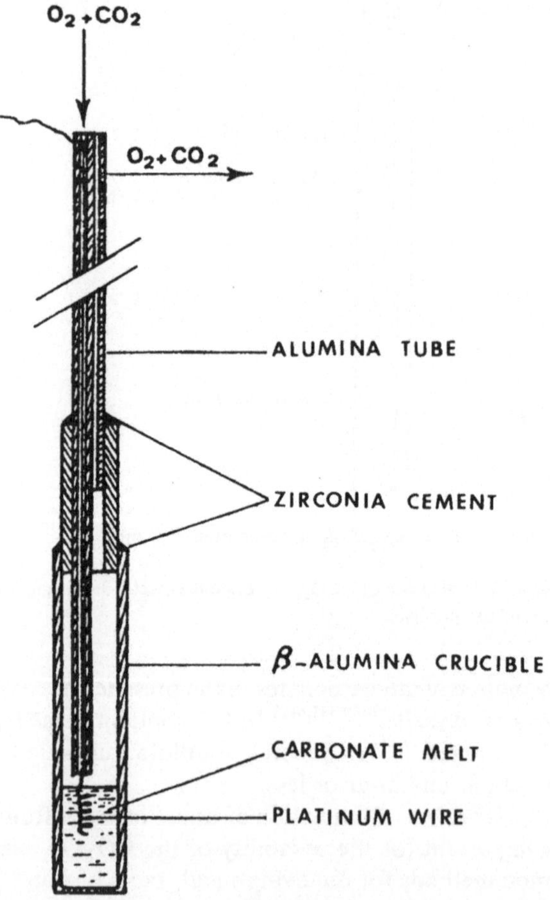

FIGURE 55. Schematic diagram of CO_2, O_2 reference electrode with β-alumina membrane (Ref. 451). (Reprinted by permission of the publisher.)

Reference electrodes with immobilized electrolyte are useful for *in situ* studies in molten carbonate fuel cells.[427,446–448]

Ionic Contact. As mentioned earlier, in the more common design of the CO_2, O_2 reference electrode, a pinhole in the alumina sheath is used for ionic contact between the reference electrode electrolyte and the electrolyte under study (ionic conduction via a restricted channel of the molten electrolyte). In molten Na_2CO_3 at 1000°C or higher, McDanel type 998 alumina has been shown to be a suitable ionic-conducting membrane[149,449] (the alumina, which contains about 0.005% Na_2O and is normally an insulator, is a Na^+-conducting material at that temperature). At lower temperatures, β-alumina may be used as the membrane[450,451] in melts containing Na_2CO_3. A reference electrode with a β-alumina membrane is shown schematically in Fig. 55. Neither alumina nor β-alumina is suitable as a membrane in Li_2CO_3-containing melts because of their reactions with the Li_2CO_3. For example, Li_2CO_3 forms $LiAlO_2$ on the surface of alumina, thus rendering it insulating even at high temperatures.

Reference Gas. Oxygen and carbon dioxide used for the reference electrode gas mixture should be dried over silica gel, P_2O_5, or molecular sieve and mixed before entering the reference electrode compartment. The flowrate of the reference gas ranges from 20 to 100 cm^3/min, with 40–50 cm^3/min the most widely used. The most accurate CO_2, O_2 reference electrode is the one with the gas composition $p_{O_2} = 1/3$ atm and $p_{CO_2} = 2/3$ atm because this electrode is least sensitive to errors in gas compositions irrespective of temperature.[452] This can be seen from Fig. 56, which shows the theoretical values of $Z_{O_2/CO_2} \times 1000$ as a function of $p_{O_2}, p_{CO_2}, p_{N_2}$ where

$$Z_{O_2/CO_2} = \frac{R}{2F}[\ln(p_{O_2}^{1/2}\, p_{CO_2})] \tag{67}$$

For an undiluted O_2 and CO_2 mixture with $p_{O_2} = 1/3$ atm and $p_{CO_2} = 2/3$ atm, the potential of the CO_2, O_2 electrode is at a maximum and at its most positive. The effect of dilution of the gas mixture with an inert component such as nitrogen is seen in the shift of the position of the maximum equilibrium potential toward the lower values of p_{O_2} and, of course, toward the lower, more negative potential. Another change with increasing dilution is the decrease in the broadness of the potential peak and hence an increased sensitivity of the value of the potential to error in the gas composition. Generally, one can expect a better potential stability if a gas reference electrode is "gas bubbling" rather than "unstirred." However, since the electrode process of the CO_2, O_2 electrode is fast, gas bubbling is unnecessary.

Studies reported in the literature[428,436,439,440,452–460] on the thermodynamic behavior of the CO_2, O_2 reference electrode all agree on their conclusions that the electrode potential is determined by the overall reaction

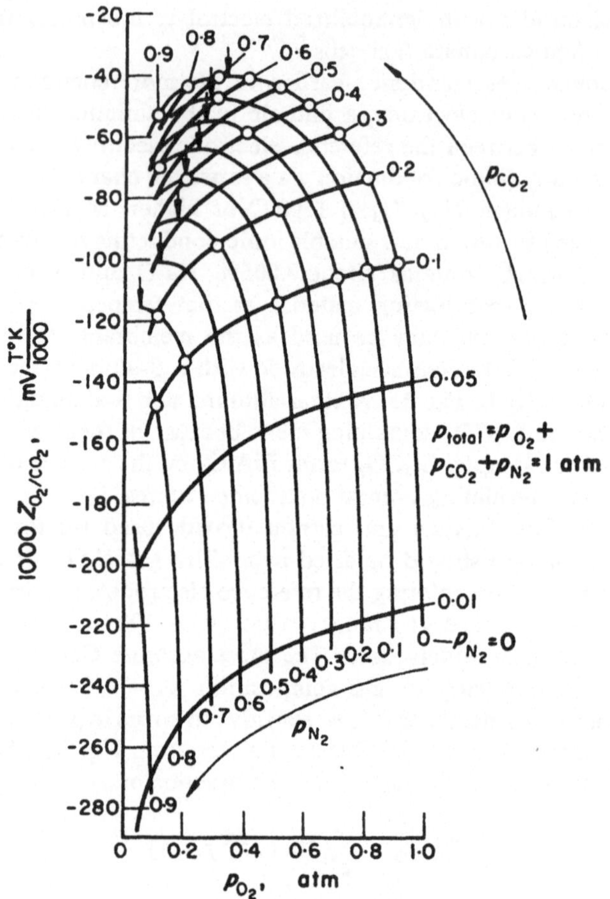

FIGURE 56. Variation of the potential of the CO_2, O_2 reference electrode with gas composition (Ref. 454). (Reprinted by permission of the publisher.)

given by Eq. (66) and that the potential (stable to within ±2 mV) obeys the Nernst equation

$$\varepsilon = \varepsilon^0 + \frac{RT}{2F}\ln\left(p_{O_2}^{1/2}p_{CO_2}\right) \tag{68}$$

This behavior is borne out by the data plotted in Fig. 57 for a cell made up of two CO_2, O_2 electrodes.

The reversibility of the CO_2, O_2 electrode has been tested by considering the exchange current density, i^0, and the polarization resistance, $R_m = (\partial\eta/\partial i)_{i\to 0}$, under unstirred and stirred conditions (simple gas bubbling was the means of agitation).[452] The test is based on the following general

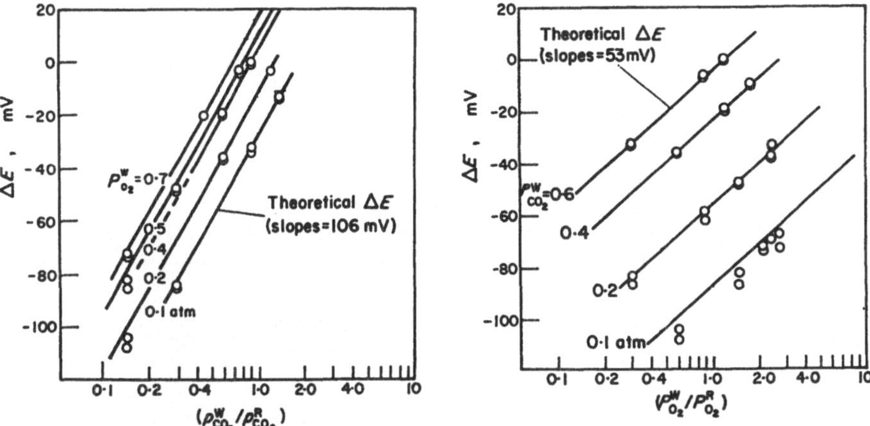

FIGURE 57. Test of the applicability of the Nernst equation for the CO_2, O_2 reference electrode in Li_2CO_3-Na_2CO_3-K_2CO_3 eutectic at 800°C (Ref. 454). Superscript W indicates test electrode, superscript R indicates reference electrode. (Reprinted by permission of the publisher.)

arguments:

$$R_m = \left(\frac{\delta\eta}{\delta i}\right)_{i\to 0} = \frac{RT}{zF}\left(\frac{1}{i^0} + \frac{1}{i_m}\right) \tag{69}$$

where i_m is the limiting current density governing the electrode process near equilibrium. Since stirring has no effect on i^0 but increases i_m, the polarization resistance under stirred conditions, ${}^sR_m$, is given as

$${}^sR_m = \frac{RT}{zF}\left(\frac{1}{i^0}\right) \tag{70}$$

Under unstirred conditions, the resistance ${}^uR_m$ and the corresponding ${}^ui_m$ are therefore related by

$${}^uR_m - {}^sR_m = \frac{RT}{zF}\left(\frac{1}{{}^ui_m}\right) \tag{71}$$

From the above two equations

$$\frac{i^0}{{}^ui_m} = \frac{{}^uR_m}{{}^sR_m} - 1 \tag{72}$$

According to Gerischer,[461] an unstirred reversible electrode is one with $i^0/i_m \geq 10$ or ${}^uR_m/{}^sR_m \geq 11$ [according to Eq. (72)] and an irreversible one with $i^0/i_m \leq 0.1$ or ${}^uR_m/{}^sR_m \leq 1.1$. However, a high value for the ratio i^0/i_m alone is not a sufficient indication of a good reversibility. Because, for

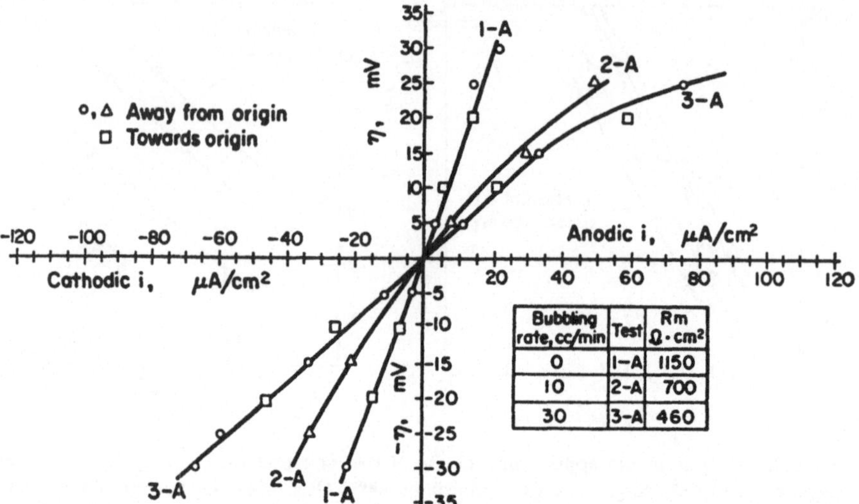

FIGURE 58. Effect of agitation in micropolarization characteristics of the CO_2, O_2 reference electrode in Li_2CO_3-Na_2CO_3-K_2CO_3 eutectic at 800°C (Ref. 455). (Reprinted by permission of the publisher.)

example, an electrode with even a high i^0 would appear to be more reversible when not stirred and less reversible when stirred.[452,455] Therefore, high i^0 and low R_m values have been suggested as criteria for the evaluation of the reversibility of these gas reference electrodes.

The results of micropolarization tests[452,455] on the CO_2, O_2 reference electrode are shown in Fig. 58. The results obtained for R_m and the calculated approximate values of i^0, $^{u}i_m$, and their ratios are given in Table XI. It can be seen from the table that above 650°C the electrode has $^{u}R_m$ values that are sufficiently low and i^0 values sufficiently high to justify its satisfactory reference electrode behavior.

The equilibrium potential of the CO_2, O_2 electrode has been correlated experimentally with that of the standard CO_2, CO reference electrode. On the standard CO_2, CO scale ($\varepsilon^0_{CO/CO_2} = 0$ for $p_{CO} = 0.382$ atm, $p_{CO_2} =$ 0.618 atm, and total pressure = 1 atm), the potential of any CO_2, O_2 reference electrode is given by the following equation:

$$\varepsilon_{O_2/CO_2} = 1200 - 0.456[T - 590] + TZ_{O_2/CO_2} \quad \text{(in mV)} \tag{73}$$

where T is in K.

4.6.2. The CO_2, CO Reference Electrode

The CO_2, CO electrode is another suitable gas reference electrode in molten carbonates. However, in spite of its accuracy in reference electrode

TABLE XI. Effect of Temperature on Reversibility of the Au/CO_2, O_2 Electrode[a]

Temperature (°C)	${}^{u}R_m$ (Ω cm^2)	${}^{s}R$ (Ω cm^2)	Approx. $i^0 \times 10^4$ (A/cm^2)	Approx. ${}^{u}i_m \times 10^4$ (A/cm^2)	$\frac{i^0}{{}^{u}i_m}$
800	1150	460	1.0	0.67	1.5
750	1580	680	0.65	0.5	1.3
700	1800?	1000	0.42	0.52?	0.8?
650	3900	1700	0.23	0.18	1.3
575	5950	2970	0.12	0.12	1.0
550	9620	3970	0.09	0.062	1.4

[a] Reference 455.

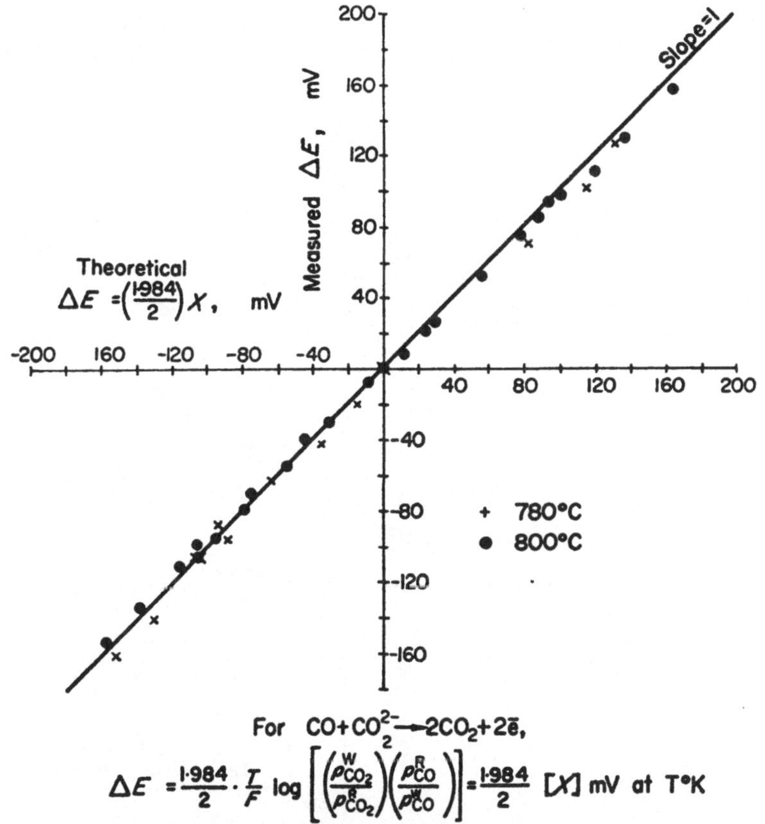

FIGURE 59. Nernstian plot for the CO_2, CO reference electrode in Li_2CO_3-Na_2CO_3-K_2CO_3 eutectic (Ref. 445). Superscript W indicates test electrode, superscript R indicates reference electrode. (Reprinted by permission of the publisher.)

applications and its convenient potential scale (to be discussed later in this section), the CO_2, CO electrode is highly toxic and therefore it is less convenient than the CO_2, O_2 reference electrode.

The construction and the operational detail of the CO_2, CO electrode are identical to those described and used for the CO_2, O_2 reference electrode. The potential of the CO_2, CO electrode is determined by the overall reaction

$$2CO_2 + 2e^- = CO + CO_3^{2-} \tag{74}$$

The measured values of the emf of the cell

$$\underset{\text{(reference)}}{(Au)CO_2, CO} | \text{molten carbonate} | \underset{\text{(test)}}{CO_2, CO(Au)}$$

Cell 33

have been found (Fig. 59) to be related very accurately to the partial pressure of CO and CO_2 by the Nernst equation.[428,445,452,455,462,463] Micropolarization

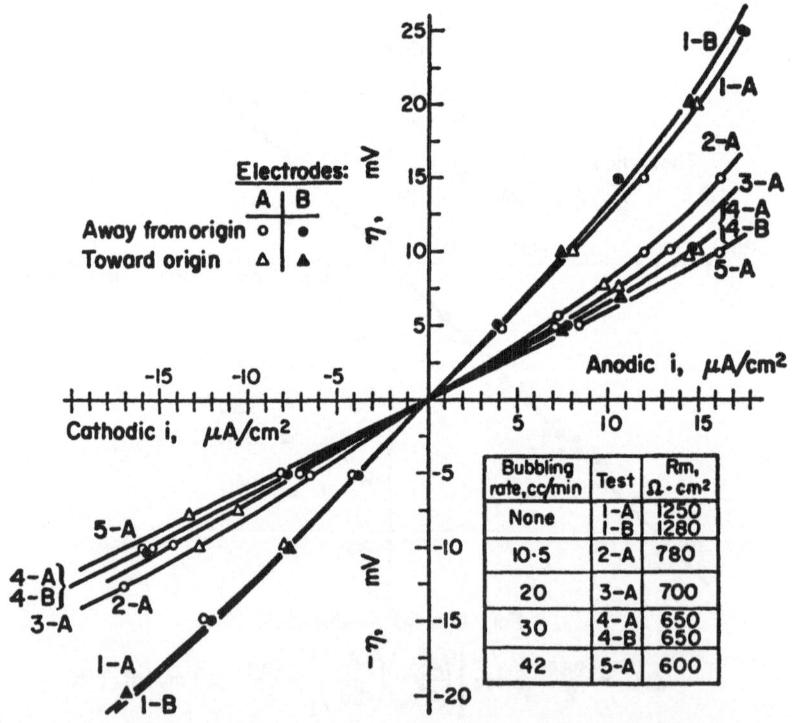

FIGURE 60. Effect of agitation on micropolarization characteristics of the CO_2, CO reference electrode in Li_2CO_3-Na_2CO_3-K_2CO_3 eutectic at 800°C (Ref. 455). (Reprinted by permission of the publisher.)

TABLE XII. **Effect of Temperature on Reversibility of the Au/CO_2, CO Electrode**[a]

Temperature (°C)	${}^uR_m$ (Ω cm^2)	${}^sR$ (Ω cm^2)	Approx. $i^0 \times 10^4$ (A/cm^2)	Approx. ${}^ui_m \times 10^4$ (A/cm^2)	$\frac{i^0}{{}^ui_m}$
800	1250	600	0.77	0.71	1.1
750	1580	595?	0.74	0.45	1.65?
700	1985	793	0.53	0.35	1.5
650	2380	950	0.42	0.28	1.5

[a] Reference 455.

tests (see discussion in the previous section on the CO_2, O_2 reference electrode) have been performed on the CO_2, CO electrode.[452,455] The experimental results obtained (Fig. 60) and the calculated values of ${}^uR_m$, i^0, ${}^ui_m$, and their ratios (Table XII) indicate a good reference electrode behavior.

The potential stability of the CO_2, CO electrode is excellent. Tests of two and three CO_2, CO electrodes against each other show that the stability of the potentials, as well as their agreement, is within ±2 mV or better. The time for the electrode to reach a new equilibrium after a change in gas composition is within 40 min (at a typical gas flowrate of 40 cm^3/min).

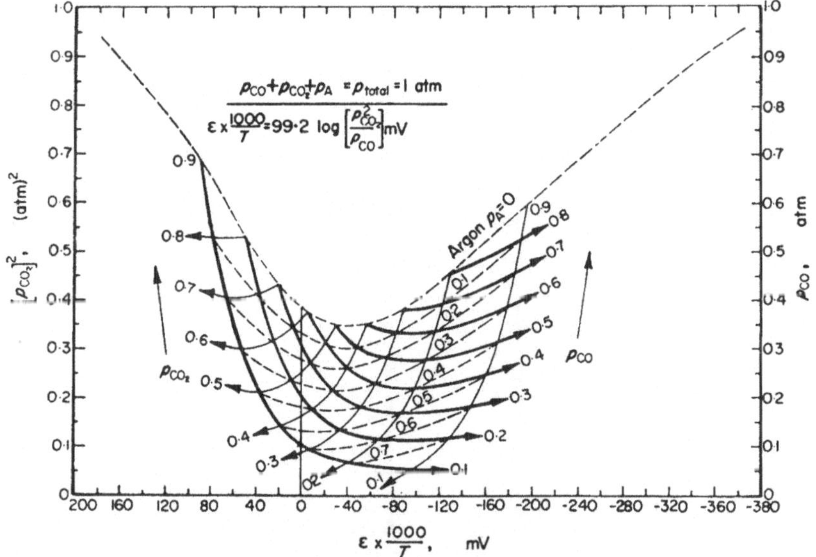

FIGURE 61. Families of curves for equipotential CO_2, CO reference electrodes as a function of gas composition (Ref. 445). (Reprinted by permission of the publisher.)

A convenient thermodynamic potential scale has been defined[445,452] in terms of the standard CO_2, CO reference electrode for which $\varepsilon^0 = 0$, at all temperatures, when $(p_{CO_2})^2/p_{CO} = 1$ atm. Figure 61 shows families of curves for equipotential CO_2, CO electrodes for the values of $(p_{CO} + p_{CO_2}) \leq$ 1 atm and when the partial pressure of inert nitrogen gas is varied from 0 to 0.7 atm. The standard CO_2, CO potential has been accurately correlated to the CO_2, O_2 electrode potential (see previous section on the CO_2, O_2 reference electrode).

The application of the CO_2, CO reference electrode is limited by the Boudouard equilibrium

$$2CO \overset{K_B}{\rightleftharpoons} CO_2 + C \tag{75}$$

which causes deposition of solid carbon unless, at the given temperature, $p_{CO} < (p_{CO_2}/K_B)^{1/2}$. To evaluate the situation of a particular CO_2, CO electrode with respect to the Boudouard equilibrium at a given temperature, one can use the alignment chart given in Fig. 62. To use the chart, it is first necessary to select the appropriate temperature on the left of the log K_B scale and hence to establish the value of log K_B. Through this point a line

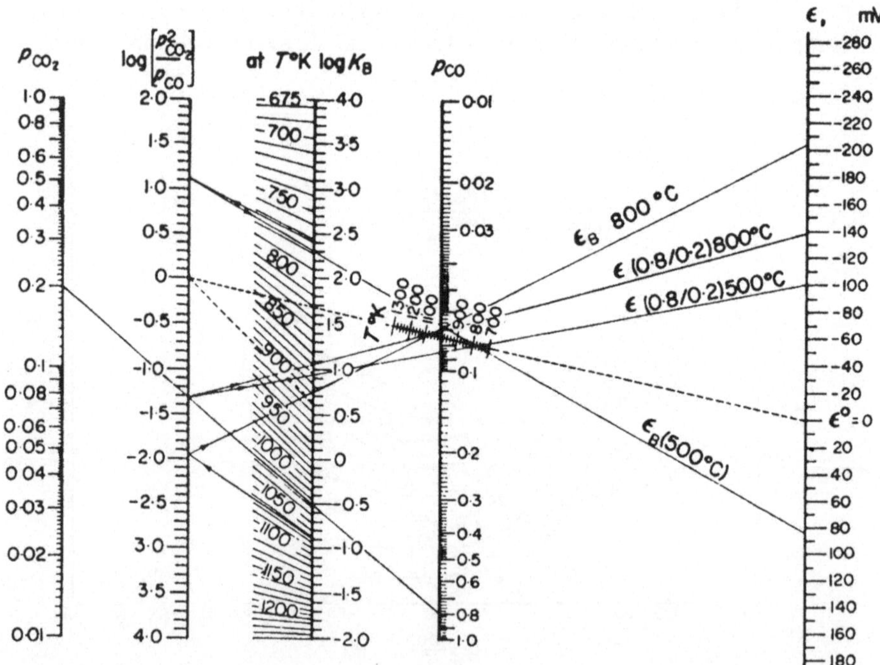

FIGURE 62. Alignment chart for evaluating the CO_2, CO reference electrode with respect to the Boudouard equilibrium (Ref. 445). (Reprinted by permission of the publisher.)

is drawn to intersect the p_{CO} and p_{CO_2} scales such that the sum of the intersects on the two scales is equal to $p_{CO} + P_{CO_2}$. This line intersects the log $p_{CO_2}^2/p_{CO}$ scale, and the point of intersection is connected with the appropriate point on the temperature scale and projected onto the potential scale on the far right-hand side of the chart. This procedure gives the potential ε_B. Secondly, the equilibrium potential ε of the electrode is obtained by drawing a line between the appropriate points on the p_{CO_2} and p_{CO} scales. This intersects the $p_{CO_2}^2/p_{CO}$ scale (giving the appropriate value of log $p_{CO_2}^2/p_{CO}$), and again connecting the intersection point on the $p_{CO_2}^2/p_{CO}$ scale with the chosen temperature on the inclined temperature scale leads to the value of ε. If ε is found to be more negative than ε_B, then carbon will be deposited. The evaluation of an CO_2, CO electrode with $p_{CO} = 0.8$ atm, $p_{CO_2} = 0.2$ atm at 500°C and 800°C is shown in Fig. 62. At 500°C, ε is -100 mV and $\varepsilon_B = 84$ mV; at 800°C, $\varepsilon = -139$ mV and $\varepsilon_B = -204$ mV. Hence, carbon will deposit at 500°C but not at 800°C.

The CO_2, CO reference electrode is recommended to be used at temperatures above 600°C. At low temperatures, the reaction $CO + 2OH^- = CO_3^{2-} + H_2$ proceeds (unless the melt is completely free of moisture, which is very difficult to attain in practice), consuming part of the incoming CO so that the electrode potential deviates from the Nernst equation. High p_{CO} is also not recommended for use at temperatures of 500°C and lower because, under these conditions, the reaction $CO_2 + CO + 2e^- = CO_3^{2-} + C$ becomes thermodynamically favored.(428)

4.6.3. Other Reference Electrodes

The Ag/Ag$^+$ Reference Electrode. This electrode, often called the Danner–Rey electrode in the literature, is a silver wire in a solution of Ag_2SO_4 in an Li_2SO_4–K_2SO_4 or Li_2SO_4–Na_2SO_4–K_2SO_4 melt contained in a closed aluminous porcelain (Pythagoras porcelain) tube (refer to the section on sulfates for a detailed discussion of this reference electrode). The Ag/Ag$^+$ electrode has been used as the reference electrode in several molten carbonate studies.(464–469) Because silica-containing Pythagoras porcelain is likely to be attacked by molten carbonates, the stability of the Danner–Rey reference electrode in carbonates, particularly for prolonged use and at high temperatures, is doubtful. Also, the electrode has never been properly evaluated in molten carbonates with respect to its membrane potential.

Miscellaneous Electrodes. Electrodes based on Pb/Pb^{2+}, Cd/Cd^{2+}, and Bi/Bi^{3+} systems have been proposed as reference electrodes in molten carbonates.(470) These electrodes, however, are not convenient for practical use. Recently, a reference electrode based on NiO has been tried in carbonate melts.(471) The reference electrode is an NiO pellet (with a gold wire for electrical contact) inside an alumina tube. The NiO sits on a mixture of

$LiAlO_2$ and Li_2CO_3–K_2CO_3 melt, which is used for gas sealing and for ionic contact. The reference gas is 30% CO_2–70% dry air. The electrode is claimed to have good stability.

A reference electrode, called the ionized air electrode, has been investigated for use in molten carbonates.[472,473] The electrode employs a low-level source of ionizing radiation (Am-241) to ionize an air gap inside the electrode housing, thereby establishing electrical contact between the reference electrode and the electrolyte under study (without introducing a liquid junction potential). The electrode is a 50-mesh platinum screen, and the electrode housing is constructed of high-temperature Vycor glass. The electrode is claimed to have the advantages of long life, high durability, low cost, simple design, and freedom from maintenance. However, the practicability of the ionized air reference electrode in molten carbonate studies is not known at present. The electrode was tested in Li_2CO_3–K_2CO_3 at 525°C and was found to give results comparable with those obtained using a platinum wire quasi-reference electrode.

4.7. Sulfates

The electrochemical behavior and the thermodynamic properties of molten sulfates, in particular the alkali metal sulfates, have been the subject of considerable study for the past two decades. Molten sulfates have been associated with hot corrosion of high-temperature alloys[474] such as those used in boiler tubes in conventional electric power generation and in superalloy components in gas turbines. (Hot corrosion is an accelerated oxidation attack of metals or alloys in contact with a thin film of fused salt.) Molten sulfates also belong to the class of oxyanionic solvents, which have interesting fundamental properties (e.g., acid–base and redox properties). The basicity of a molten sulfate is determined by the oxide activity of the melt, which is given by the equilibrium $SO_3 + O^{2-} \rightleftharpoons SO_4^{2-}$.

Several types of reference electrodes have been designed for use in sulfate systems: the Ag/Ag^+ electrode, the oxygen electrode, the SO_2, O_2 gas electrode, and the sodium electrode. Among these electrodes, the Ag/Ag^+ electrode is the most popular, owing to its long-term stability and simplicity of construction.

4.7.1. The Ag/Ag^+ Reference Electrode

The Ag/Ag^+ reference electrode is based on the Ag/Ag^+ equilibrium. The electrode consists of a silver wire that is immersed in a solution of Ag_2SO_4 in a molten alkali metal sulfate; all components are contained in a closed tube made of materials such as ceramics (mullite, Pythagoras porcelain) or glass (quartz, Supremax glass, Pyrex glass). The silver wire is normally spot-welded to a platinum wire that serves as the lead wire.

Although neither oxygen nor humidity in air has been found to have significant effect on the potential of the Ag/Ag^+ reference electrode,[475] sealing the tube under vacuum or inert gas atmosphere is advisable. Sealing is also essential to minimize any drift caused by loss of SO_3 from decomposition of the silver sulfate. It is recommended that all the chemicals used in the reference electrode be pure and free of moisture; construction materials should be dried and degassed under vacuum at high temperature before use. Figure 63 shows several designs of the Ag/Ag^+ electrode for molten sulfate studies. A summary of the types of membranes and sulfate systems employed for the Ag/Ag^+ electrode reported in the literature is listed in Table XIII.

The silver sulfate in the reference electrode melt is obtained either by *in situ* anodic dissolution of the silver wire or by addition of silver sulfate to the melt. Anodization of silver wire is normally carried out at a constant current (a few milliamperes per square centimeter) for a measured period of time; current efficiency is generally 100%. It is recommended to use an Ag_2SO_4 concentration of 1–10 mol%. Potential-time measurements on the Ag/Ag^+ electrode[475] have shown that electrodes with an Ag_2SO_4 concentration below 1 mol% exhibit considerable potential variation (Fig. 64). One of the reasons for the variation has been suggested to be due to concentration change caused by corrosion of the silver.[494] On the other hand, if the concentration of Ag_2SO_4 is too high, thermal decomposition of the silver sulfate can occur, thus adversely affecting the stability of the reference electrode. It is important to ensure the homogeneity of the reference melt

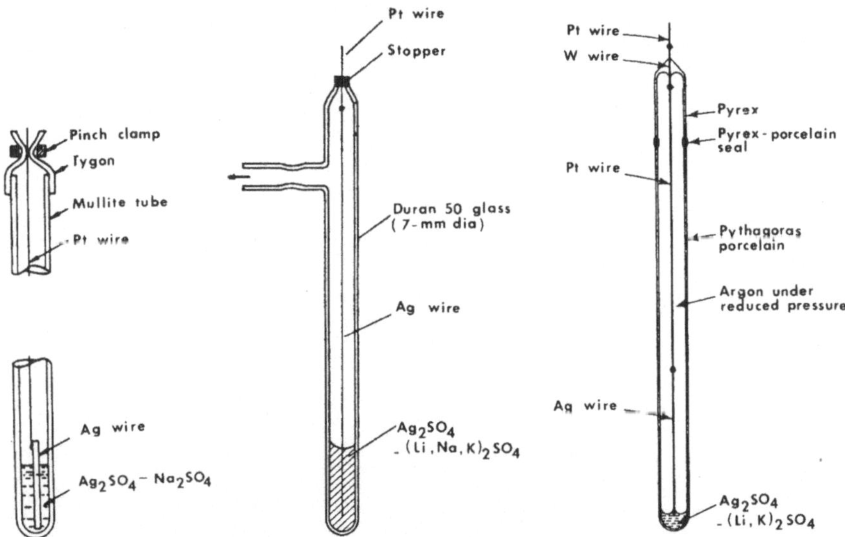

FIGURE 63. Ag/Ag^+ reference electrodes (Refs. 485, 475, 476). (Reprinted by permission of the publishers.)

TABLE XIII. Ag/Ag^+ Reference Electrodes in Molten Sulfates

Reference melt	Sulfate melt under study	Temperature (°C)	Membrane	Reference
$Ag_2SO_4 + Li_2SO_4$–K_2SO_4	—	900	Pythagoras porcelain	476
$Ag_2SO_4 + Li_2SO_4$–Na_2SO_4–K_2SO_4	Li_2SO_4–Na_2SO_4–K_2SO_4	580–625	Pyrex glass	477–480
$Ag_2SO_4 + Na_2SO_4$–K_2SO_4	Na_2SO_4–K_2SO_4	900	Mullite	481
$Ag_2SO_4 + Li_2SO_4$–Na_2SO_4–K_2SO_4	Li_2SO_4–Na_2SO_4–K_2SO_4	727	Mullite	482
$Ag_2SO_4 + Li_2SO_4$–Na_2SO_4–K_2SO_4	Li_2SO_4–Na_2SO_4	800	Supremax glass	475,
	Li_2SO_4–K_2SO_4		"	483, 484
	Li_2SO_4–Na_2SO_4–K_2SO_4		"	"
	Na_2SO_4–$MgSO_4$		"	"
	Na_2SO_4–$CaSO_4$–$MgSO_4$		"	"
$Ag_2SO_4 + Na_2SO_4$	Na_2SO_4	900–927	Mullite	485–490
$Ag_2SO_4 + Li_2SO_4$–K_2SO_4	Li_2SO_4–K_2SO_4	625	Quartz	491, 492
$Ag_2SO_4 + Li_2SO_4$–K_2SO_4	Li_2SO_4–K_2SO_4	625	Pyrex glass	493

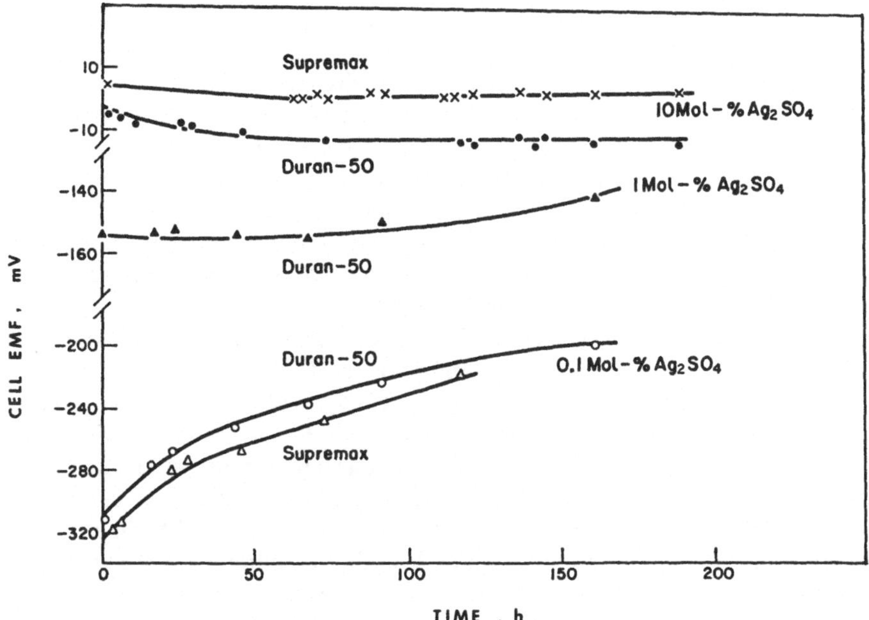

FIGURE 64. Emf versus time for cells

$$\underset{(10\,\text{mol}\%)}{\text{Ag}|\text{Ag}_2\text{SO}_4,\ \text{Li}_2\text{SO}_4,\ \text{Na}_2\text{SO}_4,\ \text{K}_2\text{SO}_4}:\text{glass}:\underset{(x\,\text{mol}\%)}{\text{Ag}_2\text{SO}_4,\ \text{Li}_2\text{SO}_4,\ \text{Na}_2\text{SO}_4,\ \text{K}_2\text{SO}_4|\text{Ag}}$$

at 625°C (Ref. 475). (Reprinted by permission of the publisher.)

in the reference electrode compartment. One cause of failure of the reference electrode may be the local recrystallizations of the silver wire arising from small temperature and concentration gradients. Gentle continuous stirring of the reference electrode melt by argon or nitrogen bubbling has been suggested to minimize inhomogeneities.[493,495]

The potential of the Ag/Ag^+ reference electrode is established by the reversible reaction

$$Ag^+ + e^- = Ag \tag{76}$$

This reaction has been shown to obey the Nernst equation in molten sulfates.[477,491] (It should be noted that the Ag/Ag^+ reaction becomes irreversible in very basic media, and it is not possible to generate Ag^+ in basic sulfate melts. However, the Ag_2SO_4-alkali sulfate melts are sufficiently acidic to ensure the reversibility of the Ag/Ag^+ reaction.) The potential of the reference electrode is stable (over a few hundred hours) and is reproducible to within ±10 mV. However, an initial period (up to 200 h) is necessary for the stabilization of the potential which can shift up to 100 mV during the stabilization period; after stabilizing, the potential is generally constant. This observed effect may be caused by ion exchange between the melt and ceramic or glass membrane, or by structural changes of the ceramic or glass tube. The reversibility, the polarizability, and the stability of an Ag/Ag^+ reference electrode with Pythagoras porcelain are shown in Fig. 65.

As shown in Table XIII, two main types of material have been used as membrane: ceramics (mullite, Pythagoras porcelain) and glasses (Pyrex and Supremax). These materials also serve as a barrier to prevent mixing the reference electrolyte and the melt under study (asbestos fiber has also been used as the junction in some studies; however, it can easily leak(496,497) and is therefore not recommended). The properties of these materials as the Ag/Ag^+ reference electrode membrane are discussed below.

Ceramics (Mullite, Pythagoras Porcelain). Mullite is a two-phase ceramic consisting of mullite grains ($2Al_2O_3 \cdot SiO_2$) enveloped by silica. At high temperatures, the silica grain boundary phase provides exclusive sodium-ion conduction. Mullite, although being slightly dissolved in molten alkali metal sulfates, strongly decreases its weight loss with time. Consequently, after aging mullite is useful as the membrane for the Ag/Ag^+ electrode. The material maintains a fairly constant membrane potential. Unfortunately, mullite tubes have shown evidence of transport of silver ions at high temperatures. The part of the mullite tube of an Ag/Ag^+ electrode that contained Ag_2SO_4–Na_2SO_4 mixture became dark, and silver could be detected in the Na_2SO_4 melt under study.(485) The silver penetration compromises the mechanical stability and thermal shock resistance of mullite. Mullite tubes often crack when quickly cooled down to room temperature owing to the solidification of the reference melt; long cooling times to ambient temperature are necessary to prevent cracking. In order to achieve a long life for mullite tubes, it is useful to keep the reference electrode at a constant temperature (for example, by changing it from one melt into another).

Pythagoras porcelain is an aluminous porcelain. The material is gas-tight and chemically stable in molten sulfates. The electrical resistance of a Pythagoras porcelain tube of about 1.5-mm wall thickness and 8-mm outer diameter is on the order of a few kilo-ohms at 900°C. Pythagoras porcelain is essentially an ionic conductor via potassium ions. Like mullite, Pythagoras porcelain is not completely impermeable to silver ions.(498) Because of the silver diffusion, the life of the Ag/Ag^+ reference electrode with Pythagoras porcelain is limited to only several hundred hours. The thermal gradient along the length of the Pythagoras porcelain and the diffusion across the material have little effect on the potential of the electrode.(476) After one week of operation, only a few millivolts of variation was observed for the Ag/Ag^+ electrode.(476)

Glass (Pyrex, Supremax). Pyrex (borosilicate glass type) has been used as a membrane at temperatures up to 625°C. Pyrex glass has a very small

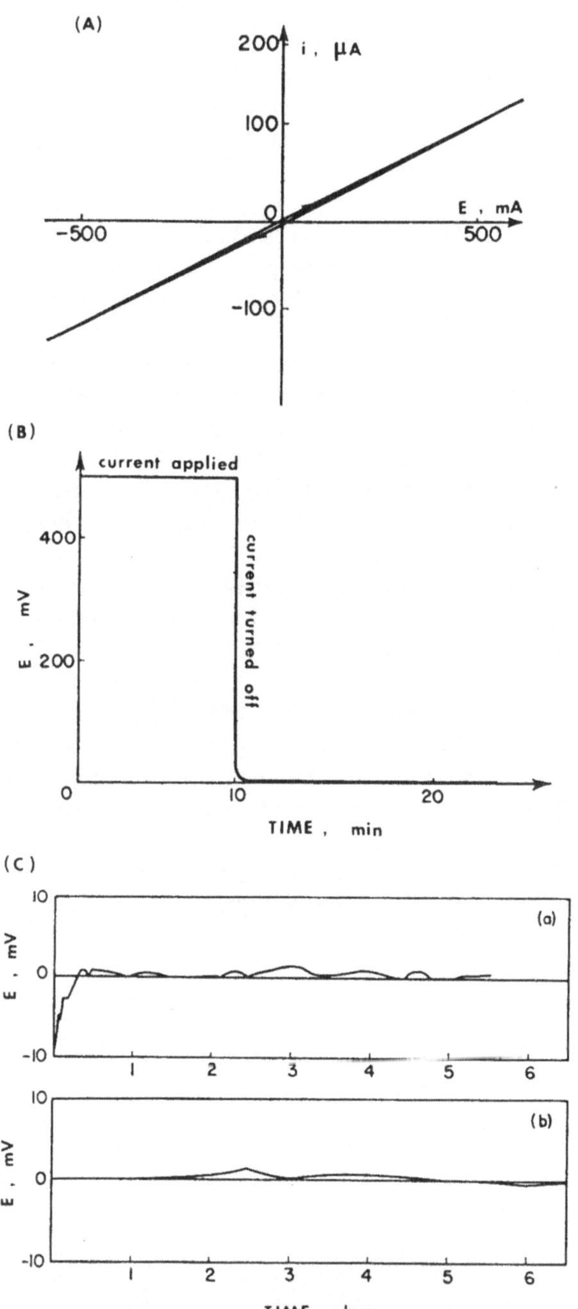

FIGURE 65. (A) Reversibility, (B) polarizability, and (C) stability of Ag/Ag^+ reference electrode (with Pythagoras porcelain membrane) in Li_2SO_4–K_2SO_4 melts at 650°C (Ref. 485). (Reprinted by permission of the publisher.)

but definite alkali metal-ion conduction and thus acts as an ionic conductor. A Pyrex sheath of 1-mm wall thickness is normally used. This thickness is thick enough to prevent diffusion of the silver ions, which have a lower permeability in the glass than alkali metal ions. The glass generally gives rise to a well-defined membrane potential. When contacted with silver sulfate solution, Pyrex glass rapidly develops a brown coloration, which is attributed to the exchange of silver ions with sodium ions of the glass. This phenomenon probably accounts for the potential drifts generally observed during the initial period of use for reference electrodes with Pyrex membrane.

Supremax glass is an alkaline-earth aluminosilicate glass. It has a softening point high enough for use at temperatures up to 800°C. A satisfactory long-term stability over a few hundred hours can be achieved up to 800°C with a Supremax membrane. Supremax is impermeable to silver ions but conductive to alkali metal ions. A Supremax glass tube of 1-mm wall thickness and 8-mm outer diameter, which is widely used, has an electrical resistance of about 400 Ω at 800°C.

Above 800°C, the use of glass membranes is impractical. At these higher temperatures, electrodes with mullite or porcelain membranes must be used. However, electrodes with ceramic membranes often fail because of the cracking of the silver wire.[481] Therefore, it was suggested that the silver wire be replaced by a platinum wire (Fig. 66). In the construction of this type of reference electrode, the platinum is insulated by a quartz tube to prevent contact with the reference melt which creeps up the wall of the mullite tube. If the platinum wire comes in contact with the mullite wall, a mixed potential may be created (depending on the temperature and concentration gradients along the tube wall), thus reducing the reproducibility of the reference electrode. Since platinum cannot be fused into quartz,

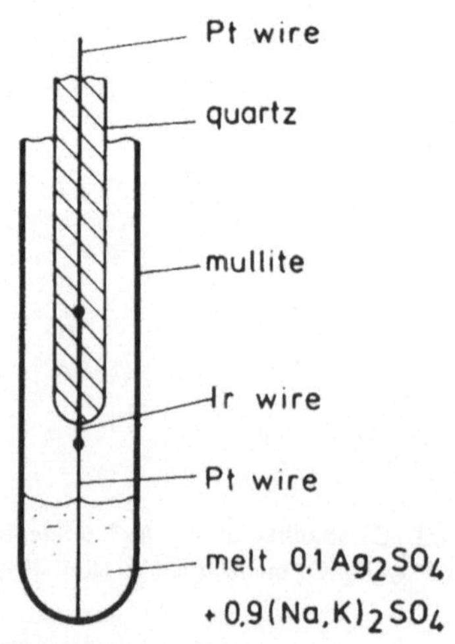

FIGURE 66. (Pt)Ag/Ag^+ reference electrode for molten sulfates (Ref. 481). (Reprinted by permission of the publisher.)

a piece of iridium wire is fused into the quartz tube to close the bottom of the tube. After an aging period of about 200 h, the potential of this type of reference electrode is stable within ±10 mV for many hundred hours. Its stable potential, which is close to the Ag/Ag^+ potential, is suggested to be established by the formation of a silver or silver-platinum alloy layer on the platinum wire.[481]

For temperatures above 960°C (the melting point of silver), a reference electrode with molten silver has been proposed.[476,499] The construction of such a reference electrode is shown schematically in Fig. 67. The platinum wire, which serves as the lead wire, does not touch the reference melt. The

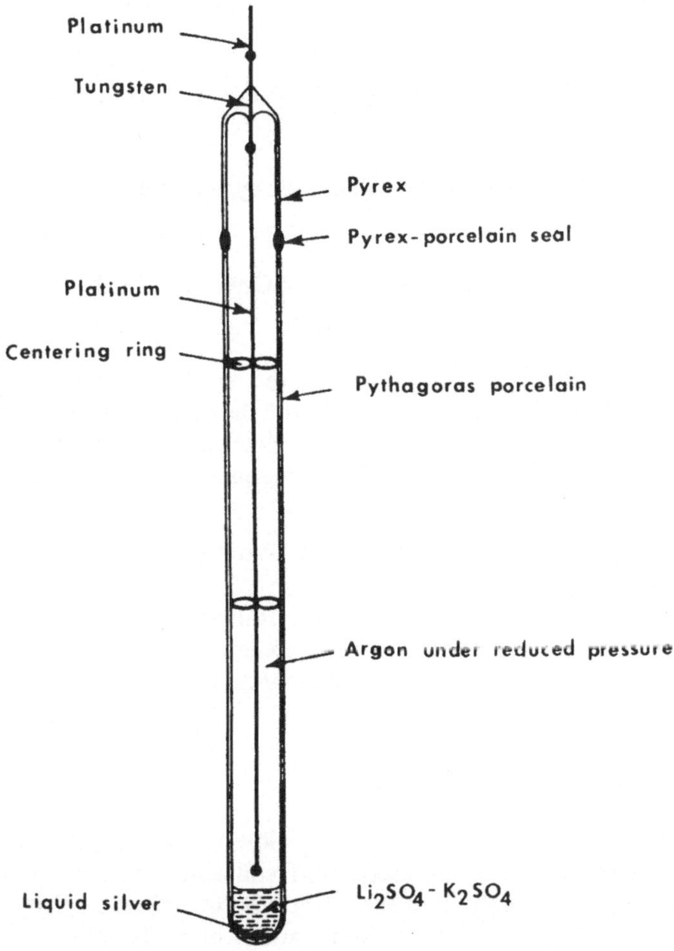

FIGURE 67. Schematic diagram of Ag/Ag^+ reference electrode for use up to 1300°C (Ref. 476). (Reprinted by permission of the publisher.)

conductivity of the argon gas at those high temperatures is suggested to be sufficient to make electrical contact between the platinum and the molten silver.[476,499] The electrode has been claimed to be stable and reproducible at temperatures up to 1200°C. The potential of this reference electrode is constant within ±25 mV.

The Ag/Ag^+ reference electrode has been compared with other reference electrodes in molten sulfates such as the zirconia oxygen electrode and the SO_2, O_2 electrode.[485] All of the emfs of the cells combined from these electrodes have been shown to obey the Nernst equation for the proposed electrochemical reactions. Several sodium oxide concentration cells have been studied, for example,

$$(Au)SO_3, O_2|Na_2SO_4 \vdots \text{mullite} \vdots 10 \text{ mol\% } Ag_2SO_4, Na_2SO_4|Ag$$

Cell 34

$$Ag|10 \text{ mol\% } Ag_2SO_4, Na_2SO_4 \vdots \text{mullite} \vdots N \text{ mol\% } Ag_2SO_4, Na_2SO_4|Ag$$

Cell 35

$$W|WS_2, Na_2S \vdots \text{mullite} \vdots 10 \text{ mol\% } Ag_2SO_4, Na_2SO_4|Ag$$

Cell 36

$$W|WS_2, Na_2S \vdots \text{mullite} \vdots Na_2SO_4|O_2, SO_3(Au)$$

Cell 37

to establish the thermodynamic basis for the Ag/Ag^+ reference electrode. Measured emfs of these cells agreed well with the calculated thermodynamic emfs. As an example, the emf of cell 37 (which is equivalent to cell 36 minus cell 34) at 900°C is calculated from free energy data and is given by[486]

$$E = 2.292 + \frac{RT}{2F} \ln (p_{SO_3} p_{O_2}^{1/2}) \tag{77}$$

Figure 68 shows the emf of cell 37 as a function of $\ln (p_{SO_3} p_{O_2})^{1/2}$. The experimental values agree with the thermodynamic voltages calculated from Eq. (77), which agreement also confirms that mullite functions as a sodium-ion membrane.

4.7.2. The Oxygen Reference Electrode

A reference electrode which is reversible to oxygen gas and oxygen ion, an oxygen electrode, is useful and convenient for thermodynamic measurements in oxyanionic melts. This type of reference electrode, for example, has been used in the determinations of oxide activity in molten sulfates.[500,501] The two common oxygen reference electrodes in molten

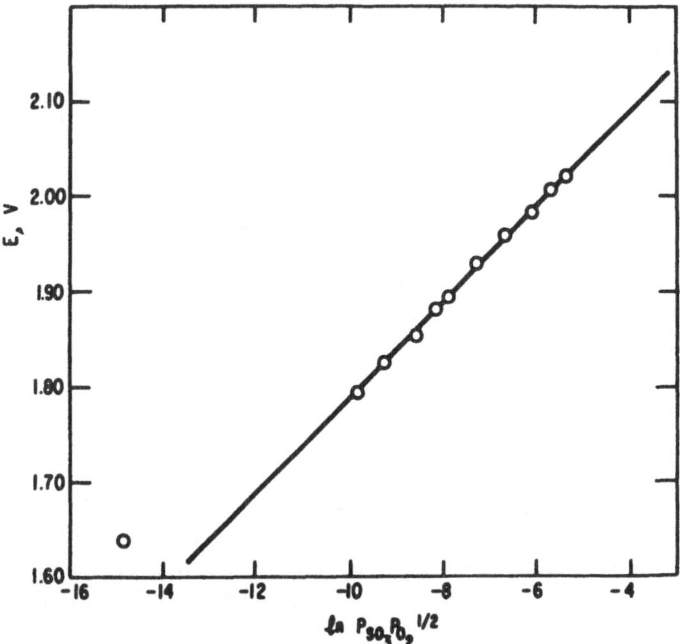

FIGURE 68. Emf of cell W|WS_2, Na_2S:mullite:Na_2SO_4|SO_3, O_2(Au) as a function of $\ln(p_{SO_3}p_{O_2}^{1/2})$ at 900°C (Ref. 486). Solid line corresponds to Eq. (77). (Reprinted by permission of the publisher.)

sulfates are the oxygen electrode on Pt or Au and the oxygen electrode with a zirconia membrane. Another type of oxygen electrode, the oxygen electrode with a β-alumina membrane, has recently been developed for melts containing sodium ions.

The Oxygen Reference Electrode on Platinum or Gold. The construction of the oxygen reference electrode on platinum or gold—called the (Pt)O_2 electrode—is quite simple. The electrode consists of a Pt (or Au) wire inserted in an alumina tube, which serves as gas delivery tube; the components are immersed in the molten sulfates. Oxygen (or air) is passed over the platinum wire, and the equilibrium,

$$\tfrac{1}{2}O_2 + 2e^- = O^{2-} \tag{78}$$

and the corresponding stable potential

$$\varepsilon = \varepsilon^0 + \frac{RT}{zF}\ln\left(p_{O_2}^{1/2}/a_{O^{2-}}\right) \tag{79}$$

are established at the electrode. Attainment of equilibrium, especially with respect to the gas phase, is rapid because platinum has the capacity to allow quick oxygen adsorption on its surface. Platinum stabilizes at a somewhat more positive potential than the gold electrode, suggesting that both platinum and gold are slightly attacked by molten sulfates. However, melt interaction with these metals is so negligible that they can be considered inert in the molten salt medium.

For the proper functioning of the $(Pt)O_2$ reference electrode, several precautions are necessary. Since the electrochemical properties of platinum depend to a large extent on its history, a standard pretreatment of the platinum electrode should be adhered to. For example, the pretreatment may consist of heating the electrode in the oxidizing portion of a Bunsen flame, dipping the electrode in concentrated HCl, and reheating it in the oxidizing flame. There is the possibility of poisoning the $(Pt)O_2$ electrode by iron deposits on the platinum wire if a small amount of dissolved iron is present in the melt. The effect of iron deposit is to raise the true apparent potential of the reference electrode and to make the potential dependent on gas stirring rate.[502] Addition of active zirconium oxide, which is insoluble in molten sulfates, helps to precipitate trace of dissolved iron.

The $(Pt)O_2$ electrode has satisfactory electrochemical behavior in acidic sulfates. In such melts, the electrode obeys the Nernstian behavior according to Eq. (79), i.e., the variation of the equilibrium potential with $\log p_{O_2}^{1/2}$ is a linear relationship with a slope of $2.3RT/4F$.[493,495] In basic melts, on the other hand, the $(Pt)O_2$ electrode is not reversible to oxide ion. The observed Nernst slopes in rather basic molten sulfates are less than the theoretical value.[503,504] This observation has been ascribed to the formation of other oxide ions in the melts such as peroxide and superoxide ions,[485,503] which are known to be stable in basic sulfates. Therefore, in basic melts, an alternative equilibrium other than Eq. (78) is potential-determining. For example, it is postulated that the potential of $(Pt)O_2$ electrodes is controlled by the reaction

$$\tfrac{1}{2}O_2 + O^{2-} = O_2^{2-} \tag{80}$$

in Na_2SO_4 at 900°C.[503] The inconsistency of the behavior of the $(Pt)O_2$ electrode with the simple interpretation based on an equilibrium of molecular oxygen and oxide ion has been observed in Na_2SO_4 melt at 927°C.[485] The presence of superoxide and peroxide ions is also suggested as the cause of the inconsistent electrode behavior.

The Oxygen Reference Electrode with a Zirconia Membrane. This electrode employs zirconia as an ionic membrane.[504-506] Doped (or stabilized) ZrO_2, such as calcia- or yttria-stabilized zirconia, is a good oxide-ion-conducting solid electrolyte at elevated temperatures. The material is impervious to gases and stable in molten sulfates. The electrode is based on the oxygen/oxide ion equilibrium at the interface between a platinum electrode and the solid electrolyte. The reference electrode is shown schematically in

Fig. 69. The platinum electrode, which is reversible to oxygen partial pressure, is prepared by platinizing the inside bottom of the zirconia tube. Platinum paste, which consists of submicron platinum particles suspended in an organic vehicle, is painted on the tube and fired to form porous platinum layers. Electrical contact with the platinum is obtained by placing a Pt/Pt–10% Rh thermocouple in an alumina protection tube and spring loading it inside the ZrO_2 tube to ensure positive contact. Air and pure oxygen have been the most frequently used reference gases because they are convenient and provide good reversibility to the reference electrode. Reversibility has been demonstrated for platinum electrodes with stabilized

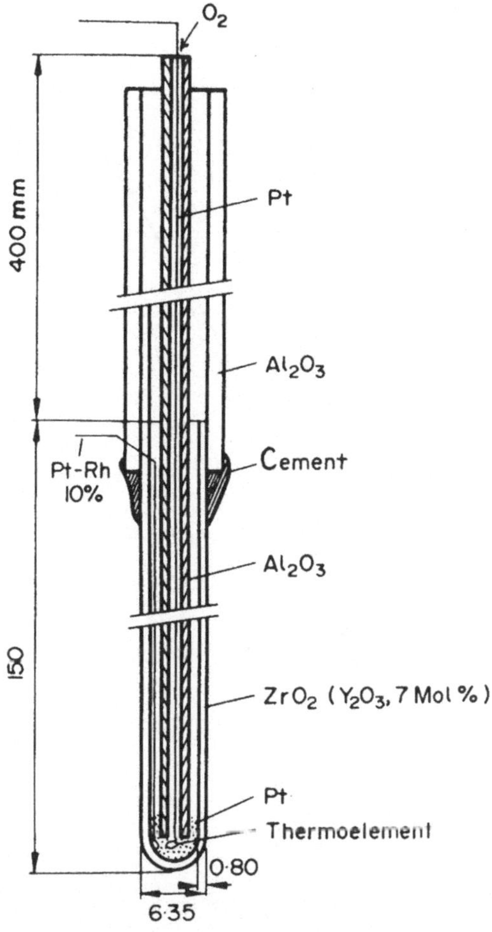

FIGURE 69. Schematic diagram of oxygen reference electrode with zirconia membrane (Ref. 505). (Reprinted by permission of the publisher.)

ZrO_2 electrolyte[507] and also for oxygen electrodes with zirconia membrane in molten sulfates at 900°C (Fig. 70). One of the advantages of the oxygen electrode with a zirconia membrane in molten sulfates is that the reference gas in the zirconia tube can be adjusted independently of the gas composition in equilibrium with the melt under study. Also, this electrode is reversible only to oxide ions in the melt. Of course, there are some inherent limitations. First, there is an operating temperature range for zirconia solid electrolyte. At low temperatures (e.g., less than 450°C) the resistivity of the ZrO_2 is very high and the electrode reaction becomes irreversible. At high temperatures (e.g., greater than 1100°C) the electronic conductivity of the solid electrolyte increases, and volatization of platinum as oxides begins. Second,

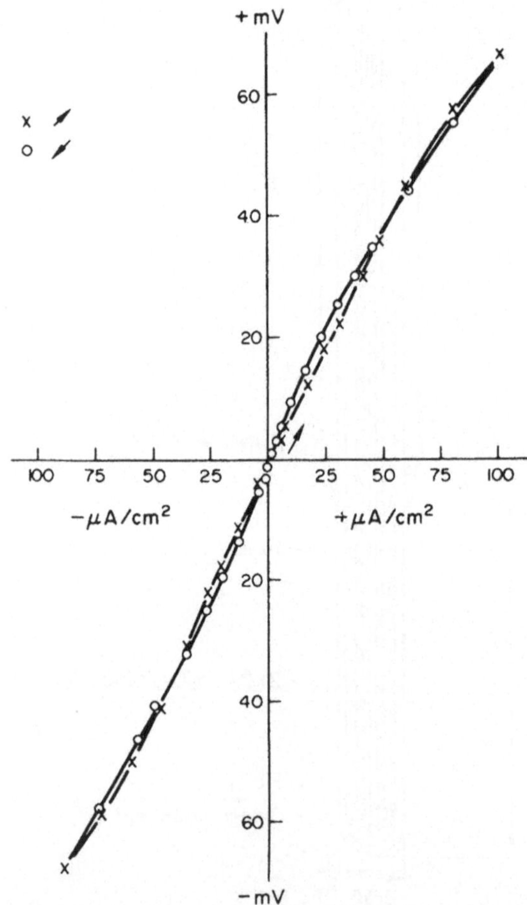

FIGURE 70. Micropolarization test for the oxygen electrode with zirconia membrane in molten sulfates at 900°C (Ref. 505). (Reprinted by permission of the publisher.)

zirconia electrolyte has very poor thermal shock resistance so great care is needed to prevent thermal cracking.

The Oxygen Reference Electrode with a β-Alumina Membrane. This oxygen electrode design consists of a platinum wire dipped in a Na_2O–WO_3 melt of known composition contained in a β-alumina tube.[508] Beta-alumina, $Na_2O{\cdot}11Al_2O_3$, is a sodium-ion-conducting solid electrolyte. This electrode is thus only useful in sulfates containing Na_2SO_4. The potential of this electrode is defined by the oxygen atmosphere inside the β-alumina tube, which is established by bubbling oxygen (at about 10 cm^3/min) in the melt via an alumina tubing. Under an oxygen atmosphere, the reference potential is fixed by the oxide activity in the Na_2O–WO_3 melt, which in turn depends on the melt composition. At 927°C, the oxide ion activity in Na_2O–WO_3 melt with composition ranging from $N_W = 0.52$ to 0.80 (where N_W is the mole fraction of WO_3) is given by the equation log(oxide activity) $= -4.14 - 15.11 N_W$.[508] The variation of oxide ion activity as a function of temperature at different melt composition also has been given.[508]

4.7.3. *The SO_2, O_2 Reference Electrode*

It is well known that the oxide activity (or basicity) of molten sulfates is given by the equilibrium

$$SO_3 + O^{2-} \rightleftharpoons SO_4^{2-} \tag{81}$$

or

$$\tfrac{1}{2}O_2 + SO_2 + O^{2-} \rightleftharpoons SO_4^{2-} \tag{82}$$

The oxide activity can thus be described quantitatively in terms of the partial pressure of SO_3 or the corresponding partial pressures of O_2 and SO_2 in the gas phase in equilibrium with the melt.

The SO_2, O_2 reference electrode[494,495,504,512,518–521] is based on the equilibrium given by Eq. (82). In this electrode, an SO_2, O_2 gas mixture is fed into a mullite tube (or quartz tube for lower-temperature applications) containing molten sulfate (Fig. 71). A platinum (or gold) wire is inserted into the molten sulfate as an electrode. To ensure the equilibrium

$$\tfrac{1}{2}O_2 + SO_2 \rightleftharpoons SO_3 \tag{83}$$

in the gas phase inside the reference electrode, the gas mixture is passed over a platinum catalyst before coming in contact with the sulfate melt.

The variation of the potential of the SO_2, O_2 reference electrode with the partial pressure of O_2 and SO_2 agrees with Eq. (82). It also appears that the reaction

$$\tfrac{3}{2}O_2 + 2SO_2 + 2e^- = S_2O_7^{2-} \tag{84}$$

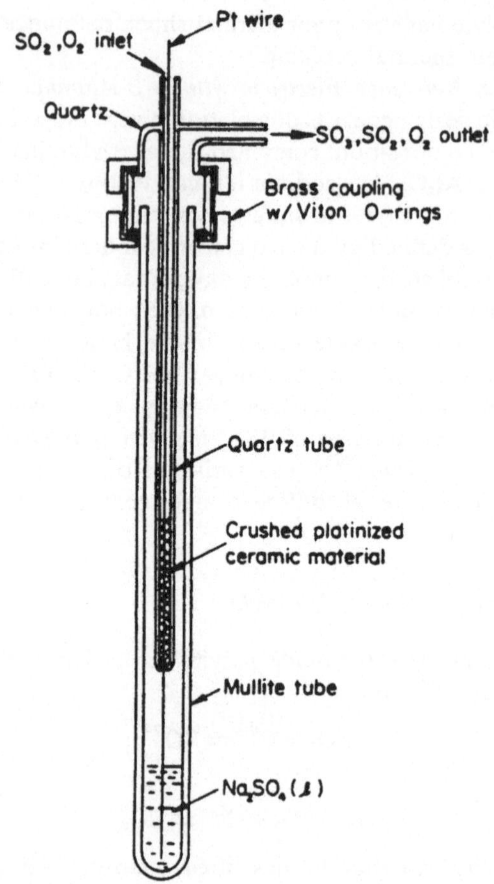

FIGURE 71. SO_2, O_2 reference electrode (Ref. 485). (Reprinted by permission of the publisher.)

has no contribution to the electrode process. The electrode with the Pt wire stabilizes at a potential somewhat more positive than that taken by the electrode with the gold wire. As mentioned earlier, this effect can be explained by the interaction of those metals with the molten salt. It has been noted that the platinum electrode approaches its equilibrium potential slightly faster than the gold electrode. When the gas composition of the SO_2, O_2 electrode is changed to another composition, the potential of the electrode normally drifts slowly in a positive direction and requires a few hours to stabilize. For a fixed gas composition, the potential generally remains constant to ±5 mV for more than 3 h.

4.7.4. The Na^+-Ion Reference Electrode

The reaction

$$\tfrac{1}{2}WS_2 + 2Na^+ + 2e^- = \tfrac{1}{2}W + Na_2S \tag{85}$$

is the basis for the Na^+-ion reference electrode.[513-516] Figure 72 shows the design of this reference electrode. As indicated in the figure, the electrode consists of analytical grade powders of W, WS_2, and Na_2S (in a weight ratio of 1 : 1 : 8, respectively) packed in the bottom of the β-alumina crucible (11 mm o.d., 6 mm i.d., 60 mm high). On top of this mixture is packed a dried powder of β-alumina. A tungsten lead is punched through the powder packings and touches the bottom of the crucible. The upper part of the tungsten lead is protected by an alumina tube. The tungsten lead is also connected to a Pt wire for electrical connection (it is necessary to correct for any measured voltages for the Pt–W thermo-emf). The alumina tube is sealed in place with fused soft glass. The seal also effectively seals the reference electrode chamber from contacting the outside atmosphere.

The emf of a cell between this electrode and the SO_2, O_2 electrode, which is known to behave reversibly, is quite stable and reproducible in the temperature range of 830–950°C. For a freshly prepared electrode, about 8 h is required for the emf of the cell to stabilize. When the temperature is changed, the emf reaches a new value within 1–2 h. If the temperature is altered and then returned to the original value, the emf of the cell is reproducible within 1 or 2 mV. Upon application of a current of 5 mA for 5 s in either direction, the cell is polarized about 500 mV above and below the steady-state voltage. The voltage returns to the previous value within 1 min after the polarizing current is turned off.

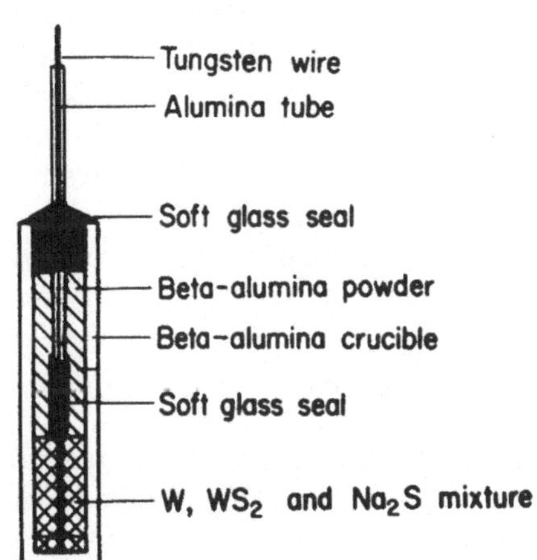

FIGURE 72. Sodium reference electrode with β-alumina membrane (Ref. 515). (Reprinted by permission of the publisher.)

The lifetime for the sodium-ion reference electrode is approximately 36 h. Failure occurs when Na_2S penetrates through the β-alumina solid electrolyte.

4.8. Hydroxides

Molten hydroxides (especially NaOH, KOH, and their eutectic mixture), besides having excellent electrical and thermal conductivities, exhibit interesting acid–base properties. Applications of hydroxides include electrosynthesis and extractive metallurgy. There has been considerable interest in recent years in the chemistry and electrochemistry of molten hydroxides.(12,517) However, hydroxides are probably the least studied among the numerous potentially interesting classes of molten salts, mainly because of the highly corrosive nature of hydroxide melts. The chemical aggressivity of hydroxides also reflects in the poor state of development for reference electrodes for use in those melts.

In molten hydroxides, the crucial problem in the development of the reference electrode is that of finding suitable insulator materials that are compatible with the molten salts under the experimental conditions involved. In general, the choice of insulating materials in hydroxides is very narrow, and the few available are not without limitations. Thus, alumina and Teflon are the only suitable insulators. However, Teflon cannot be used at temperatures higher than 280°C, and alumina is unstable in basic melt. The basicity of molten hydroxides can be described in terms of oxide activity, which in turn is determined by the self-dissociation reaction

$$2OH^- = H_2O + O^{2-} \tag{86}$$

The selection of a reversible redox couple is another problem in the development of reference electrodes in molten hydroxides. The basicity of the melt plays an important role in the selection of a redox system. The chemical reactivity of a molten hydroxide varies considerably as the condition of the melt is changed from acidic through neutral to basic. Such variation adversely affects the properties of those metals and dissolved substances which may be considered as suitable for reference electrodes. The various reference electrodes for hydroxides reported in the literature are discussed in this section.

4.8.1. The Oxygen Reference Electrode

The principle of the oxygen reference electrode in hydroxides is similar to that of the $(Pt)O_2$ electrode in molten sulfates. Again, platinum and gold are most commonly used as lead wires. It is preferable to separate the reference electrode system from the melt under study; porous alumina, alundum, and corrundum have been used for this purpose. As an example, an oxygen reference electrode with a porous alumina plug is schematically represented in Fig. 73. However, use of alumina and other refractory oxides in basic melts is limited because of their high solubility in those media.

The oxygen electrode in molten alkali hydroxides has been studied by several authors.[518-529] In acidic melts, the electrode has been shown to be reversible, nonpolarizable, stable, and reproducible (better than ±10 mV). The reaction

$$O_2 + 4e^- = 2O^{2-} \tag{87}$$

or

$$O_2 + 2H_2O + 4e^- = 4OH^- \tag{88}$$

is potential-determining, and the dependence of the potential of the electrode is given by

$$\varepsilon = \varepsilon^0 + \frac{RT}{4F} \ln \left(p_{O_2} p^2_{H_2O}\right) \tag{89}$$

At low partial pressure of oxygen (<0.05 atm), however, deviation from reversibility for the oxygen reference electrode is observed.[522] In basic melts, hydroxides are converted to peroxides. Thus, potential measurements have shown deviation from the behavior expected from the relationship described by Eq. (88). In those melts, the electrode reaction $O_2 + 2e^- = O_2^{2-}$

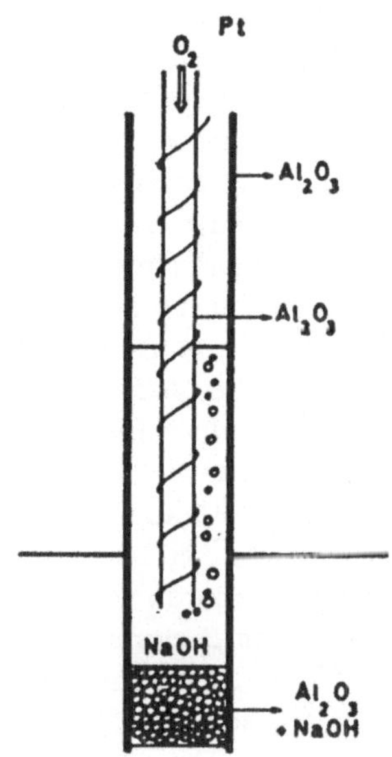

FIGURE 73. Oxygen reference electrode for molten hydroxides (Ref. 518). (Reprinted by permission of the publisher.)

takes place rapidly and is potential-determining.[521] As indicated by Eq. (88), the water content of the melt (thus its acidity) should be carefully controlled for the proper use of this reference electrode.

Another type of oxygen reference electrode for hydroxide melts is the metal/metal oxide electrode (e.g., Cu/CuO, Ni/NiO), which is similar to that commonly used in solid-state electrochemical cells.[530] The problem with this type of electrode is finding a suitable membrane to separate the reference electrode system from the hydroxide melt.

4.8.2. The Sodium Reference Electrode

The sodium electrode, which is based on the Na/Na^+ system, has been used as the reference electrode in molten NaOH at temperatures of 350–450°C.[531–534] Several versions of the sodium electrode have been described[531] and are given below.

The Driven Sodium Electrode. The potential of this electrode is determined by extrapolating the linear part of the current–voltage curve obtained for the sodium deposition at iron microelectrode. The accuracy of this potential determination thus relies on the extrapolation procedure and ignores the electrochemical reduction of water. The current–voltage curves are regular in acidic-to-neutral melts, but they are distorted in basic melts and unsuitable for extrapolation.

The Sodium Pool Electrode. The electrode is composed of molten sodium floating on the surface of molten NaOH; the molten components are contained within an open-ended alumina tube. Contact with the sodium pool is made by an iron wire. The inherent problem associated with this electrode is the dissolution of sodium in the molten hydroxide, giving a green coloration. This problem may give rise to a somewhat ill-defined potential-determining interface and the sodium contaminates the melt.

The Glass-Enclosed Sodium Electrode. In order to form this version of the sodium electrode, a pellet of sodium is sealed into a borosilicate glass test tube that has a rubber serum cap and that has been flushed with argon. An iron wire is used as the lead wire. This electrode, however, shows rapid attack of the borosilicate glass by the melt.

The Sodium Electrode with a β-Alumina Membrane. The electrode consists of either sodium contained in a fused alumina tube (5-mm bore) with a sintered β-alumina plug or sodium in a crystalline β-alumina crucible (12 mm o.d., 40 mm long, 3 mm wall thickness). The electrical contact to sodium is an iron wire. The β-alumina plug can be made by ramming moist −200 mesh β-alumina powder into the end of the alumina tube, drying at 130°C, and then sintering at about 930°C for 2 h under argon. Sodium is then placed inside the tube, which is sealed under argon with a rubber serum cap. The reproducibility of the sodium electrode with a β-alumina membrane is quite good (±1.5 mV in a wide range of conditions). The electrode, however, is subject to membrane potential associated with the transport of sodium ions across the β-alumina/melt interface. For longer periods of use, the potential of the electrode may drift. Frequent calibration by reference to a freshly prepared electrode is thus recommended. It has been recommended that the electrode be discarded when the potential difference (to a freshly prepared electrode) becomes more than 100 mV.[531] The sodium electrode with a β-alumina membrane is the most promising among the different versions of the sodium electrode, and it is suitable for thermodynamic studies if the membrane potential is stable and corrected for.

The Lead-Sodium Alloy Electrode. This reference electrode is a pool of lead-sodium alloy beneath a sodium hydroxide melt. Electrical connection is made via an iron wire sheathed in an alumina tube. This lead-sodium electrode is useful for sodium contents below 30 mol% if the melt is not so acidic as to rapidly remove sodium from the alloy.

4.8.3. The Cu^+, Cu^{2+} Reference Electrode

A schematic diagram of this reference electrode is shown in Fig. 74. It consists of a Pt wire immersed in a solution of Cu^+ and Cu^{2+} in a hydrated

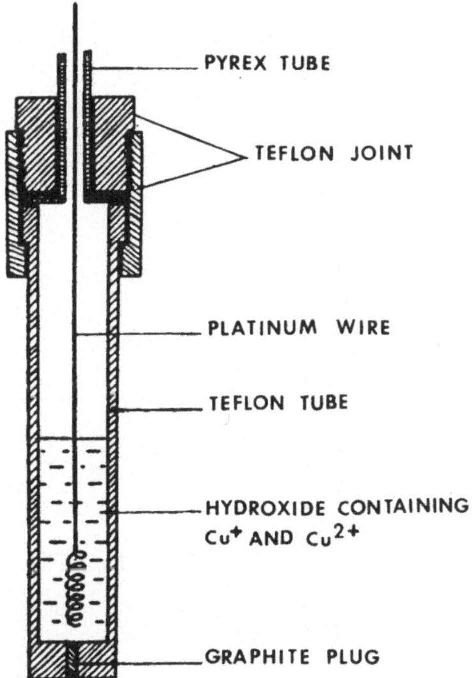

FIGURE 74. Cu^+, Cu^{2+} reference electrode (Ref. 528). (Reprinted by permission of the publisher.)

hydroxide melt.[528] The reference electrode is isolated from the bulk melt by containment within a Teflon tube that is closed at one end by a plug of porous graphite. The potential of the electrode remains constant if the pH_2O ($pH_2O = -\log[H_2O]$) remains less than 3 (H_2O concentration $>10^{-3}$ *M*). In more basic melts, the precipitation of CuO and Cu_2O gives rise to potential shifts.[535] It is necessary to control the atmosphere inside the reference electrode to prevent the variation of Cu^+ concentration due to oxidation. This electrode is designed for use at temperatures below 280°C.

4.8.4. The Cu/Cu^+ Reference Electrode

The electrode (Fig. 75) is a copper-plated platinum sheet (supported by a Pyrex tube which also serves as a gas inlet) immersed in a solution of

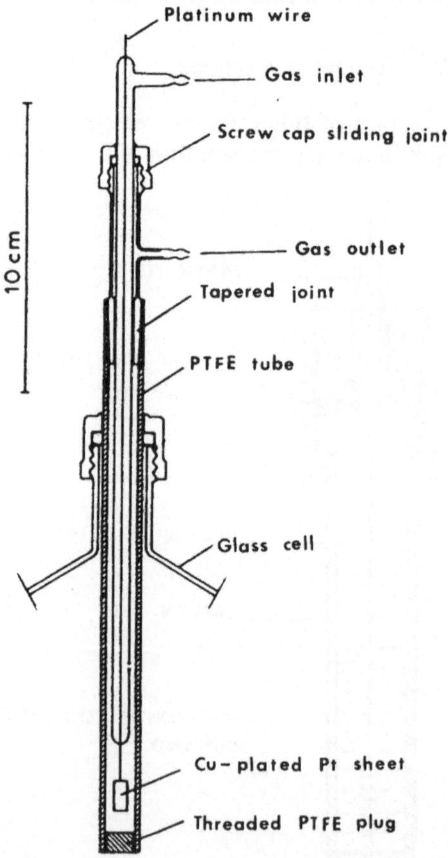

FIGURE 75. Schematic diagram of Cu/Cu^+ reference electrode (Ref. 536). (Reprinted by permission of the publisher.)

Cu^+ of known concentration (2-6 × 10^{-4} mole fraction) in an NaOH-KOH melt containing water (water content between 5 × 10^{-4} to 5 × 10^{-3} mole fraction).[536] The reference melt is contained in a PTFE tube (15 mm o.d., 5 mm thick) that is closed at the bottom by a plug of 8 mm long. The plug is prepared by threading a PTFE bar (105 mm diam) and is forced into the tube after heating the tube at about 310°C. The PTFE tube is filled with the reference melt via suction by immersing the tube in a prepared melt. A nitrogen atmosphere is maintained within the reference electrode body.

The validity of the Nernst equation for the Cu/Cu^+ system has been verified by measuring the emf of concentration cells in which the Cu^+ concentration is varied. between 1 × 10^{-4} and 2 × 10^{-3} mole fraction (Fig. 76). The stability and reproducibility of the electrode have been tested at temperatures between 217 and 237°C. The electrode is stable (±5 mV) and reproducible, even after repeated use and storage at room temperature.

4.8.5. *The Au/Au^{3+} Reference Electrode*

A gold wire dipped in a saturated solution of Au^{3+} in NaOH-KOH eutectic forms the Au/Au^{3+} electrode (Fig. 77). An alumina tube is used to contain the reference electrode melt, with a porous graphite providing the ionic contact to the melt under study. The electrode has good reversibil-

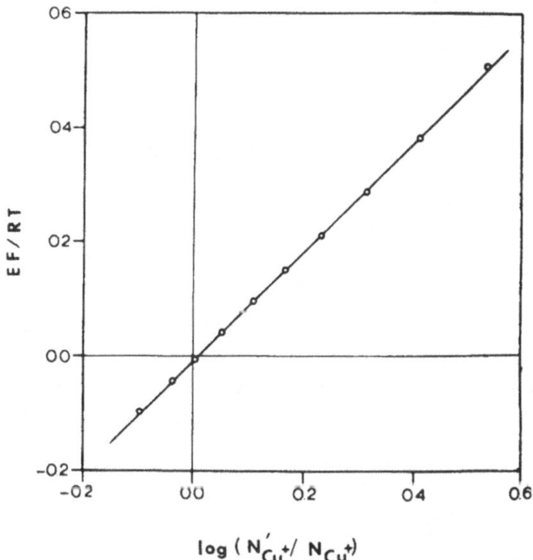

FIGURE 76. Nernst plot for concentration cell Cu|Cu^+(N), NaOH, KOH ¦¦ Cu^+(N'), NaOH, KOH|Cu (Ref. 536). (Reprinted by permission of the publisher.)

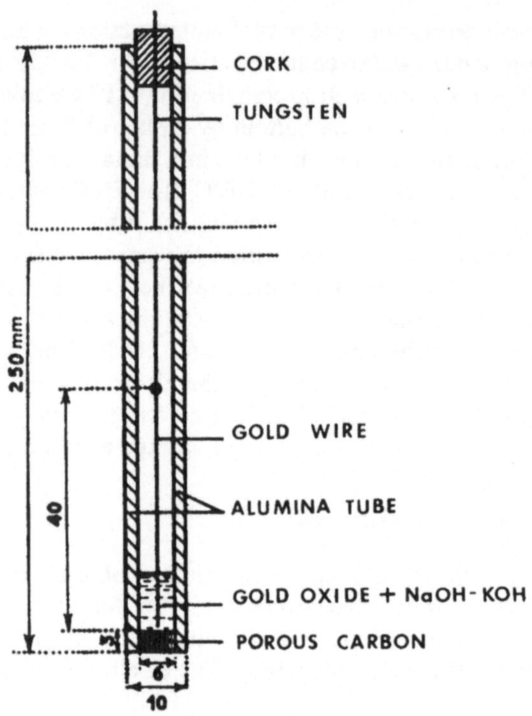

FIGURE 77. Au/Au^{3+} reference electrode (Ref. 537). (Reprinted by permission of the publisher.)

ity, low polarizability, long-term stability, and excellent reproducibility (±5 mV) at temperatures in the range of 200–500°C.[(537)] In order to use the electrode many times, it is essential to store it in the molten salt. The graphite plug may form carbonate when exposed to the air, and the pores of the graphite may remained plugged on reuse. The electrode is useful only in acidic melts,[(538)] and a constant H_2O concentration is necessary for potential stability. Also, diffusion of reducible species across the porous graphite may occur, causing potential drifts.[(539)] It has been found in one study that the electrode does not obey the Nernstian relationship with respect to Au^{3+}, and it has been suggested that the electrode is actually of the type $Au/Au_mO_n/O^{2-}$.

4.8.6. Other Reference Electrodes

The Ag/AgCl Reference Electrode. The low solubility of AgCl in molten nitrate was employed to build the reference electrode shown in Fig. 78.[(539)] The electrode consists of an Ag/AgCl electrode immersed in a solution of

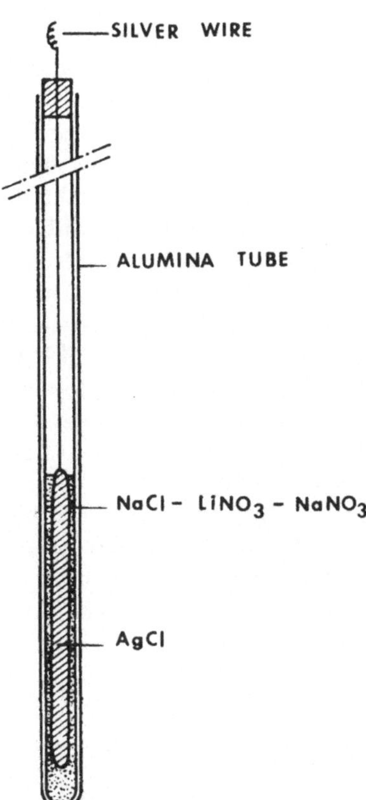

FIGURE 78. Ag/AgCl reference electrode for molten hydroxides (Ref. 539). (Reprinted by permission of the publisher.)

NaCl in $LiNO_3$-$NaNO_3$ eutectic (m.p. 198°C), all contained in a somewhat porous alumina tube. Permeability of this alumina diaphragm to molten hydroxide can be improved by abrasion or by dipping it in molten hydroxide at 1000°C.

The potential of the electrode is related to the activity of chloride in the nitrate melt through the relation

$$\varepsilon = \varepsilon^{0\prime} - \frac{RT}{F} \ln a_{\text{AgCl}} \tag{90}$$

with

$$\varepsilon^{0\prime} = \varepsilon^{0} + \frac{RT}{F} \ln K_{\text{sp(AgCl)}} \tag{91}$$

The above relation is verified for a chloride concentration greater than 5×10^{-3} mole fraction. The electrode has been reported to be reproducible (the emf between two identical electrodes returns to its equilibrium in one minute after the passage of 10 mA for 20 min), reversible, and stable (the variation of the potential between two electrodes is not more than 5 mV after 10 h). For new electrodes, a stabilization period of about 5 h is required to attain stable potential.

The H_2O, H_2 Reference Electrode. The H_2O, H_2 reference electrode is based on the diffusion of hydrogen in palladium.[540] The electrode is a closed palladium tube of 0.5-mm thickness that is immersed at a constant depth in hydroxide melt with circulation of hydrogen in the tube. Hydrogen diffuses through the metal, and the H_2/H_2O equilibrium is established at the metal/melt interface. The reversibility of the H_2/H_2O system has been demonstrated.[540–542] The stability of the potential of the electrode requires a constant concentration of H_2O in molten hydroxide, which is difficult and sometimes inconvenient to maintain.

The Liquid Metal/Metal Oxide Reference Electrode. Systems such as liquid metal/metal oxide-saturated hydroxide [e.g., Hg/HgO(saturated), Pb/PbO(saturated)][543,544] generally have stable equilibrium potentials in molten hydroxides and can be used as reference half-cells or electrodes. The Hg/HgO(saturated) system contained in a Teflon tube (with a porous Teflon plug) has been used as the reference electrode in NaOH–KOH eutectic at 227°C.[543] The potential of this electrode corresponds to the equilibrium

$$HgO_2^{2-} + 2H_2O + 2e^- \rightleftharpoons Hg + 4OH^- \tag{92}$$

The potential of the electrode depends on H_2O concentration, and this type of reference electrode is only useful when the H_2O content of the melt is carefully controlled.

Acknowledgments

We would like to thank the Chemical Technology Division of Argonne National Laboratory for its support, especially Dr. J. P. Ackerman for his encouragement, Dr. M. Blander for his valuable discussions and helpful suggestions, and Mr. J. Harmon and Mr. D. Hamrin for their editorial assistance with the manuscript.

References

1. R. W. Laity, Electrodes in fused salt systems, in: *Reference Electrodes, Theory and Practice* (D. J. G. Ives and G. J. Janz, eds.), Academic Press, New York (1961).
2. A. F. Alabyshev, M. F. Lantratov, and A. G. Morachevskii, *Reference Electrodes for Fused Salts*, Sigma Press, Washington, D.C. (1965).

3. F. Lantelme, D. Inman, and D. G. Lovering, Electrochemistry I, in: *Molten Salt Techniques*, Vol. 2 (D. G. Lovering and R. J. Gale, eds.), Plenum Press, New York (1984).
4. K. J. Vetter, *Electrochemical Kinetics*, Academic Press, New York (1967).
5. R. Parsons, *Pure Appl. Chem.* **37**, 499 (1974).
6. G. J. Janz and C. Dijkhuis, *Molten Salts*, Vol. 2; Section 1. Electrochemistry of molten salts: Gibbs free energies and excess free energies from equilibrium-type cells; Surface tension data, NSRDS-NBS 28, National Bureau of Standards, Washington, D.C. (1969).
7. JANAF Thermochemical Tables, NSRDS-NBS 37, National Bureau of Standards, Washington, D.C. (1971) and supplements: 1974 Supplement, *J. Phys. Chem. Ref. Data* **3**, 311 (1974); 1975 Supplement, *ibid.* **4**, 1 (1975); 1978 Supplement, *ibid.* **7**, 793 (1978); 1982 Supplement, *ibid.* **11**, 695 (1982); 1982 Supplement No. 2, The NBS Tables of Chemical Thermodynamic Properties, *J. Phys. Chem. Ref. Data* **11**, 1 (1982).
8. I. Barin and O. Knacke, *Thermochemical Properties of Inorganic Substances*, Springer Verlag, Berlin (1973), and I. Barin, O. Knacke, and O. Kubaschewski, Supplement (1977).
9. V. P. Glushko, L. V. Gurvich, G. A. Bergman, I. V. Veizh, V. A. Medvedjev, G. A. Hachkuruzov, and V. S. Jungman, *Termodinamicheskie svoistva individualnykh veshchestv*, Tom I (1978), Tom II (1979), Tom III (1981), Tom IV (1982), Akademia Nauk SSSR, Institut Visokih Temperatur, Izdatelstvo "Nauka," Moskva.
10. Y-T. Hsu, R. B. Escue, and T. H. Tidwell, *J. Electroanal. Chem. Interfacial Electrochem.* **15**, 245 (1967).
11. G. Mamantov and R. A. Osteryoung, in: *Characterization of Solutes in Nonaqueous Solvents* (G. Mamantov, ed.), Plenum Press, New York (1978).
12. B. Tremillon, *Pure Appl. Chem.* **25**, 395 (1971).
13. Editorial, *J. Electrochem. Soc.* **132**, 37C (1985).
14. M. Blander, Thermodynamic properties of molten salt solutions, in: *Molten Salt Chemistry* (M. Blander, ed.), Interscience, New York (1964).
15. J. A. Plambeck, *Fused Salt Systems*, Vol. 10 of *Encyclopedia of Electrochemistry of the Elements* (A. J. Bard, ed.), Marcel Dekker, New York (1976).
16. S. N. Flengas and T. R. Ingraham, *J. Electrochem. Soc.* **106**, 714 (1959).
17. W. H. Hamer, M. S. Malmberg, and B. J. Rubin, *J. Electrochem. Soc.* **103**, 8 (1956).
18. K. Grjotheim and G. Rosenblatt, in: *Selected Topics in High-Temperature Chemistry* (T. Førland *et al.*, eds.), Universitetsforlaget, Oslo (1966), p. 43.
19. C. Dijkhuis, R. Dijkhuis, and G. J. Janz, *Chem. Rev.* **68**, 253 (1968).
20. J. Newman, *Electrochemical Systems*, Prentice Hall, Englewood Cliffs, New Jersey (1973).
21. A. Klemm, Transport properties of molten salts, in: *Molten Salt Chemistry* (M. Blander, ed.), Interscience, New York (1964).
22. M. Okada and K. Kawamura, *Electrochim. Acta* **15**, 1 (1970).
23. W. K. Behl and J. J. Egan, *J. Phys. Chem.* **71**, 1764 (1967).
24. M. Temkin, *Acta Physicochim. URSS* **20**, 411 (1945).
25. S. N. Flengas and E. K. Rideal, *Proc. R. Soc. London* **233A**, 443 (1956).
26. S. N. Flengas and T. R. Ingraham, *Can. J. Chem.* **35**, 1139 (1957).
27. S. N. Flengas and T. R. Ingraham, *Can. J. Chem.* **35**, 1254 (1957).
28. H. A. Laitinen and C. H. Liu, *J. Am. Chem. Soc.* **80**, 1015 (1958).
29. J. Braunstein and R. M. Lindgren, *J. Am. Chem. Soc.* **84**, 1534 (1962).
30. H. L. Jindal, *Indian J. Chem.* **7**, 1135 (1969).
31. R. W. Laity, *J. Am. Chem. Soc.* **79**, 1849 (1957).
32. R. W. Laity and C. T. Moynihan, *J. Phys. Chem.* **67**, 723 (1963).
33. I. G. Murgulescu and D. I. Marchidan, *Russ. J. Phys. Chem.* **34**, 1196 (1960).
34. I. G. Murgulescu and S. Steinberg, *Discuss. Faraday Soc.* **32**, 107 (1961).
35. E. A. Kokh and M. V. Smirnov, in: *Electrochemistry of Molten and Solid Electrolytes* (A. N. Baraboshkin, ed.), Vol. 7, Consultants Bureau, New York (1969), p. 11.

36. E. A. Kokh and M. V. Smirnov, Ref. 35, p. 16.
37. E. A. Kokh and M. V. Smirnov, in: *Electrochemistry of Molten and Solid Electrolytes*, Vol. 8 (S. F. Pal'guev, ed.), Consultants Bureau, New York (1970), p. 1.
38. E. A. Kokh and M. V. Smirnov, Ref. 37, p. 4.
39. M. V. Smirnov and V. Ya. Kudyakov, *Electrochim. Acta* **28**, 1349 (1983).
40. G. J. Janz, *Molten Salts Handbook*, Academic Press, New York (1967).
41. G. J. Janz, F. W. Dampier, G. R. Lakshminarayan, P. K. Lorenz, and R. P. T. Tomkins, *Molten Salts*, Vol. 1, *Electrical Conductance, Density and Viscosity Data*, NSRDS-NBS 15, National Bureau of Standards, Washington, D.C. (1968).
42. J. O'M. Bockris, G. J. Hills, D. Inman, and L. Young, *J. Sci. Instrum.* **33**, 438 (1956).
43. K. H. Stern and J. A. Stiff, *J. Electrochem. Soc.* **111**, 893 (1964).
44. B. Lengyel and A. Sammt, *Z. Phys. Chem.* (*Leipzig*) **181A**, 55 (1937).
45. T. Suzuki, *Denki Kagaku* **39**, 380 (1971).
46. J. Hladik, M. Saunier, and G. Morand, *C. R. Acad. Sci. Paris Ser. C* **263**, 357 (1966).
47. G. Tamman, *Z. Anorg. Chem.* **133**, 267 (1924).
48. L. Yang and R. G. Hudson, *J. Electrochem. Soc.* **106**, 986 (1959).
49. K. Horowitz, *Z. Phys.* **15**, 369 (1923).
50. F. Conti and G. Eisenman, *Biophys. J.* **5**, 247 (1965).
51. K. Notz and A. G. Keenan, *J. Phys. Chem.* **70**, 662 (1966).
52. R. H. Doremus, *J. Phys. Chem.* **68**, 2212 (1964).
53. R. H. Doremus, *J. Electrochem. Soc.* **115**, 924 (1968).
54. T. J. van Reenen, M. van Niekerk, and W. J. de Wet, *J. Phys. Chem.* **75**, 2815 (1971).
55. K. H. Stern, *J. Phys. Chem.* **74**, 1329 (1970).
56. H. M. Garfinkel, *J. Phys. Chem.* **73**, 1766 (1969).
57. K. H. Stern, *J. Electrochem. Soc.* **114**, 1257 (1967).
58. K. Grjotheim, K. H. Stern, and L. U. Thulin, *Electrochim. Acta* **15**, 1135 (1970).
59. L. U. Thulin, *Electrochim. Acta* **15**, 1139 (1970).
60. K. Hughes and J. O. Isard, Ionic transport in glasses, in: *Physics of Electrolytes*, Vol. 1 (J. Hladik, ed.), Academic Press, London (1972).
61. K. H. Stern, *J. Phys. Chem.* **74**, 1323 (1970).
62. F. Toussaint and J. M. Jonville, *Silic. Ind.* **40**, 367 (1975).
63. K. H. Stern, *Chem. Rev.* **66**, 355 (1966).
64. H. A. Laitinen and B. B. Bhatia, *J. Electrochem. Soc.* **107**, 705 (1960).
65. L. Redey and D. R. Vissers, *J. Electrochem. Soc.* **130**, 231 (1983).
66. M. Bonnemay and R. Pineaux, *C. R. Acad. Sci. Paris* **240**, 1774 (1955).
67. F. R. Duke and H. M. Garfinkel, *J. Phys. Chem.* **65**, 1627 (1961).
68. D. Inman, *Electrochim. Acta* **10**, 11 (1965).
69. D. J. G. Ives and G. J. Janz, *Reference Electrodes, Theory and Practice*, Academic Press, New York (1961).
70. J. N. Agar and F. P. Bowden, *Proc. R. Soc. London* **169A**, 206 (1939).
71. G. B. Shilina and Yu. K. Delimarskii, *Ukr. Khim. Zh.* **30**, 1045 (1964).
72. A. D. Graves and D. Inman, *Nature* **208**, 481 (1965).
73. G. Mamantov and D. L. Manning, *Anal. Chem.* **38**, 1494 (1966).
74. Yu. K. Delimarskii, *Usp. Khim.* **23**, 766 (1954).
75. E. Black and T. DeVries, *Anal. Chem.* **27**, 906 (1955).
76. Yu. K. Delimarskii and I. D. Panchenko, *Ukr. Khim. Zh.* **19**, 47 (1953).
77. R. Narayan and D. Inman, *J. Polarogr. Soc.* **11**, 27 (1965).
78. H. A. Laitinen, C. H. Liu, and W. S. Ferguson, *Anal. Chem.* **30**, 1266 (1958).
79. L. N. Antipin and N. G. Tyurin, *Zh. Fiz. Khim.* **32**, 640 (1958).
80. H. Bloom, *The Chemistry of Molten Salts*, Benjamin, New York (1967).
81. H. Bloom and J. O'M. Bockris, Structural aspects of ionic liquids, in: *Fused Salts* (B. S. Sundheim, ed.), McGraw-Hill, New York (1964).

82. G. N. Papatheodorou, Structure and thermodynamics of molten salts, in: *Comprehensive Treatise of Electrochemistry*, Vol. 5 (B. E. Conway *et al.*, eds.), Plenum Press, New York (1983).
83. M. Blander, M.-L. Saboungi, and A. Rahman, Extended Abstracts, 168th Electrochem. Soc. Meeting, Las Vegas, Vol. 85-2, The Electrochem. Soc., Pennington, New Jersey (1985), p. 701.
84. D. Inman and D. G. Lovering, Electrochemistry in molten salts, in: *Comprehensive Treatise of Electrochemistry*, Vol. 7 (B. E. Conway *et al.*, eds.), Plenum Press, New York (1983).
85. M. A. Bredig, Mixtures of metals with molten salts, in: *Molten Salt Chemistry* (M. Blander, ed.), Interscience, New York (1964).
86. J. D. Corbett, The solution of metals in their molten salts, in: *Fused Salts* (B. S. Sundheim, ed.), McGraw-Hill, New York (1964).
87. N. H. March and M. P. Tosi, Liquid metals and their mixtures with molten salts, in: *Coulombic Liquids*, Academic Press, London (1984), p. 38.
88. E. A. Ukshe and N. G. Bukun, *Russ. Chem. Rev.* **30**, 90 (1961).
89. W. W. Warren Jr., Electronic properties of solutions of liquid metals and ionic melts, in: *Advances in Molten Salt Chemistry*, Vol. 4 (G. Mamantov *et al.*, eds.), Plenum Press, New York (1981).
90. W. W. Warren Jr., Extended Abstracts, 168th Electrochem. Soc. Meeting, Las Vegas, Vol. 85-2, The Electrochem. Soc., Pennington, New Jersey (1985), p. 688.
91. H. Bloom and J. O'M. Bockris, Structural aspects of ionic liquids, in *Fused Salts* (B. S. Sundheim, ed.), McGraw-Hill, New York (1964).
92. Yu. K. Delimarskii and B. F. Markov, *Electrochemistry of Fused Salts*, Sigma, Washington, D.C. (1961), p. 201.
93. G. J. Janz, *Molten Salts Handbook*, Academic Press, New York (1967), p. 109.
94. R. A. Sharma and R. N. Seefurth, *J. Electrochem. Soc.* **123**, 1763 (1976).
95. H. R. Bronstein and M. A. Bredig, *J. Phys. Chem.* **65**, 1220 (1961).
96. J. L. Settle and Z. Nagy, *J. Electrochem. Soc.* **132**, 1619 (1985).
97. *Phase Diagrams for Ceramists* (E. M. Levin *et al.*, eds.), American Ceramists Society, Columbus, Ohio (1964, 1965, 1975).
98. U. Gesenhues and H. Wendt, in *Proc. of 4th International Symp. on Molten Salts*, Vol. 84-2 (M. Blander *et al.*, eds.), The Electrochem. Soc., Pennington, New Jersey (1984), p. 74.
99. H. H. Emons and W. Holbeck, in *Proc. of 4th International Symp. on Molten Salts*, Vol. 84-2 (M. Blander *et al.*, eds.), The Electrochem. Soc., Pennington, New Jersey (1984), p. 411.
100. K. H. Stern and E. L. Weise, *High Temperature Properties and Decomposition of Inorganic Salts*, National Standard Reference Data Series, National Bureau of Standards, Washington, D.C. Part 1. Sulfates, NSRDS-NBS 7 (1966). Part 2. Carbonates. NSRDS-NBS 30 (1969).
101. K. H. Stern, *High Temperature Properties and Decomposition of Inorganic Salts*, National Standard Reference Data Series, Part 3. Nitrates and Nitrites, *J. Phys. Chem. Ref. Data* **1**, 747 (1972).
102. *Molten Salt Techniques*, Vol. 1 (D. G. Lovering and R. J. Gale, eds.), Plenum Press, New York (1983), p. 5.
103. C. L. Hussey, Room temperature molten salt systems, in: *Advances in Molten Salt Chemistry*, Vol. 5 (G. Mamantov, ed.), Elsevier, Amsterdam (1983).
104. H. A. Laitinen and C. H. Liu, *J. Am. Chem. Soc.* **80**, 1015 (1958).
105. *Physics of Electrolytes*, Vol. 2, *Thermodynamics and Electrode Processes in Solid State Electrolytes* (J. Hladik, ed.), Academic Press, London (1972).
106. W. H. Hamer, M. S. Malmberg, and B. J. Rubin, *J. Electrochem. Soc.* **112**, 750 (1965).

107. D. G. Lovering and R. M. Oblath, Water in molten salts: Industrial and electrochemical consequences, in: *Ionic Liquids* (D. Inman and D. G. Lovering, eds.), Plenum Press, New York (1981).
108. S. H. White, The role of water during the purification of high-temperature ionic salts, Ref. 107.
109. R. Combes, The solution chemistry of water in melts, Ref. 107.
110. S. H. White and U. M. Twardoch, in *Proc. of the Fourth International Symp. on Molten Salts*, Vol. 84-2 (M. Blander *et al.*, eds.), The Electrochem. Soc., Pennington, New Jersey (1984) p. 392.
111. A. A. El Hosary, D. H. Kerridge, and A. M. Shams El Din, Oxide species in molten salts, in: *Ionic Liquids* (D. Inman and D. G. Lovering, eds.), Plenum Press, New York (1981).
112. Anderson Physics Laboratory, 406 Busey Ave., Urbana, Illinois 61801.
113. *Ionic Liquids* (D. Inman and D. G. Lovering, eds.), Plenum Press, New York (1981), p. 187.
114. G. J. Janz, Physical Properties Data Compilations Relevant to Energy Storage, Molten Salts, U.S. Department of Commerce, National Bureau of Standards, Washington, D.C., NSRDS-NBS 61. G. J. Janz, C. B. Allen, J. R. Donney, and R. P. T. Tomkins, Part I. (1978); G. J. Janz, C. B. Allen, N. P. Bansal, R. M. Murphy, and R. P. T. Tomkins, Part II. (1979); G. R. Miller and D. G. Paquette, Part III. (1979); G. J. Janz and R. P. T. Tomkins, Part IV. (1981).
115. J. G. Gibson and J. L. Sudworth, *Specific Energies of Galvanic Reactions*, Chapman and Hall, London (1973).
116. G. J. Janz, C. B. Allen, J. R. Downey Jr., and R. P. T. Tomkins, *Eutectic Data, Safety, Hazard, Corrosion, Melting Points, Compositions and Bibliography*, Vol. 1-2, Molten Salts Data Center, Rensselaer Polytechnic Institute, Troy, New York (1976).
117. G. J. Janz, *High Temp. Sci.* **19,** 173 (1985).
118. *Instrumental Methods in Electrochemistry*, Southampton Electrochemistry Group, R. Greef *et al.*, Ellis Horwood Ltd., Chichester (1985), p. 368.
119. R. Piontelli, G. Bianchi, and R. Aletti, *Z. Elektrochem.* **56,** 86 (1952).
120. R. Piontelli, G. Bianchi, U. Bertocci, C. Guerci, and B. Rivolta, *Z. Elektrochem.* **58,** 54 (1954).
121. R. Piontelli, B. Rivolta, and G. Montanelli, *Z. Elektrochem.* **59,** 64 (1955).
122. K. Tokuda, T. Gueshi, K. Aoki, and M. Matsuda, *J. Electrochem. Soc.* **132,** 2390 (1985).
123. Development of Molten Carbonate Fuel Cells for Power Generation, Final Report, SRD-80-055, General Electric Co., Schenectady, New York (April 1980), p. 5-3.
124. L. Redey, D. R. Vissers, J. Newman, and S. Higuchi, in *Proc. of the Symp. on Porous Electrodes: Theory and Practice*, Vol. 84-8 (H. C. Maru *et al.*, eds.), The Electrochem. Soc., Pennington, New Jersey (1984), p. 322.
125. W. E. Haupin, *J. Met.* (46) (October 1971).
126. L. Redey, unpublished data.
127. H. Taniguchi and G. J. Janz, *J. Electrochem. Soc.* **104,** 123 (1957).
128. J. Plambeck, *J. Chem. Eng. Data* **12,** 77 (1967).
129. G. A. Sacchetto, G. A. Mazzocchin, and G. G. Bombi, *J. Electroanal. Chem.* **20,** 435 (1969).
130. Z. Tomczuk, L. Redey, and D. R. Vissers, *J. Electrochem. Soc.* **130,** 1074 (1983).
131. I. Epelboin, C. Gabrielli, and M. Keddam, Analysis of the electrochemical noise, Ref. 165, p. 160.
132. A. Bezegh and J. Janata, Extended Abstracts, 168th Electrochem. Soc. Meeting, Las Vegas, Vol. 85-2, The Electrochem. Soc., Pennington, New Jersey (1985), p. 772.
133. G. Blanc, I. Epelboin, C. Gabrielli, and M. Keddam, *J. Electroanal. Chem. Interfacial Electrochem.* **75,** 97 (1977).

134. K. Tachibana and G. Okamoto, Rev. on Coatings and Corrosion, Sci. Publ. Division, Freund Publ. House, Tel Aviv, Vol. 4, p. 229 (1981).
135. A. U. Seybolt and J. E. Burke, *Experimental Metallurgy*, Chap. 4, Wiley, New York (1953).
136. D. T. Livey and P. Murray, The stability of refractory materials, in: *Physicochemical Measurements at High Temperatures* (J. O'M. Bockris *et al.*, eds.), Academic Press, New York (1959).
137. R. A. Sharma and T. G. Bradley, *J. Electrochem. Soc.* **128**, 1835 (1981).
138. Atomergic Chemetals Co., 584 Mineola Ave., Carle Place, Long Island, New York 11514. Coors Porcelain Co., 600 Ninth Street, Golden, Colorado 80401. McDanel Refractory Porcelain Co., 510 Ninth Ave., Beaver Falls, Pennsylvania 15010. National Beryllia Corp., Haskell, New Jersey 07420. 3M Co., Technical Ceramic Products Division, Chattanooga, Tennessee 37405. Union Carbide Corp., Carbon Products Division, 120 S. Riverside Plaza, Chicago, Illinois 60606, BN and graphite. AIRCO Carbon Co., P.O. Box 387, St. Marys, Pennsylvania 15857. Pure Carbon Co., 441 Hall Ave., St. Marys, Pennsylvania 15857.
139. A. U. Seybolt and J. E. Burke, *Procedures in Experimental Metallurgy*, Wiley, New York (1953).
140. I. E. Campbell, *High Temperature Technology*, Wiley, New York (1956).
141. J. E. Battles, Materials for high-temperature Li-Al/FeS_x secondary batteries, in: *Critical Materials Problems in Energy Production*, Academic Press, New York (1976).
142. J. D. Mackenzie, Container materials for melts, Appendix 1 in: *Physicochemical Measurements at High Temperatures* (J. O'M. Bockris *et al.*, eds.), Academic Press, New York (1959).
143. *Molten Salt Techniques*, Vol. 2 (D. G. Lovering and R. J. Gale, eds.), Plenum Press, New York (1984), p. 186.
144. G. J. Janz, *Molten Salts Handbook*, Academic Press, New York (1967), p. 388.
145. H. V. Venkatasetty and D. J. Saathoff, in *Proc. of the International Symp. on Molten Salts*, Vol. 76-4 (J. P. Pemsler *et al.*, eds.), The Electrochem. Soc., Pennington, New Jersey (1976), p. 575.
146. G. J. Janz and R. P. T. Tomkins, *Corrosion—NACE* **35**, 485 (1979).
147. A. A. Bauer and J. L. Bates, Batelle Mem. Inst. Report 1930 (July 1974).
148. W. D. Kingery, H. K. Bowen, and D. R. Uhlmann, *Introduction to Ceramics*, Chap. 7, Electrical Conductivity, Wiley, New York (1976).
149. K. Y. Kim and O. F. Devereux, in *Proc. of 2nd International Symp. on Molten Salts*, Vol. 81-10 (J. Braunstein and J. R. Selman, eds.), The Electrochem. Soc., Pennington, New Jersey (1981), p. 145.
150. BC 1200 High-Performance Potentiostat Manual, Stonehart Associates, 17 Cottage Rd., Madison, Connecticut 06443.
151. M. C. H. McKubre and D. D. MacDonald, Electronic Instrumentation for Electrochemical Studies, in: *Comprehensive Treatise of Electrochemistry*, Vol. 8 (R. E. White *et al.*, eds.), Plenum Press, New York (1984).
152. *Instrumental Methods in Electrochemistry*, Southampton Electrochemistry Group, R. Greef *et al.*, Ellis Horwood Ltd., Chichester (1985), p. 370.
153. C. C. Herrmann, G. G. Perrault, and A. A. Pilla, *Anal. Chem.* **40**, 1173 (1968).
154. K. Doblehofer and A. A. Pilla, *J. Electroanal. Chem. Interfacial Electrochem.* **68**, 147 (1976).
155. B. D. Cahan, Z. Nagy, and M. A. Gershaw, *J. Electrochem. Soc.* **119**, 64 (1972).
156. J. E. Harrai and C. L. Pomernacki, *Anal. Chem.* **45**, 57 (1973).
157. U. Bertocci, *J. Electrochem. Soc.* **127**, 1931 (1980).
158. R. Littlewood, *Electrochim. Acta* **3**, 270 (1961).
159. Signal Conditioning, Applications Booklet No. 101, Gould Inc., Instrument Systems Division, 3631 Perkins Ave., Cleveland, Ohio 44114.

160. R. Morrison, *Grounding and Shielding Techniques in Instrumentation*, Wiley, New York (1977).
161. *Electrochemistry: Calculations, Simulation, and Instrumentation* (J. S. Mattson *et al.*, eds.), Marcel Dekker, New York (1972).
162. H. V. Malmstadt, *Electronics and Instrumentation for Scientists*, Benjamin/Cummings Publishers, Menlo Park, California (1981).
163. T. H. Ridgway and H. B. Mark, Computerization in electroanalytical chemistry, in: *Comprehensive Treatise of Electrochemistry*, Vol. 8 (R. E. White *et al.*, eds.), Plenum Press, New York (1984).
164. *Instrumental Methods in Electrochemistry*, Southampton Electrochemistry Group, R. Greef *et al.*, Ellis Horwood Ltd., Chichester (1985).
165. *Comprehensive Treatise of Electrochemistry*, Vol. 8, *Experimental Methods in Electrochemistry* (R. E. White *et al.*, eds.), Plenum Press, New York (1984).
166. *Comprehensive Treatise of Electrochemistry*, Vol. 9, *Electrodics: Experimental Techniques* (E. Yeager *et al.*, eds.), Plenum Press, New York (1984).
167. J. Greenberg and B. R. Sundheim, *J. Chem. Phys.* **29**, 1029 (1958).
168. W. D. Treadwell, A. Ammann, and Th. Zurrer, *Helv. Chim. Acta* **19**, 1255 (1936).
169. W. D. Treadwell and L. Terebesi, *Helv. Chim. Acta* **18**, 103 (1935).
170. H. A. Laitinen and J. W. Pankey, *J. Am. Chem. Soc.* **81**, 1053 (1959).
171. S. Senderoff and G. W. Mellors, *Rev. Sci. Instrum.* **29**, 151 (1958).
172. R. Lorenz and M. Fox, *Z. Phys. Chem.* **63**, 109 (1908).
173. R. Lorenz and H. Velde, *Z. Anorg. Allg. Chem.* **183**, 81 (1921).
174. J. H. Hildebrand and G. C. Rule, *J. Am. Chem. Soc.* **49**, 422 (1927).
175. A. Wachter and J. H. Hildebrand, *J. Am. Chem. Soc.* **52**, 4655 (1930).
176. E. J. Salstrom and J. H. Hildebrand, *J. Am. Chem. Soc.* **52**, 4641, 4650 (1930).
177. S. Smirnov, L. Palguyev, and L. E. Ivanovski, *Zh. Fiz. Khim.* **29**, 772 (1955).
178. J-P. Millet, H. Pham, G. Pourcelly, and M. Rolin, *J. Appl. Electrochem.* **12**, 575 (1982).
179. K. H. Stern, *J. Phys. Chem.* **60**, 679 (1956).
180. M. F. Lantratov and A. F. Alabyshev, *Zh. Prikl. Khim.* **26**, 263, 353 (1953); **27**, 722 (1954).
181. K. Hagemark and D. Hengstenberg, *J. Chem. Eng. Data* **11**, 596 (1966).
182. P. Drossbach, *J. Electrochem. Soc.* **103**, 700 (1956).
183. D. L. Maricle and D. N. Hume, *J. Electrochem. Soc.* **107**, 354 (1960).
184. A. Yazawa and M. Kameda, *Can. Metall. Q.* **6**, 263 (1967).
185. B. P. Artamonov, *Tr. Gos. Issled. Proektn. Inst. Khim.* **33**, 39 (1940).
186. G. Grube and E. A. Rau, *Z. Elektrochem.* **40**, 352 (1934).
187. J. D. Corbett and S. Winbuch, *J. Am. Chem. Soc.* **77**, 3964 (1955).
188. J. A. Plambeck, in *Proc. of 4th International Symp. on Molten Salts*, Vol. 84-2 (M. Blander, ed.), The Electrochem. Soc., Pennington, New Jersey (1984), p. 707.
189. M. Takahashi, *Denki Kagaku* **25**, 432 (1957).
190. S. Senderoff and A. Brenner, *J. Electrochem. Soc.* **101**, 31 (1954).
191. B. Eichler, *Z. Phys. Chem.* (*Leipzig*) **249**, 195 (1972).
192. J. Guion, M. Blander, D. Hengstenberg, and K. Hagemark, *J. Phys. Chem.* **72**, 2086 (1968).
193. S. M. Selis and L. P. McGinnis, *J. Electrochem. Soc.* **108**, 191 (1961).
194. R. J. Labrie and V. A. Lamb, *J. Electrochem. Soc.* **106**, 895 (1959).
195. D. Inman and M. J. Weaver, *J. Electroanal. Chem. Interfacial Electrochem.* **51**, 45 (1974).
196. A. DeHaan and H. Van der Poorten, *C. R. Acad. Sci. Ser. C* **261**, 5462 (1965).
197. M. Takahashi, Y. Katsuyama, and Y. Kanzaki, *J. Electroanal. Chem. Interfacial Electrochem.* **62**, 363 (1975).
198. D. R. Vissers, Z. Tomczuk, L. Redey, and J. Battles, High-temperature lithium-alloy/iron sulfide batteries, in: *Lithium: Current Applications in Science, Medicine, and Technology* (O. Bach, ed.), Wiley, New York (1985), p. 121.

199. K. M. Myles, F. C. Mrazek, J. A. Smaga, and J. L. Settle, in *Proc. of Symp. and Workshop on Adv. Battery R&D*, March 1976, U.S. ERDA Report No. ANL-76-8, p. B-50.
200. C. J. Wen, W. Weppner, B. A. Boukamp, and R. A. Huggins, *Metall. Trans.* **11B**, 131 (1980).
201. C. J. Wen, B. A. Boukamp, R. A. Huggins, and W. Weppner, *J. Electrochem. Soc.* **126**, 2258 (1979).
202. W. Weppner and R. A. Huggins, *J. Electrochem. Soc.* **125**, 7 (1978).
203. N-P. Yao, L. A. Heredy, and R. C. Saunders, *J. Electrochem. Soc.* **118**, 1039 (1971).
204. R. A. Sharma and R. N. Seefurth, *J. Electrochem. Soc.* **23**, 1763 (1976).
205. R. N. Seefurth and R. A. Sharma, *J. Electrochem. Soc.* **122**, 1049 (1975).
206. G. J. Reynolds, M. C. Y. Lee, and R. A. Huggins, in *Proc. of 4th International Symp. on Molten Salts*, Vol. 84-2 (M. Blander *et al.*, eds.), The Electrochem. Soc., Pennington, New Jersey (1984), p. 519.
207. J. R. Selman, D. K. DeNuccio, C. Sy, and R. K. Steunenberg, *J. Electrochem. Soc.* **124**, 1160 (1977).
208. L. Redey and D. R. Vissers, *J. Electrochem. Soc.* **128**, 2703 (1981).
209. C. A. Melendres, *J. Electrochem. Soc.* **124**, 650 (1977).
210. C. H. Liu, A. J. Zielen, and D. M. Gruen, *J. Electrochem. Soc.* **120**, 67 (1973).
211. L. Redey, to be published.
212. L. Redey and D. R. Vissers, Extended Abstracts, Meeting of The Electrochem. Soc., Denver, Colorado, Vol. 81-2, The Electrochem. Soc., Pennington, New Jersey (1981), p. 1412.
213. D. Warin, Z. Tomczuk, and D. R. Vissers, *J. Electrochem. Soc.* **130**, 64 (1983).
214. Z. Tomczuk, D. R. Vissers, and M-L. Saboungi, in *Proc. of 4th International Symp. on Molten Salts*, Vol. 84-2 (M. Blander *et al.*, eds.), The Electrochem. Soc., Pennington, New Jersey (1984), p. 352.
215. A. F. Alabyshev, M. F. Lantratov, and A. G. Morachevskii, *Usp. Khim.* **27**, 921 (1958).
216. A. F. Alabyshev and A. G. Morachevskii, *Zh. Neorg. Khim.* **2**, 669 (1957).
217. F. Halla and R. Herdy, *Z. Elektrochem.* **56**, 213 (1952).
218. Yu. K. Delimarskii and A. A. Kolotii, *Zh. Fiz. Khim.* **28**, 1169 (1954).
219. Yu. K. Delimarskii and A. A. Kolotii, *Ukr. Khim. Zh.* **14**, 124 (1948).
220. Yu. K. Delimarskii and R. S. Khaimovich, *Ukr. Khim. Zh.* **15**, 77, 360 (1969).
221. Yu. K. Delimarskii and A. A. Kolotii, *Ukr. Khim. Zh.* **16**, 1, 438, 599 (1950).
222. Yu. K. Delimarskii and A. A. Kolotii, *Zh. Fiz. Khim.* **28**, 1169 (1956).
223. A. A. Kolotii, Yu. K. Delimarskii, and V. F. Grishchenko, Physical Chemistry of Molten Salts and Slags, Academy of Sciences USSR, The Ural Branch, Institute of Electrochemistry (1960). English translation AEC-tr-5948.
224. Yu. K. Delimarskii and A. A. Kolotii, *Ukr. Khim. Zh.* **24**, 146 (1958).
225. A. F. Alabyshev, M. F. Lantratov, and A. G. Morachevskii, *Reference Electrodes for Fused Salts*, Sigma Press, Washington D.C. (1965), p. 118.
226. M. Takahashi, *Denki Kagaku* **25**, 481 (1957).
227. Yu. K. Delimarskii and Yu. G. Roms, *Ukr. Khim. Zh.* **29**, 781 (1963).
228. L. S. Leonova, Yu. M. Ryabukhin, and E. A. Ukshe, *Sov. Electrochem.* **4**, 1126 (1968).
229. R. Combes, J. Vedel, and B. Tremillon, *Anal. Lett.* **3**, 523 (1970).
230. R. Combes, J. Vedel, and B. Tremillon, *Electrochim. Acta* **20**, 191 (1975).
231. N. S. Wrench and D. Inman, *J. Electroanal. Chem. Interfacial Electrochem.* **17**, 319 (1968).
232. Y. Kanzaki and M. Takahashi, *J. Electroanal. Chem. Interfacial Electrochem.* **58**, 339, 349 (1975).
233. R. Littlewood and E. J. Argent, *Electrochim. Acta* **4**, 114 (1961).
234. Yu. K. Delimarskii, L. S. Berenblum, and I. N. Sheiko, *Zh. Fiz. Khim.* **25**, 398 (1951).
235. Yu. K. Delimarskii and R. S. Khaimovich, *Ukr. Khim. Zh.* **15**, 340 (1949).
236. Yu. K. Delimarskii and A. A. Kolotii, *Zh. Fiz. Khim.* **23**, 339 (1949).

237. Yu. K. Delimarskii, *Ukr. Khim. Zh.* **16,** 414 (1950).
238. M. F. Lantratov and T. N. Shevlyakova, *Zh. Neorg. Khim.* **4,** 1153 (1959), **6,** 193 (1961).
239. E. J. Salstrom, *J. Am. Chem. Soc.* **58,** 1848 (1936).
240. K. Frøyland, T. Førland, N. H. Lundberg, and T. Østvold, in: *Selectred Topics in High-Temperature Chemistry* (T. Førland *et al.*, eds.), Universitetsforlaget, Oslo (1966), p. 27.
241. N. P. Kulik, M. V. Smirnov, and V. P. Stepanov, *Elektrokhimiya* **21,** 1681 (1985).
242. Yu. K. Delimarskii, E. M. Skobets, and V. D. Ryabokon, *Zh. Fiz. Khim.* **21,** 843 (1947).
243. T. Førland and T. Østvold, in: *Selected Topics in High-Temperature Chemistry* (T. Førland *et al.*, eds.), Universitetsforlaget, Oslo (1966), p. 36.
244. N. G. Chovnyk, *Dokl. Akad. Nauk SSSR* **100,** 495 (1955).
245. A. A. Kolotii, *Zh. Fiz. Khim.* **30,** 508 (1956).
246. S. Karpachev and O. Poltoratskaya, *Zh. Fiz. Khim.* **6,** 967 (1935).
247. K. Grjotheim, *Z. Phys. Chem. N. F.* **11,** 150 (1957).
248. R. J. Gale and R. A. Osteryoung, Haloaluminates, in: *Molten Salt Techniques,* Vol. 1 (D. G. Lovering and R. Gale, eds.), Plenum Press, New York (1983), p. 55.
249. H. L. Chum and R. A. Osteryoung, Chemical and electrochemical studies in room temperature aluminum-halide-containing melts, in: *Ionic Liquids* (D. Inman and D. G. Lovering, eds.), Plenum Press, New York (1981).
250. H. Linga, Z. Stojek, and R. A. Osteryoung, *J. Am. Chem. Soc.* **103,** 3754 (1981).
251. L. G. Boxall, H. L. Jones, and R. A. Osteryoung, *J. Electrochem. Soc.* **120,** 223 (1973).
252. C. L. Hussey, Room temperature molten salt systems, in: *Advances in Molten Salt Chemistry,* Vol. 5 (G. Mamantov, ed.), Elsevier, Amsterdam (1983), p. 197.
253. R. J. Gale and R. A. Osteryoung, *Inorg. Chem.* **18,** 1603 (1979).
254. L. G. Boxall, H. L. Jones, and R. A. Osteryoung, *J. Electrochem. Soc.* **121,** 212 (1974).
255. A. Bjøgrum, A. Sterten, V. B. Sørensen, J. Thonstad, and R. Tunold, *Electrochim. Acta* **26,** 487 (1981).
256. C. L. Hussey and T. B. Scheffler, in *Proc. of 4th International Symp. on Molten Salts,* Vol. 84-2 (M. Blander *et al.*, eds.), The Electrochem. Soc., Pennington, New Jersey (1984), p. 133.
257. G. T. Cheek and R. B. Herzog, Ref. 256, p. 163.
258. H. Wendt and K. Reuhl, Ref. 256, p. 418.
259. L. Redey, I. Molnar, and I. Porubszky, in *Power Sources 5* (D. H. Collins, ed.), Academic Press, London (1975), p. 559.
260. T. M. Laher and C. L. Hussey, *Inorg. Chem.* **21,** 4079 (1982).
261. J. S. Wilkes, J. A. Levisky, R. A. Wilson, and C. L. Hussey, *Inorg. Chem.* **21,** 1263 (1982).
262. M. Blander and M-L. Saboungi, in *Proc. of 3rd International Symp. Molten Salts,* Vol. 81-9 (G. Mamantov *et al.*, eds.), The Electrochem. Soc., Pennington, New Jersey (1981), p. 212.
263. R. Wehrman and L. F. Yntema, *J. Phys. Chem.* **48,** 259 (1944).
264. R. G. Verdieck and L. F. Yntema, *J. Phys. Chem.* **46,** 344 (1942).
265. R. G. Verdieck and L. F. Yntema, *J. Phys. Chem.* **48,** 268 (1944).
266. S. Senderoff, G. W. Mellors, and W. J. Reinhart, *J. Electrochem. Soc.* **112,** 840 (1965).
267. H. W. Jenkins, G. Mamantov, and D. L. Manning, *J. Electroanal. Chem. Interfacial Electrochem.* **19,** 385 (1968).
268. H. W. Jenkins, G. Mamantov, and D. L. Manning, *J. Electrochem. Soc.* **117,** 183 (1970).
269. H. W. Jenkins, G. Mamantov, D. L. Manning, and J. P. Young, *J. Electrochem. Soc.* **116,** 1712 (1969).
270. P. Chardard, Note CEA-N-2090, Centre d'Etudes Nucléaires de Fontenay-aux-Roses (1979).
271. T. Suzuki, *Denki Kagaku* **41,** 795 (1973).

272. P. Taxil and Zhiyu Qiao, *J. Chim. Phys.* **82**, 83 (1985).
273. Zhiyu Qiao and P. Taxil, *J. Appl. Electrochem.* **15**, 259 (1985).
274. R. Winand, *C. R. Acad. Sci. Paris Ser. C* **264**, 649 (1967).
275. R. Winand, *Electrochim. Acta* **17**, 251 (1972).
276. C. Decroly, A. Mukhtar, and R. Winand, *J. Electrochem. Soc.* **115**, 905 (1968).
277. H. R. Bronstein and D. L. Manning, *J. Electrochem. Soc.* **119**, 125 (1972).
278. C. E. Bamberger, in *Advances in Molten Salt Chemistry*, Vol. 3 (J. Braunstein, G. Mamantov, and G. P. Smith, eds.), Plenum Press, New York (1975), p. 177.
279. F. R. Clayton, G. Mamantov, and D. L. Manning, *High Temp. Sci.* **5**, 358 (1973).
280. F. R. Clayton, G. Mamantov, and D. L. Manning, *J. Electrochem. Soc.* **120**, 1193 (1973).
281. J. de Lepinay and P. Paillere, *Electrochim. Acta* **29**, 1243 (1984).
282. P. Paillere, thesis, Grenoble (1980).
283. M. Brigaudeau and J. F. Wagner, Journées d'Etudes sur les Sels Fondus, Liège, Belgium, June 6–8, 1979, CEA-CONF 4732.
284. H. Coriou, J. Dirian, and J. Huré, *J. Chim. Phys.* **52**, 479 (1955).
285. B. F. Hitch and C. F. Baes, Jr., *J. Inorg. Nucl. Chem.* **34**, 163 (1972).
286. F. R. Clayton, G. Mamantov, and D. L. Manning, *J. Electrochem. Soc.* **120**, 1199 (1973).
287. D. L. Manning and G. Mamantov, *J. Electroanal. Chem. Interfacial Electrochem.* **7**, 102 (1964).
288. S. Pizzini and R. Morlotti, *Electrochim. Acta* **10**, 1033 (1965).
289. G. Dirian, K. A. Romberger, and C. F. Baes, Jr., USAEC Report, ORNL-3789, January (1965).
290. B. F. Hitch and C. F. Baes, Jr., *Inorg. Chem.* **8**, 201 (1969).
291. K. A. Romberger, J. Braunstein, and R. E. Thoma, *J. Phys. Chem.* **76**, 1154 (1972).
292. K. A. Romberger and J. Braunstein, *Inorg. Chem.* **9**, 1273 (1970).
293. D. L. Manning, J. M. Dale, and G. Mamantov, in *Polarography 1964* (G. J. Hills, ed.), MacMillan, London (1966), p. 1143.
294. J. S. Hammond and D. L. Manning, *High Temp. Sci.* **5**, 50 (1973).
295. A. Fontana and R. Winand, *Electrochim. Acta* **16**, 257 (1971).
296. R. Boen and J. Bouteillon, *J. Appl. Electrochem.* **13**, 277 (1983).
297. D. Elwell and G. M. Rao, *Electrochim. Acta* **27**, 673 (1982).
298. Y. Ito, T. Takenaka, K. Ema, and J. Oishi, *Denki Kagaku* **51**, 864 (1983).
299. N. Q. Minh and N. P. Yao, *J. Electrochem. Soc.* **130**, 1025 (1983).
300. D. L. Manning and G. Mamantov, *High Temp. Sci.* **8**, 219 (1976).
301. D. L. Manning and G. Mamantov, *J. Electrochem. Soc.* **124**, 480 (1977).
302. G. Mamantov, in *Molten Salts: Characterization and Analysis* (G. Mamantov, ed.), Marcel Dekker, New York (1969), p. 537.
303. U. Cohen, *J. Electrochem. Soc.* **128**, 731 (1981).
304. K. Grjotheim, C. Krohn, M. Malinovský, K. Matiašovský, and J. Thonstad, *Aluminium Electrolysis, the Chemistry of the Hall–Héroult Process*, Aluminium-Verlag, Düsseldorf (1977).
305. P. Drossbach, *Z. Elektrochem.* **40**, 605 (1934).
306. Yu. V. Baimakov, V. P. Mashovets, and I. G. Kil, *Legk. Met.*, No. 1, 22 (1937).
307. A. I. Belyaev, *Legk. Met.*, No. 1, 7 (1938).
308. V. P. Mashovets and A. A. Revazyan, *J. Appl. Chem. USSR* **30**, 1069 (1957).
309. M. M. Vetyukov and R. G. Chuvilyaev, *Izv. Vyssh. Ucheb. Zaved., Tsvetn. Metall.* **8**(2), 65 (1965).
310. M. Rey, *C. R. Acad. Sci. Paris* **260**, 5528 (1965).
311. A. Sternten, S. Haugen, and K. Hamberg, *Electrochim. Acta* **21**, 589 (1976).
312. A. Sternten and K. Hamberg, in *Light Metals 1976*, Vol. I (S. R. Leavitt, ed.), The Metallurgical Society of AIME, New York (1976), p. 203.

313. J. Thonstad, *Electrochim. Acta* **13,** 449 (1968).
314. V. P. Mashovets and A. A. Revazyan, in *Soviet Electrochemistry,* Vol. II, Consultants Bureau, New York (1961), p. 185.
315. J. Thonstad and E. Hove, *Can. J. Chem.* **42,** 1542 (1964).
316. N. E. Richards and B. J. Welch, in *Proceedings of the First Australian Conference on Electrochemistry,* Pergamon, London (1965), p. 901.
317. J. Thonstad, *Electrochim. Acta* **15,** 1569 (1970).
318. V. P. Mashovets and A. A. Revazyan, *J. Appl. Chem. USSR* **31,** 560 (1958).
319. H. T. Shiver, J. L. Dewey, and W. E. Campbell, French Patent No. 1,457,746 (1966).
320. J. Thonstad, F. Nordmo, and J. K. Rødseth, *Electrochim. Acta* **19,** 761 (1974).
321. J. Thonstad, F. Nordmo, and K. Vee, *Electrochim. Acta* **18,** 27 (1973).
322. P. Drossbach, *Z. Elektrochem.* **42,** 65 (1936).
323. N. E. Richards, cited in Ref. 315.
324. L. N. Antipin and A. N. Khudyakov, *J. Appl. Chem. USSR* **29,** 985 (1956).
325. S. I. Rempel and L. P. Khodak, *J. Appl. Chem. USSR* **26,** 857 (1953).
326. R. A. Lewis, French Patent No. 1,592,315 (1968).
327. S. I. Rempel, N. A. Anisheva, and L. P. Khodak, *Dokl. Akad. Nauk SSSR* **97,** 859 (1954).
328. S. I. Rempel and L. P. Khodak, *Dokl. Akad. Nauk SSSR* **75,** 833 (1950).
329. L. N. Antipin, *Dokl. Akad. Nauk SSSR* **99,** 1019 (1954).
330. L. N. Antipin and A. N. Khudyakov, *Dokl. Akad. Nauk SSSR* **100,** 93 (1955).
331. L. N. Antipin, *Zh. Fiz. Khim.* **30,** 1425 (1956).
332. L. N. Antipin, *Zh. Fiz. Khim.* **30,** 1767 (1956).
333. S. Karpachev, S. I. Rempel, and E. Yordan, *Zh. Fiz. Khim.* **23,** 422 (1949).
334. M. Rey, *Electrochim. Acta* **14,** 991 (1969).
335. R. Piontelli, *Ann. N.Y. Acad. Sci.* **79,** 1025 (1960).
336. R. Piontelli and G. Montanelli, *Alluminio* **22,** 672 (1953).
337. R. Piontelli, *Acad. Nat. Lincei, C. R. Sc. Fis. Mat. Nat.* **26,** 18 (1959).
338. J. P. Sarget, P. Homsi, V. Plichon, and J. Badoz-Lambling, *Electrochim. Acta* **20,** 819 (1975).
339. K. Yoshida and E. W. Dewing, *Metall. Trans.* **3,** 683 (1972).
340. J. Thonstad and S. Rolseth, *Electrochim. Acta* **23,** 223 (1978).
341. J. Thonstad and S. Rolseth, *Electrochim. Acta* **23,** 233 (1978).
342. J. Thonstad, *J. Appl. Electrochem.* **3,** 315 (1973).
343. R. Piontelli, B. Mazza, and P. Pedeferri, *Electrochim. Acta* **10,** 1117 (1965).
344. J. Thonstad, A. Solbu, and A. Larsen, *J. Appl. Electrochem.* **1,** 261 (1971).
345. A. Kerouanton and V. Plichon, *J. Electroanal. Chem. Interfacial Electrochem.* **79,** 337 (1977).
346. S. Zuca, C. Herdlicka, and M. Terzi, *Electrochim. Acta* **25,** 211 (1978).
347. P. Homsi, V. Plichon, and J. Badoz-Lambling, *J. Electroanal. Chem. Interfacial Electrochem.* **81,** 141 (1977).
348. J. Badoz-Lambling and J. P. Sarget, *C. R. Acad. Sci. Paris* **271,** 250 (1970).
349. V. P. Mashovets and R. V. Svoboda, *J. Appl. Chem. USSR* **32,** 2210 (1959).
350. P. Homsi, V. Plichon, and J. Badoz-Lambling, *J. Inorg. Nucl. Chem. Suppl.* 193 (1976).
351. E. W. Dewing and E. Th. van der Kouwe, *J. Electrochem. Soc.* **122,** 358 (1975).
352. M. M. Vetyukov and F. Akgva, *Sov. Electrochem.* **6,** 1806 (1970).
353. D. Dumas and J. Brenet, *Rev. Roum. Chim.* **14,** 1339 (1969).
354. D. Dumas and J. Brenet, *C. R. Acad. Sci. Paris Ser. C* **265,** 1395 (1967).
355. K. Grjotheim and B. J. Welch, *Aluminium Smelter Technology,* Aluminium-Verlag, Düsseldorf (1980).
356. J. Thonstad, *Electrochem. Technol.* **6,** 346 (1968).
357. R. Piontelli and G. Montanelli, *Alluminio* **25,** 79 (1956).
358. F. Lantelme, D. Damianacos, J. Chevalet, and M. Chemla, *Electrochim. Acta* **22,** 261 (1977).

359. M. Rolin and J. J. Gallay, *Bull Soc. Chim. Fr.* 2096 (1960).
360. M. Rolin and J. J. Gallay, *Electrochim. Acta* **7**, 153 (1962).
361. M. Rolin and A. Ducouret, *Bull. Soc. Chim. Fr.* 790 (1964).
362. M. Rolin and A. Ducouret, *Bull. Soc. Chim. Fr.* 794 (1964).
363. M. Rolin, *Rev. Int. Hautes Temp. Refract.* **9**, 333 (1972).
364. H. Pham, M. Rolin, and P. Revin, *Rev. Int. Hautes Temp. Refract.* **10**, 37 (1973).
365. M. Feinleib and B. Porter, *J. Electrochem. Soc.* **103**, 231 (1956).
366. J. C. Mitchell and C. S. Samis, *Trans. Metall. Soc. AIME* **245**, 1227 (1969).
367. H. Ginsberg and S. Wilkening, *Metall* **27**, 787 (1973).
368. M. Rolin and J. J. Gallay, *Bull. Soc. Chim. Fr.* 2093 (1960).
369. M. L. Kronenberg, *J. Electrochem. Soc.* **116**, 1160 (1969).
370. C. Kubík, K. Matiašovský, M. Malinovský, and J. Zeman, *Electrochim. Acta* **9**, 1521 (1964).
371. Yu. K. Delimarskii, N. A. Pavlenko, and N. V. Vlasyuk, *Sov. Prog. Chem.* **32**, 511 (1964).
372. N. Q. Minh and N. P. Yao, *J. Electrochem. Soc.* **131**, 2279 (1984).
373. A. Kerouanton and V. Plichon, *C. R. Acad. Sci. Paris. Ser C* **280**, 497 (1975).
374. J. Hinden, J. Augustynski, and R. Monnier, *Electrochim. Acta* **21**, 459 (1976).
375. G. Charlot, J. Badoz-Lambling, P. Homsi, V. Plichon, J. P. Sarget, and A. Kerouanton, *J. Electroanal. Chem. Interfacial Electrochem.* **75**, 665 (1977).
376. P. G. Zambonin and J. Jordan, *J. Am. Chem. Soc.* **89**, 6365 (1967).
377. P. G. Zambonin and J. Jordan, *Anal. Lett.* **1**, 1 (1967).
378. P. G. Zambonin and J. Jordan, *J. Am. Chem. Soc.* **91**, 2225 (1969).
379. L. G. Boxall and K. E. Johnson, *Trans. Faraday Soc.* **67**, 1433 (1971).
380. S. N. Flengas, *J. Chem. Soc.* 534 (1956).
381. M. Blander, F. F. Blankenship, and R. F. Newton, *J. Phys. Chem.* **63**, 1259 (1959).
382. M. Bakes, J. Guion, and J. P. Brenet, *Electrochim. Acta* **10**, 1001 (1965).
383. L. G. Boxall and K. E. Johnson, *Anal. Chem.* **40**, 831 (1968).
384. F. R. Duke and H. M. Garfinkel, *J. Phys. Chem.* **65**, 461 (1961).
385. D. G. Hill and M. Blander, *J. Phys. Chem.* **65**, 1866 (1961).
386. J. Braunstein and M. Blander, *J. Phys. Chem.* **64**, 10 (1960).
387. H. S. Swofford and H. A. Laitinen, *J. Electrochem. Soc.* **110**, 814 (1963).
388. M. S. Zakharyevsky and T. V. Permyakova, *J. Gen. Chem. USSR* **26**, 3275 (1956).
389. D. G. Hill, J. Braunstein, and M. Blander, *J. Phys. Chem.* **64**, 1038 (1960).
390. H. M. Garfinkel and F. R. Duke, *J. Phys. Chem.* **65**, 1629 (1961).
391. C. Thomas and J. Braunstein, *J. Phys. Chem.* **68**, 957 (1964).
392. A. Alvarez-Funes, J. Braunstein, and M. Blander, *J. Am. Chem. Soc.* **84**, 1538 (1962).
393. D. L. Manning, R. C. Bansal, J. Braunstein, and M. Blander, *J. Am. Chem. Soc.* **84**, 2028 (1962).
394. M. H. Miles, D. A. Fine, and A. N. Fletcher, *J. Electrochem. Soc.* **125**, 1209 (1978).
395. K. Notz and A. G. Keenan, *J. Phys. Chem.* **70**, 662 (1966).
396. I. G. Murgulescu and D. I. Marchidan. *J. Phys. Chem.* **68**, 3086 (1964).
397. F. R. Duke, R. W. Laity, and B. B. Owens, *J. Electrochem. Soc.* **104**, 299 (1957).
398. F. Lantelme and M. Chemla, *Electrochim. Acta* **10**, 663 (1965).
399. D. Inman, *J. Sci. Instrum.* **39**, 391 (1962).
400. A. G. Keenan, K. Notz, and F. L. Wilcox, *J. Phys. Chem.* **72**, 1085 (1968).
401. R. H. Doremus, *J. Phys. Chem.* **72**, 2877 (1968).
402. R. H. Doremus, *J. Phys. Chem.* **72**, 2665 (1968).
403. A. G. Keenan and W. H. Duewer, *J. Phys. Chem.* **73**, 212 (1969).
404. J. Braunstein and R. M. Lindgren, *J. Am. Chem. Soc.* **84**, 1534 (1962).
405. D. L. Manning, M. Blander, and J. Braunstein, *Inorg. Chem.* **2**, 345 (1963).
406. J. Braunstein and A. S. Minano, *Inorg. Chem.* **3**, 218 (1964).
407. J. Braunstein and R. E. Hagman, *J. Phys. Chem.* **67**, 2881 (1963).
408. J. B. Lesourd and C. Vallet, *C. R. Acad. Sci. Paris Ser. C* **273**, 1213 (1971).

409. J. B. Lesourd, C. Vallet, and Y. Doucet, *J. Electroanal. Chem. Interfacial Electrochem.* **48,** 99 (1973).
410. R. A. Osteryoung, C. Kaplan, and D. L. Hill, *J. Phys. Chem.* **65,** 1951 (1961).
411. J. E. B. Randles and W. White, *Z. Elektrochem.* **59,** 666 (1955).
412. M. Steinberg and N. H. Nachtrieb, *J. Am. Chem. Soc.* **72,** 3558 (1950).
413. G. W. Harrington and H. T. Tien, *J. Phys. Chem.* **66,** 173 (1962).
414. M. F. Lantratov and A. G. Morachevskii, *Russ. J. Phys. Chem.* **33,** 417 (1959).
415. H. Tien and G. W. Harrington, *Inorg. Chem.* **2,** 369 (1963).
416. H. Tien, *Anal. Chem.* **36,** 929 (1964).
417. H. Tien, *J. Phys. Chem.* **69,** 3763 (1965).
418. J. A. A. Ketelaar and A. Dammers de Klerk, *Rec. Trav. Chim. Pays-Bas Belg.* **83,** 322 (1964).
419. W. Triaca and A. Arvia, *Electrochim. Acta* **9,** 919 (1964).
420. M. Okada and K. Kawamura, *Electrochim. Acta* **15,** 1 (1970).
421. A. F. J. Goeting and J. A. A. Ketelaar, *Electrochim. Acta* **19,** 267 (1974).
422. J. M. de Jong and G. H. J. Broers, *Electrochim. Acta* **21,** 605 (1976).
423. J. M. de Jong and G. H. J. Broers, *Electrochim. Acta* **21,** 893 (1976).
424. J. R. Selman and L. G. Marianowski, in *Molten Salt Technology* (D. G. Lovering, ed.), Plenum Press, New York (1982), p. 323.
425. J. R. Selman and H. C. Maru, in *Advances in Molten Salt Chemistry,* Vol. 4 (G. Mamantov and J. Braunstein, eds.), Plenum Press, New York (1981), p. 159.
426. J. Dubois, *Ann. Chim.* **10,** 145 (1965).
427. E. S. Argano and J. Levitan, *J. Electrochem. Soc.* **116,** 153 (1969).
428. P. K. Lorenz and G. J. Janz, *J. Electrochem. Soc.* **116,** 1061 (1969).
429. G. J. Janz and F. Saegusa, *Electrochim. Acta* **7,** 393 (1962).
430. G. K. Stepanov and A. M. Trunov, in *Electrochemistry of Molten and Solid Electrolytes,* Vol. 3 (A. N. Baraboshkin, ed.), Consultants Bureau, New York (1966), p. 73.
431. M. V. Smirnov, L. A. Tsiovkina, and V. A. Oleinikova, in *Electrochemistry of Molten and Solid Electrolytes,* Vol. 3 (A. N. Baraboshkin, ed.), Consultants Bureau, New York (1966), p. 61.
432. G. G. Arkhipov and G. K. Stepanov, in *Electrochemistry of Molten and Solid Electrolytes,* Vol. 3 (A. N. Baraboshkin, ed.), Consultants Bureau, New York (1966), p. 67.
433. P. K. Lorenz and G. J. Janz, *Electrochim. Acta* **15,** 1025 (1970).
434. A. V. Silakov, G. S. Tyurikov, and N. V. Vasilistov, *Sov. Electrochem.* **1,** 541 (1965).
435. A. M. Trunov, *Sov. Electrochem.* **1,** 1336 (1965).
436. G. K. Stepanov and A. M. Trunov, *Dokl. Akad. Nauk SSSR* **142,** 866 (1962).
437. G. J. Janz and F. Saegusa, *J. Electrochem. Soc.* **108,** 663 (1961).
438. G. J. Janz, F. Colom, and F. Saegusa, *J. Electrochem. Soc.* **107,** 581 (1960).
439. I. N. Ozeryanaya, N. A. Krasil'nikova, M. V. Smirnov, and V. N. Danilin, in *Electrochemistry of Molten and Solid Electrolytes,* Vol. 4 (A. N. Baraboshkin and S. F. Pal'guev, eds.), Consultants Bureau, New York (1967), p. 85.
440. A. M. Trunov and G. K. Stepanov, *Tr. Inst. Elektrockhim., Ural' Fil., Akad. Nauk SSSR,* No. 2, 97 (1961).
441. J. J. Scheer, A. E. van Arkel, and R. D. Heyding, *Can. J. Chem.* **33,** 683 (1955).
442. G. J. Janz, A. Conte, and E. Neuenschwander, *Corrosion* **19,** 292t (1963).
443. B. K. Andersen, Thermodynamic Properties of Molten Alkali Carbonates, Ph.D. thesis, The Technical University of Denmark, Lyngby, Denmark (1975).
444. A. J. Appleby and S. B. Nicholson, *J. Electroanal. Chem. Interfacial Electrochem.* **53,** 105 (1974).
445. A. Borucka, *Electrochim. Acta* **13,** 295 (1968).
446. R. L. Moss and H. R. Gibbens, *Cobalt* **28,** 115 (1965).
447. I. Trachtenberg, *J. Electrochem. Soc.* **111,** 110 (1964).

448. G. H. J. Broers and M. Schenke, in *Hydrocarbon Fuel Cell Technology* (B. S. Baker, ed.), Academic Press, New York (1965), p. 225.
449. K. Y. Kim and O. F. Devereux, *J. Electrochem. Soc.* **125**, 1796 (1978).
450. J. Braunstein, H. R. Bronstein, S. Cantor, D. Heatherly, and C. E. Vallet, Molten Carbonate Fuel Cell Research at ORNL, Oak Ridge National Laboratory Report, ORNL/TM-5886, May (1977).
451. M. Azzi and J. J. Rameau, *Corros. Sci.* **24**, 935 (1984).
452. A. Borucka, in *Fuel Cell Systems—II*, Advances in Chemistry Series, American Chemical Society, Washington, D.C. (1969), p. 242.
453. N. Busson, S. Palous, R. Buvet, and J. Millet, *C. R. Acad. Sci. Paris* **260**, 6097 (1965).
454. A. Borucka and C. M. Sugiyama, *Electrochim. Acta* **13**, 1887 (1968).
455. A. Borucka and C. M. Sugiyama, *Electrochim. Acta* **14**, 871 (1969).
456. H. E. Bartlett and K. E. Johnson, *J. Electrochem. Soc.* **114**, 457 (1967).
457. N. Busson, S. Palous, R. Buvet, and J. Millet, *C. R. Acad. Sci. Paris* **261**, 720 (1965).
458. L. P. Klevtsov, G. G. Arkhipov, and G. K. Stepanov, *Sov. Electrochem.* **1**, 1170 (1965).
459. Yu. M. Pashkov and G. S. Tyurikov, *Sov. Electrochem.* **3**, 386 (1967).
460. N. Busson, S. Palous, J. Millet, and R. Buvet, *Electrochim. Acta* **12**, 1609 (1967).
461. H. Gerischer, *Z. Elektrochem.* **59**, 604 (1955).
462. A. Borucka, *J. Electrochem. Soc.* **124**, 972 (1977).
463. A. Borucka and A. J. Appleby, *J. Chem. Soc. Faraday Trans.* **73**, 1420 (1977).
464. P. Degobert and O. Bloch, *Bull. Soc. Chim. Fr.*, 1887 (1962).
465. G. J. Janz and A. Conte, *Electrochim. Acta* **9**, 1267 (1964).
466. G. J. Janz and A. Conte, *Electrochim. Acta* **9**, 1279 (1964).
467. M. D. Ingram, B. Baron, and G. J. Janz, *Electrochim. Acta* **11**, 1629 (1966).
468. S. H. White and U. M. Twardoch, *Electrochim. Acta* **27**, 1599 (1982).
469. S. H. White and M. M. Bower, in *Proceedings of the Third International Symposium on Molten Salts* (G. Mamantov, M. Blander, and G. P. Smith, eds.), The Electrochemical Society, Pennington, New Jersey (1981), p. 334.
470. P. Brossbach and D. Sauermann, *Electrochim. Acta* **9**, 1373 (1964).
471. K. Sato, H. Urushibata, T. Murahashi, and I. Hirata, Extended Abstracts, The Electrochemical Society Fall Meeting, Las Vegas, Nevada, October 13–18 (1985), p. 766.
472. F. R. Foulkes, W. F. Graydon, and M. Garamszeghy, *J. Electrochem. Soc.* **131**, 1325 (1984).
473. D. W. Kirk and F. R. Foulkes, *J. Electrochem. Soc.* **131**, 1332 (1984).
474. J. Stringer, in *Proceedings of the Symposium on Properties of High Temperature Alloys* (Z. A. Foroulis and F. S. Pettit, eds.), The Electrochemical Society, Pennington, New Jersey (1976), p. 513.
475. A. Rahmel, *Electrochim. Acta* **15**, 1267 (1970).
476. G. Danner and M. Rey, *Electrochim. Acta* **4**, 274 (1961).
477. K. E. Johnson and H. A. Laitinen, *J. Electrochem. Soc.* **110**, 314 (1963).
478. A. Rahmel, *Chem. Ing. Tech.* **41**, 169 (1969).
479. A. Rahmel and B. Vollmer, *J. Electroanal. Chem. Interfacial Electrochem.* **48**, 93 (1973).
480. A. J. B. Cutler, *J. Appl. Electrochem.* **1**, 19 (1971).
481. W. T. Wu, A. Rahmel, and M. Schmidt, *Electrochim. Acta* **28**, 1487 (1983).
482. J. A. Goebel, F. S. Pettit, and G. W. Goward, *Metall. Trans.* **4**, 261 (1973).
483. E. Tatar-Moisescu and A. Rahmel, *Electrochim. Acta* **20**, 479 (1975).
484. A. Rahmel, M. Schmidt, and M. Schorr, *Oxid. Met.* **18**, 195 (1982).
485. G. W. Watt, R. E. Andersen, and R. A. Rapp, in *Proceedings of the Second International Symposium on Molten Salts* (J. Braunstein and J. R. Selman, eds.), The Electrochemical Society, Pennington, New Jersey (1981), p. 81.
486. D. A. Shores and R. C. John, *J. Appl. Electrochem.* **10**, 275 (1980).
487. D. A. Shores and W. C. Fang, *J. Electrochem. Soc.* **128**, 346 (1981).

488. C. A. C. Sequeira and M. G. Hocking, *J. Appl. Electrochem.* **8,** 145 (1978).
489. C. A. C. Sequeira, *Rev. Port. Quim.* **21,** 117 (1979).
490. C. T. Brown, N. S. Bornstein, and M. A. DeCrescente, in *High Temperature Metallic Corrosion of Sulfur and its Compounds* (Z. A. Foroulis, ed.), The Electrochemical Society, Pennington, New Jersey (1970), p. 170.
491. C. H. Liu, *J. Phys. Chem.* **66,** 164 (1962).
492. C. H. Liu, *Anal. Chem.* **33,** 1477 (1961).
493. B. W. Burrows and G. J. Hills, *Electrochim. Acta* **15,** 445 (1970).
494. A. Rahmel, *Corros. Sci.* **13,** 833 (1973).
495. O. E. Abdel-Salam and M. A. Soliman, *Electrochim. Acta* **28,** 347 (1983).
496. H. Guion, *Inorg. Chem.* **6,** 1882 (1967).
497. E. S. Woolner, Jr., and D. G. Hill, *J. Phys. Chem.* **67,** 1571 (1963).
498. M. Petit, *C. R. Acad. Sci. Paris* **258,** 6143 (1964).
499. M. Rey, G. Danner, and M. Abraham, *C. R. Acad. Sci. Paris* **248,** 2868 (1959).
500. H. Lux, *Z. Elektrochem.* **45,** 303 (1939).
501. H. Flood, T. Forland, and K. Motzfeldt, *Acta Chem. Scand.* **6,** 257 (1952).
502. D. G. Hill, B. Porter, and A. S. Gillespie, Jr., *J. Electrochem. Soc.* **105,** 408 (1948).
503. C. A. C. Sequeira and M. G. Hocking, *Electrochim. Acta* **22,** 1161 (1977).
504. D. A. Shores, *Corrosion* **31,** 434 (1975).
505. E. Erdoes and H. Altorfer, *Electrochim. Acta* **20,** 937 (1975).
506. D. K. Gypta and R. A. Rapp, *J. Electrochem. Soc.* **127,** 2194 (1980).
507. T. H. Etsell and S. N. Flengas, *J. Electrochem. Soc.* **118,** 1890 (1971).
508. R. Y. Lin, D.Sc. thesis, M.I.T., Cambridge, Massachusetts (1980).
509. H. Flood and N. C. Boye, *Z. Elektrochem.* **66,** 184 (1962).
510. L. G. Boxall and K. E. Johnson, *J. Electrochem. Soc.* **118,** 885 (1971).
511. F. J. Salzano and L. Newman, *J. Electrochem. Soc.* **119,** 1273 (1972).
512. W. P. Stroud and R. A. Rapp, in *High Temperature Metal Halide Chemistry* (D. L. Hildenbrand and D. D. Cubicciotti, eds.), The Electrochemical Society, Pennington, New Jersey (1978), p. 574.
513. W. W. Liang, H. K. Bowen, and J. F. Elliott, in *Metal-Slag-Gas Reactions and Processes* (Z. A. Foroulis and W. W. Smeltzer, eds.), The Electrochemical Society, Pennington, New Jersey (1978), p. 574.
514. W. W. Liang and J. F. Elliott, in *Proceedings of the Symposium on Properties of High Temperature Alloys* (Z. A. Foroulis and F. S. Pettit, eds.), The Electrochemical Society, Pennington, New Jersey (1976), p. 557.
515. W. W. Liang and J. F. Elliott, *J. Electrochem. Soc.* **123,** 617 (1976).
516. W. W. Liang and J. F. Elliott, *J. Electrochem. Soc.* **125,** 572 (1978).
517. O. G. Zarubitskii, *Russ. Chem. Rev.* **49,** 536 (1980).
518. C. Grange and L. Heerman, *J. Appl. Electrochem.* **4,** 279 (1974).
519. B. A. Rose, G. J. Davis, and H. J. T. Ellingham, *Discuss. Faraday Soc.* **4,** 154 (1948).
520. A. S. Afanas'ev and V. P. Gamasov, *Russ. J. Phys. Chem.* **38,** 1537 (1964).
521. H. J. Krüger, A. Rahmel, and W. Schwenk, *Electrochim. Acta* **13,** 625 (1968).
522. V. P. Gamasov, *Sov. Electrochem.* **4,** 1216 (1968).
523. V. P. Gamasov, *Sov. Electrochem.* **4,** 1218 (1968).
524. Yu. K. Delimarskii, O. G. Zarubitskii, O. A. Omel'chenko, and V. G. Budnik, *J. Appl. Chem. USSR* **41,** 718 (1968).
525. L. I. Antropov and D. A. Tkalenko, *Sov. Electrochem.* **6,** 583 (1970).
526. L. I. Antropov and D. A. Tkalenko, *Sov. Electrochem.* **6,** 1500 (1970).
527. J. Goret, *Bull. Soc. Chim. Fr.* 1074 (1964).
528. J. Goret and B. Tremillon, *Bull. Soc. Chim. Fr.* 67 (1966).
529. R. G. Doisneau and B. Tremillon, *J. Chim. Phys.* **71,** 1445 (1974).

530. *Solid Electrolytes and Their Applications* (E. C. Subbarao, ed.), Plenum Press, New York (1980).
531. R. Dorin, H. J. Gardner, and L. J. Rogers, *J. Electroanal. Chem. Interfacial Electrochem.* **77**, 385 (1977).
532. R. Dorin, H. J. Gardner, and L. J. Rogers, Extended Abstracts, 30th Meeting of the International Society of Electrochemistry, Trondheim, Norway, August 26–31 (1979), p. 136.
533. H. Hayashi, S. Yoshizawa, and Y. Ito, *J. Electroanal. Chem. Interfacial Electrochem.* **124**, 229 (1981).
534. H. Hayashi, S. Yoshizawa, and Y. Ito, *Electrochim. Acta* **28**, 149 (1983).
535. A. Eluard and B. Tremillon, *J. Electroanal. Chem. Interfacial Electrochem.* **18**, 277 (1968).
536. G. Schiavon, S. Zecchin, and G. G. Bombi, *J. Electroanal. Chem. Interfacial Electrochem.* **38**, 473 (1972).
537. G. Kern and O. Bloch, *C. R. Acad. Sci. Paris* **258**, 5431 (1964).
538. J. Goret and B. Tremillon, *Electrochim. Acta* **12**, 1065 (1967).
539. M. Bonnemay, L. Lacourcelle, and J. Royon, *Electrochim. Acta* **14**, 1061 (1969).
540. G. Kern, P. Degobert, and O. Bloch, *C. R. Acad. Sci. Paris* **256**, 1500 (1963).
541. S. G. Ogryz'ko-Zhukovskaya, N. A. Fedotov, and V. P. Belokopytov, *Sov. Electrochem.* **7**, 1491 (1971).
542. V. P. Belokopytov, S. G. Ogryz'ko-Zhukovskaya, and N. A. Fedotov, *Sov. Electrochem.* **14**, 421 (1978).
543. A. Eluard and B. Tremillon, *J. Electroanal. Chem. Interfacial Electrochem.* **30**, 323 (1971).
544. A. Eluard and B. Tremillon, *J. Electroanal. Chem. Interfacial Electrochem.* **27**, 117 (1970).

5

Neutron Diffraction

Susan Biggin

1. Introduction

Scattering techniques include neutron, x-ray, light, and Raman scattering, and EXAFS studies: these obviously enable a wide variety of measurements to be made on molten salts. Of comparable power in some cases are neutron and x-ray methods, which will therefore be our focus. The general principles of neutron scattering can be found in texts such as Bacon[1] or Squires,[2] while x-ray techniques are covered, for example, by Warren.[3] Structural information—in particular ion–ion separations and coordination numbers—is readily obtained from such experiments, as exemplified by the studies of Enderby and Biggin[4] and Enderby.[5] The principal work on molten salts using x-rays and neutrons has been on static structure, and this chapter therefore will be restricted to this field. Other studies include inelastic neutron scattering to investigate the dynamical structure factor, $S(k, \omega)$, where the structure factor is mapped out in energy space; dispersion relations; and the measurement of diffusion coefficients. Here k is wave vector and $\hbar\omega$ is energy. For example, McGreevy and Mitchell recently have carried out inelastic neutron scattering studies of the dynamics of molten alkali halides, evaluating mass and charge correlations.[6]

In order to choose whether to use neutrons or x-rays, the technique must be matched to the problem. In general, x-rays provide a simple, laboratory-based method for finding a particular melt structure, whereas a neutron experiment requires considerable advanced arrangement but yields *more specific* information.

Susan Biggin • H. H. Wills Physics Laboratory, University of Bristol, Bristol BS8 1TL, United Kingdom.

In the technique of neutron or x-ray scattering, data are collected as a function of angle. These angles are readily converted to wave vector or momentum transfer, k [Eq. (2)]. A Fourier transform converts the k-space data to real-space, r, data, there being an inverse relationship between k and r. Hence features at low-k values tend to represent long-range features. Therefore, a point to note is that *long-range* order of inhomogeneities can be investigated, to some extent, using small-angle scattering. In small-angle neutron scattering, the wave vector, k, values are very low, typically 0.003–0.6 $Å^{-1}$.

The organization of this chapter is given briefly as follows: First, we shall consider what information is available from experiment, by discussing the background of structure factors, radial distribution functions, and coordination number, in the context of neutron and x-ray scattering.

We will turn our attention then solely to neutron methods, mentioning x-rays only when comparison is relevant. Dwelling on practical details, we ask such questions as *which* samples can be studied, how are samples prepared, what types of furnace are required, where can the experiments be performed, and how long does data analysis require? Neutron beam facilities, methods of applying for beam time, as well as descriptions of available diffractometers are given, and a brief description of data analysis is provided. Some results of melts studied to date are given as examples of the scope of the technique. Finally, a brief summary is made of the technique.

2. Molten Salt Structure

2.1. Structure Factors

A liquid static structure is defined in terms of radial distribution functions (rdf's), this structure being simply the Fourier transform of the structure factors of the liquid. The static structure factor for a liquid, $S_{ij}(k)$, is analogous to that for a crystal, which is usually given as F_{hkl} for the structure factor of the unit cell for the hkl reflection

$$F_{hkl} = \left| \sum_{\substack{\text{all}\\ \text{atoms}\\ \text{in unit cell}}} b_{xyz} \exp 2\pi i(hx/a_0 + ky/b_0 + lz/c_0) \right|$$

where b_{xyz} is the scattering length for each atomic position; a_0, b_0, c_0 represent the unit cell dimensions; x, y, z denote position. In this way, the structure factor F_{hkl} is equivalent to the Fourier transform of the lattice. So, too, for a liquid the structure factor $S(k)$ is related to the Fourier transform of the atom positions in the liquid.

In terms of neutron scattering cross section, we may write

$$S(k) = (Nb^2)^{-1}\frac{d\sigma}{d\Omega} = 1 + \int dr\, e^{ikr} g(r)$$

where $d\sigma/d\Omega$ is the coherent differential scattering cross section per unit solid angle, N is number of scattering centers, b is scattering length, $g(r)$ is a radial distribution function giving the probability of finding a particular atom at a distance r from an origin atom, and r is the distance from any arbitrarily chosen origin atom, in Ångstroms. From a single x-ray or neutron experiment a total structure factor $F(k)$ is obtained. This is the Fourier transform of the total radial distribution function $G(r)$:

$$G(r) = \frac{1}{2\pi^2 \rho r}\int_0^\infty F(k) k \sin kr\, dk \qquad (1)$$

where

$$k = \frac{4\pi}{\lambda}\sin\frac{\theta}{2} \qquad (2)$$

Here, k is a wave vector ($Å^{-1}$), λ is neutron wavelength, θ is the scattering angle, ρ is number density ($Å^{-3}$), and r is the ion–ion separation in Ångstroms. On the other hand, from a *series* of experiments, *partial* structure factors, $S_{ij}(k)$, and so partial rdf's, $g_{ij}(r)$, are extracted:

$$F(k) = \sum_i \sum_j c_i c_j b_i b_j [S_{ij}(k) - 1] \qquad (3)$$

$$g_{ij}(r) = 1 + \frac{1}{2\pi^2 \rho r}\int_0^\infty [S_{ij}(k) - 1] k \sin kr\, dk \qquad (4)$$

where c_i, b_i are concentration and scattering length of component i, respectively.

These $S(k)$ relate to *specific* ion–ion interactions and transform to *partial* radial distribution functions (rdf's), $g_{ij}(r)$. Ion separation can be deduced directly (from these partial rdf's). As mentioned above, the transform of $F(k)$ gives a total rdf, $G(r)$, Eq. (1), which represents a *combination* of ion–ion correlation functions. Examples of $S(k)$ and $g(r)$ for molten $CaCl_2$ are given in Figs. 1 and 2. Figures 3 and 4 show $F(k)$ and $G(r)$ for molten $MnCl_2$.

In some cases, knowledge of $G(r)$ alone is adequate to make some definite statements about the liquid structure. This pertains when one ion is much smaller than the other in a simple melt of the type MX or MX_2 (M, cation; X, anion), for example Biggin and Enderby.[7] It results in there being very little overlap of partial radial distribution functions around the

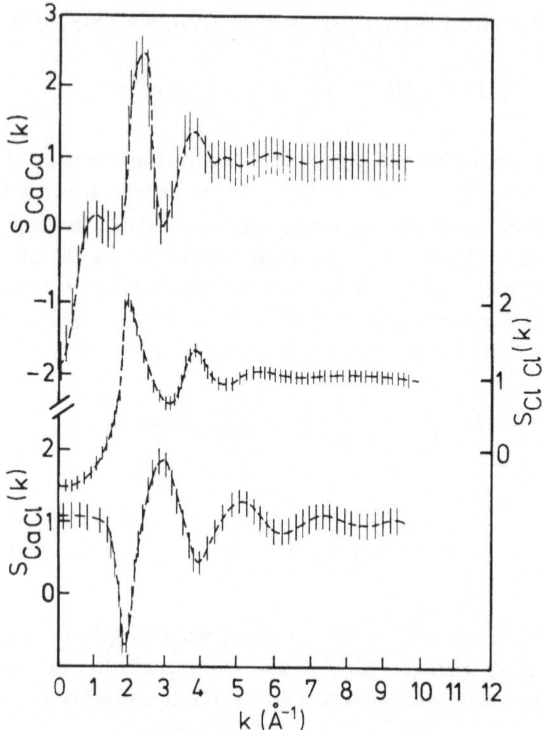

FIGURE 1. **Partial structure factors, $S_{ij}(k)$, for $CaCl_2$.**

first peak: the cross-term, Zn–Cl, shown in Fig. 5, is distinct from the other two partials, Zn–Zn and Cl–Cl, so that deduction of the first peak position and area is quite valid. This is not the case when the anion and cation are closer in their values of ion radius, e.g., $BaCl_2$, Fig. 6. This is important,

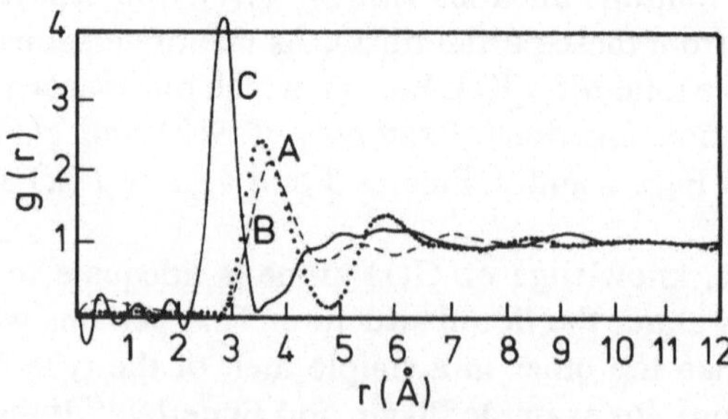

FIGURE 2. **Partial radial distribution functions, $g_{ij}(r)$, for $CaCl_2$. A, CaCa; B, ClCl; C, CaCl.**

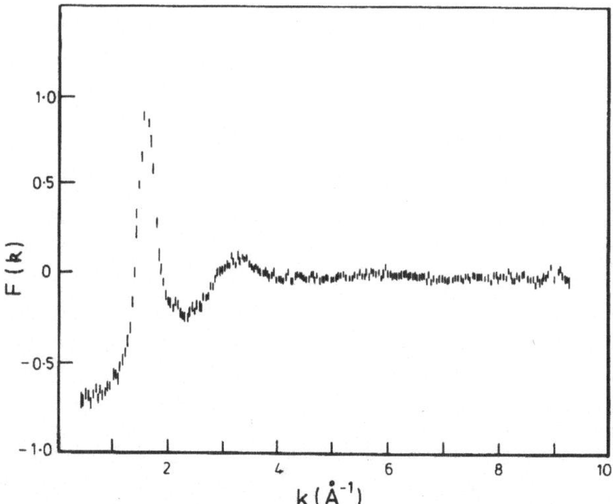

FIGURE 3. Total structure factor, $F(k)$, for $MnCl_2$.

since the area under a peak leads to a determination of the coordination number.

In general, however, specific ion–ion interaction is required, such as the Cl–Cl interaction in $ZnCl_2$. For such cases, partial rdf's are essential.

2.2. Small-Angle Neutron Scattering (SANS)

In certain systems, such as the binary melt $LiAlCl_4$, it is expected that some long-range order should be observable as a low-k feature in the partial

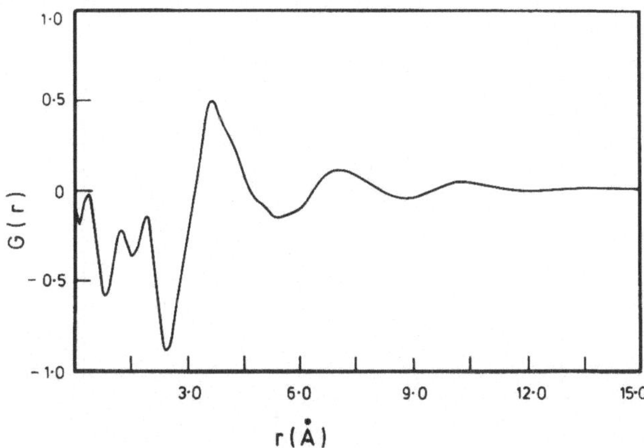

FIGURE 4. Total radial distribution function, $G(r)$, for $MnCl_2$.

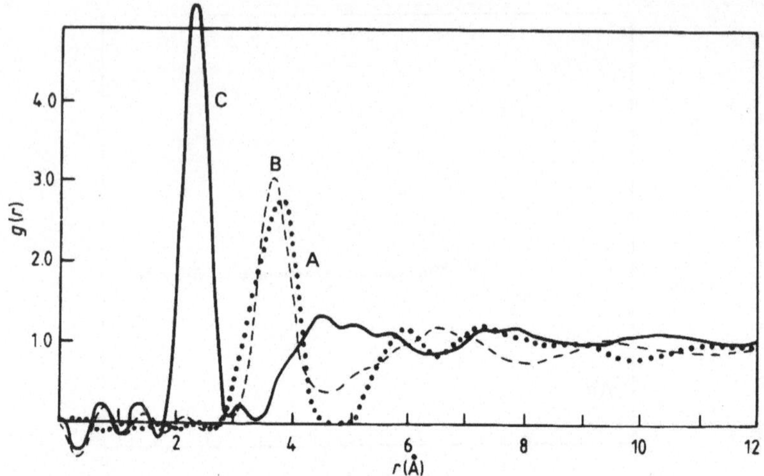

FIGURE 5. Partial radial distribution functions, $g_{ij}(r)$, for $ZnCl_2$. A, ZnZn; B, ClCl; C, ZnCl.

structure factors. In this particular melt, it is believed that bridging occurs, forming entities such as the heptachloroaluminate anion,

$$2AlCl_4^- \rightleftharpoons \mathrm{Cl{-}\overset{\overset{\Large Cl}{|}}{\underset{\underset{\Large Cl}{|}}{Al}}{-}Cl{-}\overset{\overset{\Large Cl}{|}}{\underset{\underset{\Large Cl}{|}}{Al}}{-}Cl^{(-)}} + Cl^-$$

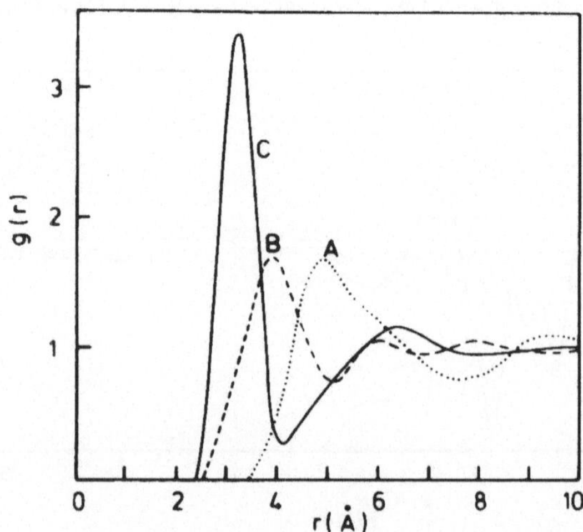

FIGURE 6. Partial distribution functions, $g_{ij}(r)$, for $BaCl_2$. A, BaBa; B, ClCl; C, BaCl.

If long-range information of this nature is required, then a special diffractometer which only covers a low-k range must be used. Both ORNL and ILL have good facilities for this type of experiment (Section 8.1).

3. Neutron or X-Ray Diffraction

In order to extract the full structural information from partial rdf's, the series of experiments to be performed involves the use of isotopes in a neutron experiment, or variations of energy in the x-ray case. The latter technique has not been used, since for resolution of the partial rdf's the energy ranges at conventional sources have been too small. However, calculations have shown (Narten[8]) that when the National Synchrotron Light Source at the Brookhaven National Laboratory comes on-line fully, the energy differences available there will be adequate. This source is dedicated to multidisciplinary experiments, using both ultraviolet synchrotron light (700 MeV electron storage ring) and x-rays (2.5-GeV electron storage ring).

To explain why isotopes are essential in the determination of *partial* rdf's, consider Eqs. (3) and (5):

$$G(r) = \sum_i \sum_j c_i c_j b_i b_j [g_{ij}(r) - 1] \tag{5}$$

where b_i is the neutron scattering length of component i.

In the neutron case, the scattering length, b, is a function of isotope, and hence by suitable choice of isotopes, Eq. (3) can be solved for $S_{ij}(k)$. For a two-component melt, e.g., NaCl, three samples are needed, i.e., $n(n+1)/2$; for three components, six are needed, etc. For x-rays, the form factor f (equivalent of b) does not depend on the isotope, but varies systematically with atomic number and is a function of k. Consequently, when using x-rays a different approach is required in order to solve Eq. (3) for $S_{ij}(k)$. The technique evaluated by Narten[8] is to determine the x-ray scattering at different energies (wavelengths) near an element's absorption edge. Synchrotron radiation has a wide energy spectrum and brilliance, and can provide high flux, from which a narrow energy band can be selected to interact strongly with specific elements and indicate their arrangements in matter. By taking differences between spectra, interactions of the element of interest alone will be highlighted. This method should enable most elements to be studied.

Hence, in either neutron or x-ray cases, sufficient experiments could be performed to provide simultaneous versions of Eq. (3) for solution for $S_{ij}(k)$. In practice, to date, isotope experiments have been the main route to $S_{ij}(k)$.

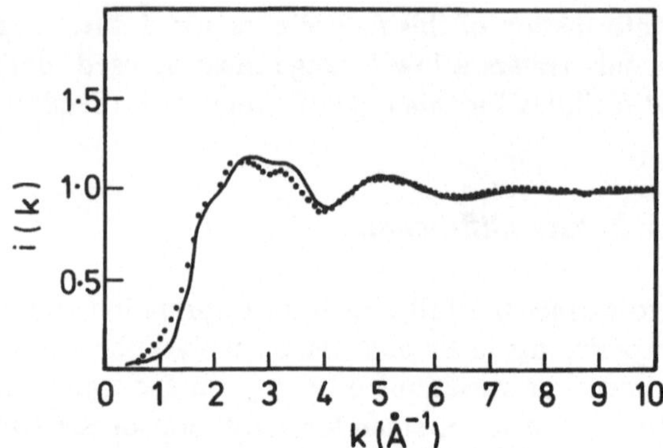

FIGURE 7. Coherent reduced scattering x-ray intensity, $i(k)$, for NaCl—derived from the structure factors shown in Fig. 13; . . . , obtained experimentally by Ohno and Furukawa.[9]

3.1. Combination of Neutron and X-Ray Data

In some cases, sets of x-ray data can be combined with neutron data sets, to extract the partial structure factors and radial distribution functions. For example, one x-ray set and two neutron sets should enable the determination of partial structure factors for a simple melt of the type AX_n.

To test the compatibility of data sets, the neutron-deduced structure factors for NaCl melt were weighted according to *x-ray* form factors, and summed to give an artificial x-ray total scattering function. This function was then compared with actual x-ray data obtained by Ohno and Furukawa.[9] The degree of agreement is perfectly acceptable, as shown in Fig. 7, which is evidence that both neutron and x-ray experiments can give $S_{ij}(k)$'s devoid of systematic (e.g., instrument- or technique-dependent) errors.

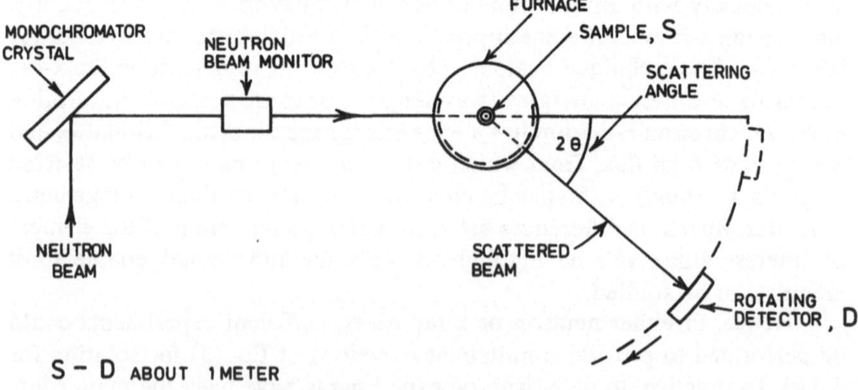

FIGURE 8. Geometry of a neutron scattering experiment.

TABLE I. Typical Experimental Runs and Time Scales for Studying a Salt at a Low-Flux Reactor

Run	Time (days)	Conditions
Background	1	
Vanadium standard	1	Rod dimensions as for sample, to eliminate beam profile artefacts
Empty furnace	1	At elevated temperature and under vacuum (Empty furnace through Sample 3)
Empty container	3	
Sample 1 $Nd^{35}Cl_3$	6	
Sample 2 $Nd^{37}Cl_3$	6	
Sample 3 $Nd^{mix}Cl_3$	6	

3.2. Scattering Experiment

In a neutron scattering experiment, as for x-ray scattering, the sample is held in its sealed container within a heat-shielded furnace. (The container typically is quartz tubing.) The sample lies at the center of a circle on which lie neutron detectors at about a 1-m radius. In some cases, the detectors move around part of the circle, typically covering 5–120°. Figure 8 represents a schematic of the sample/beam/detector geometry. The detectors are usually the ^{3}He type and detect protons produced from the scattered neutrons ($^3\mathrm{He} + n \rightarrow {}^3\mathrm{H} + p$).

A neutron experiment on three samples takes about one month of beam time at a low-flux reactor, or about 5 days at a high-flux facility.

At the Dido reactor, Harwell, for example, a salt such as $NdCl_3$ can be investigated according to the schedule shown in Table I.

4. Samples Amenable to Neutron Diffraction Study

Appropriate isotopes must be available, affordable, and have suitable scattering characteristics. Obviously, the coherent, incoherent, and absorption cross sections must be known. Table II contains examples of these factors. It can immediately be seen that there are limitations to the technique. First of all, more than one scattering length is required in order to carry out a multiple-pattern series of experiments (commonly three). Chlorine, for example, is an ideal melt anion since it has two very different scattering

TABLE II. Scattering Parameters of Various Elements, from a Compilation Produced in 1982 by Koester and Yelon[a,b]

Element	Scattering length, b (10^{-12} cm)	σ_{abs} at 1.8 Å (barns)	σ_{inc} (barns)
^{35}Cl	1.17	44.1	4.418
^{37}Cl	0.308	0.433	0.008
^{nat}Cl	0.958	33.5	5.10
H	−0.374	0.333	79.91
D	0.667	0.001	2.033
^{6}Li	0.187 − 0.026i	940.0	0.541
^{7}Li	−0.220	0.045	0.832
^{nat}Li	−0.203	70.5	0.882
V	−0.038	5.08	4.935
Ti	−0.344	6.1	2.575
Zr	0.716	0.185	0.118
^{nat}Ni	1.03	4.5	4.23
^{58}Ni	1.44	4.6	0
^{62}Ni	−0.87	14.5	0.09

[a] Reference 10.
[b] Total scattering cross section = coherent plus incoherent scattering cross sections, i.e., $\sigma_{tot} = \sigma_{coh} + \sigma_{inc}$, where $\sigma_{coh} = 4\pi\bar{b}^2$. Absorption cross sections are quoted at 1.8 Å. Complex scattering length values correspond to a neutron wavelength of 1 Å.

lengths. Thus adequate differences in the scattering contributions from the labeled anions will arise—note the values of scattering lengths for chlorine. To produce the three samples required to solve Eq. (3) for a simple melt such as NaCl,(7,11) two isotopes of sufficiently different scattering length are used, and their mixture will provide the third cross section value. Many successful experiments have been performed using ^{35}Cl, ^{37}Cl, and ^{mix}Cl (where the isotope mixture consists of 50% of each). Nickel is also very suitable because of the positive and negative scattering lengths available; note Table II: this element lends itself to elegant manipulation of scattering cross sections. It is clear that a null scatterer can be produced by mixing ^{62}Ni and ^{nat}Ni. This approach simplifies solution of Eq. (3) for specific $S_{ij}(k)$. It is also available for lithium.

Another neutron parameter that affects the choice of melt sample that can be studied is incoherent scattering cross section, σ_{inc}. Table II contains some σ_{inc} values for various elements: it is seen that hydrogen has an unusually high cross section. This type of scattering masks the coherent signal. The remedy is deuteration, which is necessary when using ammonium salts, for example.

Furthermore, the sample absorption must not be too large. Again referring to Table II, some representative absorption cross sections, σ_{abs}, can be seen. For example, ^{6}Li has a very high value of 940 barns. A compromise has to be made, therefore, when using a zero scattering length mixture of ^6Li and ^7Li, denoted "^0Li", in an experiment where null scattering from high absorbers is desired, since the absorption of neutrons by the sample will compete with the coherent scattering. Typically, samples are held in 6 mm i.d. tubes, but the sample size can be reduced to cut down absorption. This may also be necessary to bring multiple-scattering corrections under better control. Absorption occurs in accordance with the equation

$$I = I_0 \exp(-\mu x) \tag{6}$$

where I is intensity, x is the sample diameter (neutron path length), and μ is the attenuation coefficient defined by

$$\mu = \frac{\rho \sigma_T N}{A} \tag{7}$$

In Eq. (7), N is Avogadro's number, ρ is the melt density, σ_T is total cross section (including coherent, incoherent, absorption), and A is the mean atomic weight. However, the scattering signal-to-noise ratio decreases if less sample is used, so a compromise must be reached.

A final point is that nuclear resonances at particular energies will have a dramatic effect on the capture cross section. Thus, it is necessary to become acquainted with the resonance energies and widths of each isotope that may be used, most particularly when work at a pulsed source is involved (refer to Section 8). For example, the energy range of neutrons at the pulsed source at Argonne National Laboratory, where 0.1–5.6-Å neutrons are used, is about 3–8180 meV. Resonances for all neodymium isotopes lie outside this range, except for ^{145}Nd, which has a narrow resonance at 4350 meV (0.137 Å). This information is available in the extensive publication commonly called "The Barn Book."[12]

5. Sample Preparation

Samples to be used in neutron scattering usually occupy about 4 cm^3 at the most; the sample volume is around 1.5 cm^3 when isotopes are used.

Many salt samples studied by the Bristol group contain chlorine isotopes, and for this reason, drying with standard techniques[13] using HCl or Cl_2 gas is not possible. Instead, lengthier methods, such as drying the salt gently over an oil bath and under vacuum, are necessary. To check for water content, it is possible in some cases to weigh the salt as it is dried, monitor how many water molecules of crystallization are being pumped off, and, eventually, to observe at which stage the salt has become anhydrous. Hydrolysis is always a possibility, however, when this method of drying has been used (Section 5.2).

Infrared tests are used to obtain the value of residual water content, although usually just a maximum possible content can be quoted, since the values are small and at the limit of the instrument resolution.

5.1. Isotope Exchange

To describe how a sample containing a *particular* isotope is prepared, consider the preparation of one example, namely, $ZnCl_2$. Magnesium and calcium chlorides may be prepared in the same way.

Chlorine isotopes usually are obtained in the form of Na*Cl. The first step is to convert the Na*Cl to H*Cl with an H^+ ion exchange resin. Using an IR120(H) resin column, the following exchange reaction occurs:

$$\text{Na*Cl} + \text{H}^+ \text{ resin} \rightarrow \text{H*Cl} + \text{Na}^+ \text{ resin} \qquad \text{(in water)}$$

Commonly, a weighed amount of the metal oxide (ZnO in this example) is then dissolved in the H*Cl to produce the required salt:

$$\text{ZnO} + 2\text{H*Cl} \rightarrow \text{Zn*Cl}_2 + \text{H}_2\text{O}$$

Because of the presence of the Cl^- isotopes in the samples, all of the above reactions must be carried out in solution. This is not the case, of course, if metal isotopes are being used. As mentioned, the Zn^*Cl_2 cannot be dried easily, since HCl or Cl_2 gas may not be used.

5.2. Sample Drying

The usual method of drying is simply to apply heat. The problems involved with this lie in hydrolytic decomposition: very often, oxychlorides, oxides, or hydroxides are produced. To test for stability, a salt is dried with gentle heat until anhydrous, and then redissolved in H_2O. Quite often insoluble material remains—an indication of the amount of hydrolysis.

Standard texts, such as *Inorganic Syntheses*,[14] often recommend drying techniques for salts, but usually these are only suitable for cation-substituted samples.

If the *cation* in a salt is to be isotopically replaced, then preparation and drying generally are easier. For example, we have studied $NdCl_3$, with ^{144}Nd, ^{146}Nd, and natural chlorine; plus natural neodymium with ^{37}Cl. *Inorganic Syntheses*[14] describes a preparation technique:

$$\text{Nd}_2\text{O}_3 + 6\text{NH}_4\text{Cl} \rightarrow 2\text{NdCl}_3 + 3\text{H}_2\text{O} + 6\text{NH}_3$$

However, a large excess of chlorine is needed for the preparation, which is too expensive for the Cl-isotope samples, so a different method was needed for the anion-substituted salts. This involved preparation of the hexahydrate $NdCl_3 \cdot 6H_2O$ and drying with thionyl chloride. This procedure works well for $^*NdCl_3$, and $^{nat}Nd\,^{nat}Cl_3$, but again, if thionyl chloride is used in the drying process, there is a possibility of exchange of the isotopic chloride ions for natural chloride ions. Mass spectroscopy must be performed to check on the amount of exchange.

Neodymium chloride, with ^{37}Cl isotope, was dried from the hexahydrate by spreading out the salt and allowing evaporation under a vacuum at ambient laboratory temperature. The salt was weighed at regular intervals. In this method the salt does not decompose, but the water of crystallization is drawn off, leaving the last replaceable molecule in the hydrate $NdCl_3 \cdot H_2O$. This will eventually be removed if the salt is left for 1–2 weeks. Weighing gives an indication of how much water is present and how many water molecules of crystallization are left. The removal of the last hydrating water molecule is checked using infrared spectroscopy (as mentioned above).

6. *Sample Containers*

Table II shows elements that are ideal container materials for neutron scattering studies, bearing in mind that as little coherent scattering (structure) as possible is desired from any materials in the beam except from the sample itself. Those commonly used are vanadium, titanium–zirconium, and quartz. Vanadium offers almost zero contribution to the coherent scattered beam (and incidentally is used for this reason as a reference material to normalize data). Furthermore, it is resistant to many molten salts, as are titanium–zirconium, and quartz. Titanium and zirconium can be combined in a 68–32 alloy (approx) (Table II) to give a zero coherent scattering material. Quartz also has very favorable properties. All elements mentioned here have relatively small total scattering and absorption cross sections.

Vanadium and titanium–zirconium are machinable into suitable cells, approximately 5 cm in height, 6 mm i.d., and 7 mm o.d. Silica is obtained as thin-walled tubes (wall 0.6–0.7 mm).

Alternatively, standard wall size silica can be machined to reduce the diameter to an acceptable value.

Samples are sealed into the containers under vacuum, or under a reduced pressure of an inert gas. First, the volume above the sample is evacuated, and then typically about a quarter of an atmosphere of argon is backfilled into the tube, which is then sealed.

7. *Furnaces*

At most sites for neutron experiments, furnaces are available to the user, although occasionally a furnace must be designed by the user to suit special needs. For example, at the intense pulsed neutron source, Argonne, both low- and high-temperature furnaces are available for liquid experiments (Section 8 describes facilities for neutron diffraction). These have been built especially to meet the requirements of melt studies at a neutron source.

There are several specifications peculiar to use with neutrons. These include the following.

- Water cooling out-of-the-beam is required.
- All parts of the furnace that are seen by either the incident or the scattered beam should be as thin as possible, to minimize the contribution of the furnace to the scattered beam.
- Neither the furnace body, heat shields, nor heating element should have high absorption or incoherent cross sections, if at all possible.
- It should be possible to insert and change samples in the furnace without moving or dismantling the furnace.
- A "window" is machined in the furnace wall to allow neutrons to pass through a minimum amount of material, usually "Dural" (Al-40% Cu). The wall thickness is about 0.1 mm. Tantalum normally is used for the heating element.

The choice of heating-element material results from consideration of several factors: operating temperature, power supply (i.e., material resistance), stability against embrittlement, stability against thermal cycling.

Candidate materials for heating elements are tantalum, molybdenum, tungsten, vanadium, niobium, these being refractory or high-melting materials suitable for operation up to 2000°C. Unfortunately, molybdenum and tungsten have low specific resistivities, which eliminates them for high-temperature work, despite their being relatively inert (less tendency to form oxides, for example). In contrast, vanadium and niobium have high specific resistances, but are unstable in the presence of oxygen or nitrogen at high temperatures. The remaining material, tantalum, is preferred, since it also has a high specific resistance, but is not so active chemically. Nevertheless, it does become brittle sooner than the other materials, and care should be taken to achieve high vacuum (around 10^{-5} Torr) before applying heat.

The neutron scattering characteristics are not as critical for heater and furnace materials as they are for the sample, since the material can be very thin. Typical heating-element thickness is about 0.1 mm. Also, the scattering is "diffused," by which is meant that the detectors do not "see" all scattering events at the furnace and its components because of the geometry and so-called soller slits in front of the detectors. This is illustrated in Figure 9, which shows a typical slit arrangement.

Samples sealed in their containers normally are inserted from the top of the furnace. Neutron radiography reveals the position of the sample in the beam and can be used for adjustment. Polaroid film is used in conjunction with a neutron-activated scintillator located within a standard Polaroid camera.

Furthermore, very often melting point can be detected by a sudden darkening of the sample area on the film at the melting temperature, when a series of exposures is taken as the temperature is brought up to the melting temperature. Photographs can therefore be used as sample-temperature indicators.

Finally, it should be noted that it is recommended that the furnace design include a crucible positioned below the sample and out of the neutron beam to catch the sample should a rare breakage of the container occur (isotope samples can cost up to $2000 each).

A photograph of a typical furnace for use to temperatures of ~1000°C is given in Fig. 10. Figure 11 is a photograph of the furnace in position on the diffractometer.

8. Neutron Beam Facilities

8.1. Diffractometers

The principal user-oriented locations for carrying out neutron diffraction studies include the reactors and linear accelerator (LINAC) at Harwell; the high-flux reactor at the Institute Laue-Langevin (ILL), Grenoble; the intense pulsed neutron source (IPNS) at Argonne National

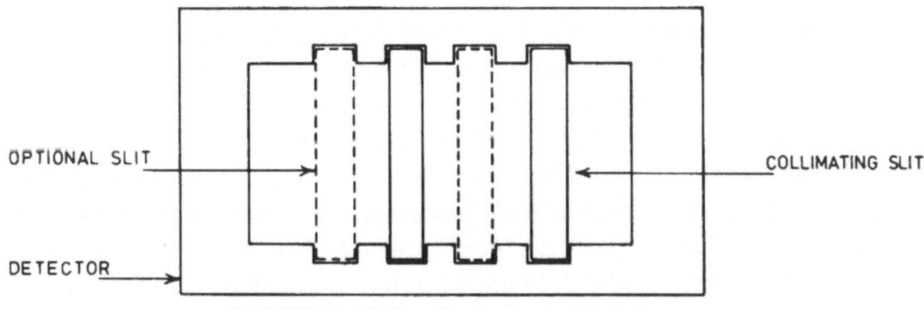

FIGURE 9. Soller slits in front of a detector.

FIGURE 10. Photograph of the furnace used on the DIDO reactor, Harwell. Temperatures up to about 1000°C can be reached. Note the long quartz sample tube, which is loaded from the top of the furnace when the latter has been assembled.

Laboratory; the high-flux reactor (HFIR) at Oak Ridge National Laboratory; in Japan the Tohoku LINAC, the JAERI reactor, and the KENS pulsed neutron source at the KEK National Laboratory for High-Energy Physics; and, near Paris at Orphee, the SACLAY reactor. Examples of diffractometers at some of the sources mentioned are listed in Table III. The spallation neutron source (ISIS) at the Rutherford–Appleton Laboratory is now also available for use, and promises well since it is designated as the world's most powerful neutron source.

At Harwell, U.K., there is a low-flux reactor, DIDO, which has diffractometers suitable for liquid structure study. There is also a linear accelerator, the LINAC, which produces neutrons at diffractometers also suitable for liquid structure determination. In March 1984, the SERC (Science and Engineering Research Council, U.K.) withdrew direct support for experimental work of University Users on both the reactor and the LINAC. Thus, in Europe, there remains for general use only the ILL reactor, and ISIS at the Rutherford–Appleton Laboratory, which is becoming routinely available. On the other hand, the diffractometer D4 at the ILL enables collection of top-quality data with excellent statistics, having been improved considerably when the diffractometer was rebuilt during 1981–1983. In Europe, this is at present the source of data with the highest statistics. At the SACLAY reactor, there are facilities for carrying out liquid structure, but this reactor

FIGURE 11. Photograph of the Harwell furnace in place on the Curran diffractometer. (Courtesy of the MPD, AERE Harwell, U.K.) Note the adjustable slit size at the neutron beam exit; the five rotatable neutron detectors at the rear; the cooling water pipes; and the vacuum pump set at the rear above the detectors.

TABLE III. Neutron Sources and Some Available Diffractometers

Site	Source	Diffractometer	k range (Å^{-1})
Harwell	25-MW reactor DIDO	Curran	0.8–10
		10HPD	0.8–11
	136-MeV LINAC	TSS (total scattering spectrometer)	0.8–60
Rutherford–Appleton	ISIS (spallation neutron source)	LAD (from Harwell LINAC)	0.5–100
ILL (Grenoble)	High flux reactor	D4	0.4–24
IPNS (Argonne)	500-MeV accelerator (pulsed neutrons)	GPPD (general purpose powder diffractometer)	0.5–100
		SEPD (special environment powder diffractometer)	0.5–40
ORNL (Oak Ridge)	HFIR (high flux reactor)	HB2A beam line	0.8–13

is neither equipped nor organized to provide an important user service. Experimental proposals are, in the main, only accepted from French Institutions.

Let us consider one facility (the IPNS) in more detail in order to gain a feel for the processes involved in selecting a diffractometer for an experiment.

At IPNS, few liquid experiments have been carried out yet, since there is not a dedicated liquids diffractometer there at present. The usable diffractometers are currently set up for powder analysis, which requires optimization of quite different parameters from those critical in amorphous studies. However, the next few years should see design and commission of a suitable instrument. The current layout of instruments at the IPNS is shown in Fig. 12. Diffractometers are labeled. The SEPD is more successful as a liquid diffractometer than the GPPD, Table III, for two reasons: (i) it is positioned nearer to the neutron source than the GPPD (14 m cf. 20 m, hence less loss of flux); (ii) neutron counters are available at lower scattering angles, thus exploiting more lower-wavelength neutrons for a given k. This is an advantage since absorption increases with wavelength and can be an insurmountable problem (e.g., ^{6}Li). Furthermore, long-wavelength neutrons are more subject to inelastic effects (studied in particular by Placzek,[15] and hence termed "Placzek corrections"), for which adequate corrections are more complicated (Section 9).

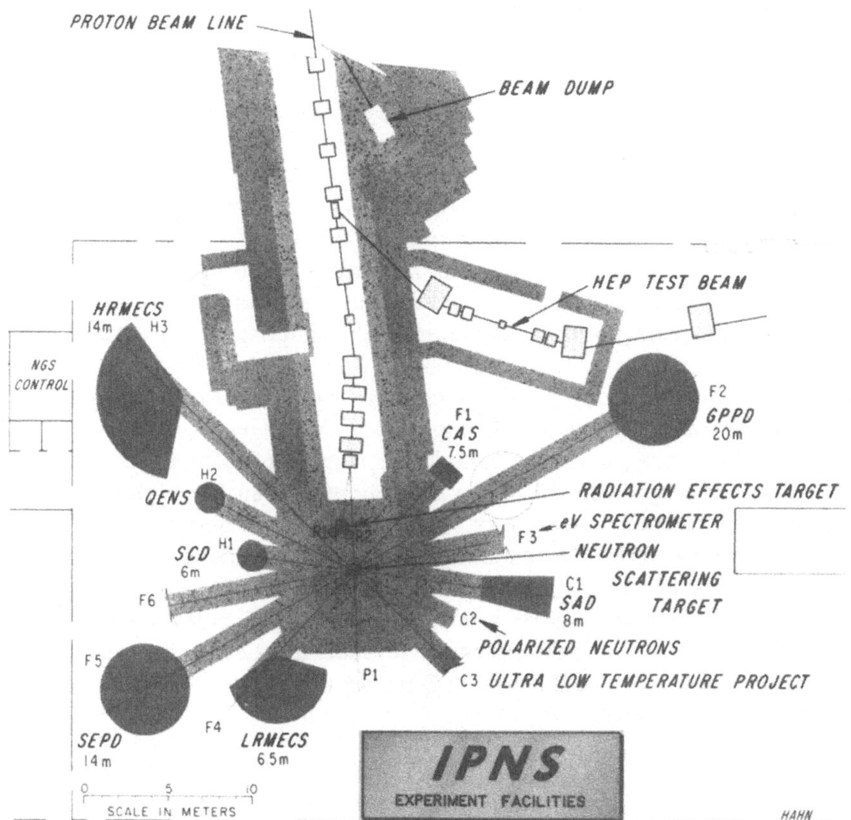

FIGURE 12. Layout of diffractometers around the intense pulsed neutron source (IPNS) monolith. The acronyms HRMECS, QENS, SCD, SEPD, LRMECS, SAD, and CAS represent, respectively, the high resolution medium energy chopper spectrometer, quasielastic neutron scattering (proposed), single-crystal diffractometer, special environment powder diffractometer, low-resolution medium-energy chopper spectrometer, small-angle diffractometer, and crystal analyzer spectrometer. (Courtesy of IPNS, Argonne National Laboratory.)

At Oak Ridge National Laboratory (ORNL), liquids have been studied extensively on the HB2A beam line, by Narten and co-workers (see Ref. 16, for example), but few melts have been investigated there to date. However, the program is expanding somewhat to include high-temperature studies. Fairly recently, a new position-sensitive detector has been installed, which will reduce counting time dramatically. Once commissioned, this detector will be available for counting neutrons from molten salts, as well as other liquids. Unfortunately, ORNL does not support a neutron user community in the manner that ILL, IPNS, and Rutherford-Appleton do, except for small-angle neutron scattering (SANS), for which it is the national (U.S.) center. However, approaches can be made for beam time.

8.2. Experiment Time Scales

In order to carry out an experiment at ILL or IPNS, proposals are submitted approximately twice yearly to the scientific secretary, for approval by a Program Committee. Beam time (around 5 days at either location for a simple melt structure experiment) is allocated, on merit, 1–3 months in advance of the experiment date. Around 6–12 months elapse between submitting a proposal and carrying out an experiment.

Sample preparation is always the key element in a neutron scattering experiment. When working to a fixed schedule, often announced with little notice, it is best to begin sample preparation ahead of time, in the expectation of being allocated beam time. Certainly it is expedient to have prepared and carry a spare set of samples (a different melt), in order to be able to face contingencies such as the chosen melt not being ready, isotopes being delayed at Customs, loss of a sample, or delay during final preparation (dehydration) stages. Since neutron beam time is at such a high premium and the experimenter has to work to someone else's schedule, all aspects of the experiment must be available and in perfect working order for the allocated few days. In this respect, it is therefore wise to run the furnace at the required temperature for a few days in advance. Furthermore, since all runs, including container and empty furnace, must be carried out at the melt temperature, one must ensure that the heating element will survive the entire experiment time.

9. Data Analysis

9.1. Introduction

Data reduction can often be carried out, at least in part, on the computer at the neutron source institution. The reduction to total scattering function $F(k)$, from intensity as a function of angle $I(\theta)$, or angle and wavelength in the case of a pulsed source, involves correcting the data for absorption, incoherent and multiple scattering, and inelastic effects. These latter are the Placzek corrections. They are most prominent, being a recoil effect, for low-atomic-weight scattering nuclei such as deuterium or lithium. It is usually convenient for the user to progress at least this far in the data reduction before leaving the neutron collection site, so that $F(k)$ can be taken away from each sample, and these data treated more conveniently on the home institution's computer.

Computer code is normally available at the neutron source institution, or can be developed by the user. Once $F(k)$ is obtained, it is a simple matter to Fourier transform to obtain the r-space data. Combination of, say, three $F(k)$ to obtain three $S(k)$, the partial structure factors, is also a straightforward procedure. From Eq. (3), we have

$$\begin{aligned} F_1(k) &= a_{11}S_{\alpha\alpha}(k) + a_{12}S_{\beta\beta}(k) + a_{13}S_{\alpha\beta}(k) \\ F_2(k) &= a_{21}S_{\alpha\alpha}(k) + a_{22}S_{\beta\beta}(k) + a_{23}S_{\alpha\beta}(k) \\ F_3(k) &= a_{31}S_{\alpha\alpha}(k) + a_{32}S_{\beta\beta}(k) + a_{33}S_{\alpha\beta}(k) \end{aligned} \tag{8}$$

where the a_{ij} involve concentrations and scattering lengths. Simple inversion of the matrix $[a]$ where

$$[F(k)] = [a][S_{ij}(k)] \tag{9}$$

yields the $S_{ij}(k)$. The matrix $[a]$ should be optimized so that solution for $S_{ij}(k)$ is best conditioned. This is done by choosing the scattering lengths of the isotopic mixtures used. Furthermore, it is inevitable that the errors involved in each $S_{ij}(k)$ are different from each other: one can choose to obtain highest accuracy in a particular $S_{ij}(k)$ at the expense of the others. When deciding to minimize errors on a particular $S_{ij}(k)$, one must bear in mind the isotopes available, and their cost. Sometimes, very much smaller errors could be obtained on the data by using a better choice of isotope [Eq. (9)], but the price of the isotope is prohibitive. Consequently, compromises using natural materials are often necessary.

The transform to r space can involve problems resulting from truncation of the data, when the data run to only around 10 Å^{-1}, rather than, say 20 Å^{-1} or even higher k values. This truncation can also result in broadening and slight shift of the main peak in $g_{ij}(r)$, i.e., the first ion–ion separation. High-k data are one of the promises of the pulsed sources, since the neutron wavelength spread permits large values of k to be reached [Eq. (2)].

Nevertheless, the first peak position is a very reliable parameter obtainable from neutron scattering experiments. Errors are typically 0.03 Å. A great deal can be deduced about ion sites (tetrahedral, octahedral, . . .) using this information, together with a knowledge of coordination number, n. This number if obtained from $g_{ij}(r)$ by integrating under the relevant peak, through one of several similar definitions. A more detailed discussion of this procedure is given by Waseda.[17]

Integration can be carried out to the first minimum in $g(r)$ or $r^2g(r)$. Alternatively, the first peak in $rg(r)$ or $r^2g(r)$ can be made symmetrical about the peak position using the low-r edge and integration performed under this artificial peak. Waseda describes and justifies each technique.

Some contention surrounds the definition of coordination shells. This is necessarily so, since it is integrals of $g_{ij}(r)$ over *all* r that have significance and relate to thermodynamic functions. From superimposed plots of all partial structure factors from one melt, one can see immediately the degree of penetration of certain ions into other coordination shells.[18] As an example, compare the amount of penetration of chloride ions into the Zn–Cl and Ba–Cl first coordination shells in Figs. 5 and 6—it is a function of ion size. This penetration obviously affects the integrations made, and therefore coordination number.

9.2. Magnetic Scattering

Certain elements contribute to the total scattering from a sample by virtue of their unpaired electrons. Manganese, neodymium, and iron, for example, show this effect. The result is that the k-space data are somewhat high at low k values and at higher values there is a gradual drop in the scattering data. This is consistent with a magnetic (m) scattering contribution of the form $b_m \exp(-ak^2)$, where b_m is the zero-k value cross section, a is a constant, and k is a wave vector. A magnetic correction using this magnetic form factor can be applied, thus correcting $F(k)$ data to the absolute values.

10. Melts Studied Using Neutron Diffraction

The technique of neutron diffraction, using isotopes, and applied to the structure of melts, has been used by four groups in particular—the Bristol group, the Leicester group, the Oxford group, and the French group, efforts led by Enderby, Howe, Mitchell, and Dupuy, respectively. X-ray diffraction of a very wide variety of salts, on the other hand, has to a large extent been carried out by Japanese workers, principally Ohno, Furukawa, and colleagues (see Ref. 9, for example). The melts fall mainly into three categories. These are 1:1 salts, 2:1 salts (both single melts), and binary melts.

A full review of all melts studied to date has been given by Enderby and Biggin.[4]

10.1. Melts of the Type MX

Most of the 1:1 salts studied by neutron diffraction have been melts having the NaCl crystal structure in the solid state: NaCl,[7,9,11] KCl,[19] RbCl,[20] and AgCl.[21] The structure of solid cuprous chloride, CuCl, however, is quite different, being the zinc blende structure which has tetrahedral symmetry in the solid (as does $ZnCl_2$). The NaCl-type salts change cation–anion coordination on melting from 6 (octahedral) to about 4. Figures 13 and 14 show the $S_{ij}(k)$ and $g_{ij}(r)$ for NaCl, typical curves for a NaCl-type melt. Cuprous chloride[22] has quite a different structure (Fig. 15). The lack of features in the CuCu curve is believed to reflect the electronic configuration of the cation, but interpretation of the curves for CuCl is not yet satisfactorily complete.

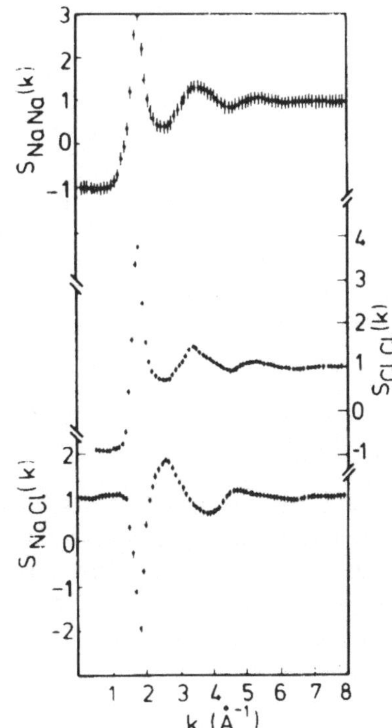

FIGURE 13. Partial structure factors, $S_{ij}(k)$, for NaCl.

10.2. Melts of the Type MX_2

The various melts of the 2:1 series studied by neutron scattering fall into two groups: (i) the solid-state layer lattices such as $MgCl_2$,[(23)] $MnCl_2$,[(23)] $NiCl_2$,[(24)] and $ZnCl_2$,[(25)] and (ii) the others, $CaCl_2$,[(18)] $BaCl_2$,[(26)] and $SrCl_2$.[(27)] With the layer lattices ($CdCl_2$ type, except for $ZnCl_2$, which has several forms), the coordination-number changes upon melting from six

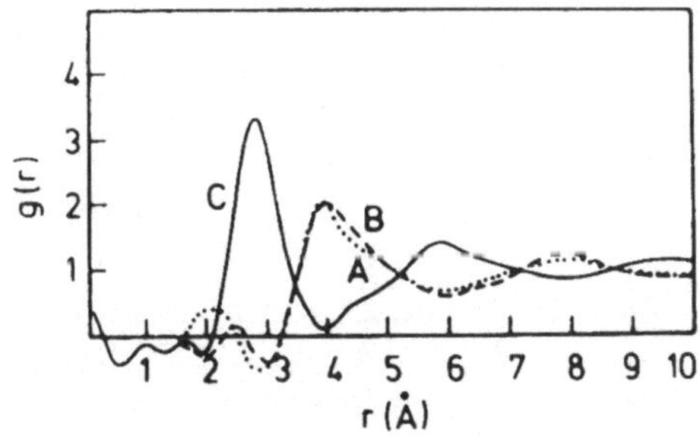

FIGURE 14. Partial radial distribution functions, $g_{ij}(r)$, for NaCl. A, NaNa; B, ClCl; C, NaCl.

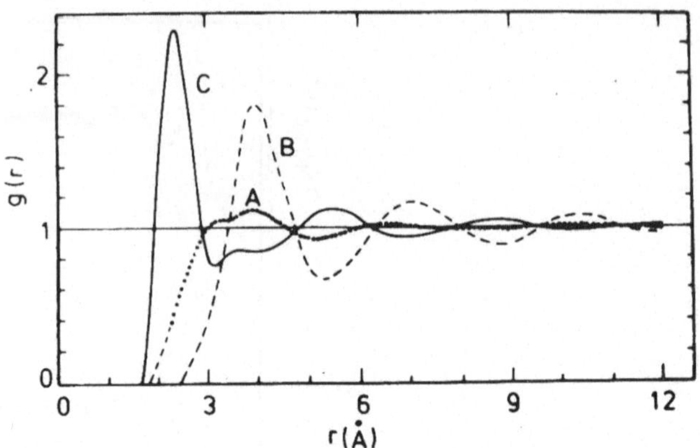

FIGURE 15. Partial radial distribution functions, $g_{ij}(r)$, for CuCl. A, CuCu; B, ClCl; C, CuCl.

(octahedral) to about four. Further general results are that the Cl–Cl correlation curves, shown in Fig. 16, show a dependency on cation size: the peak remains in a similar position, while the peak shape broadens and becomes lower with increase in cation size. The Cl–Cl coordination number of about 8 varies little with the various melts. Figures 1–6 show examples of $S_{ij}(k)$, $F(k)$, $g_{ij}(r)$, and $G(r)$ for these salts.

10.3. Binary Melts

A few binary melts have been planned for study using neutron scattering with isotopes. Mentioned in Section 2.2, the chloroaluminate, $LiAlCl_4$, is of great interest in aluminum production and advanced batteries, since it is a basic melt similar to but simpler than those actually used in industry.

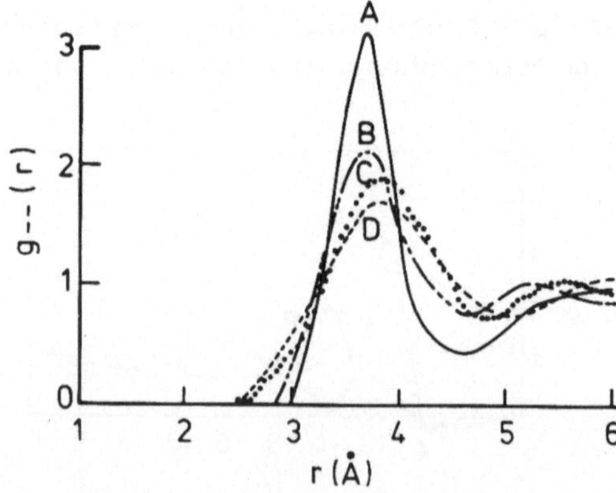

FIGURE 16. Chlorine-chlorine partial radial distribution functions for various 2:1 melts: A, $ZnCl_2$; B, $CaCl_2$; C, $SrCl_2$; D, $BaCl_2$.

A study of $LiAlCl_4$ on standard diffractometers—D4 at ILL and the GPPD at IPNS—was carried out by Biggin, Enderby, and Blander.[28,29] Absorption from the ^6Li salt used was predictably high, increasing linearly with wavelength, and this made the GPPD analysis very difficult. The ILL data were of excellent quality, however. It was confirmed that the aluminum ion is tetrahedrally coordinated by chloride ions, at a separation of 2.13 ± 0.03 Å. Comparison with molecular dynamics of Saboungi *et al.*[30] showed overall agreement, but with a few slight differences. For example, the peak height of the Al–Cl rdf was a little lower in the experiment, indicating that the effective pair potential in the real system is softer than that used in the MD calculations.

An x-ray study of this melt has also been carried out by S. Takahashi *et al.*[31] These x-ray results agree with those from neutron scattering but, in the latter case, they are more extensive since the isotope method was used.

Other plans include study of binary melts $AX_m x(BY_n)$ to observe changes in coordination number as x is varied.

10.4. Melts Discussion

As a general rule for simple *ionic* melts, the first peak position in $g_{+-}(r)$, indicating the cation–anion separation, occurs at *approximately* a distance equal to the sum of the cationic and anionic radii. This is borne out by the figures taken from Table IV for ionic radii; a comparison of ionic radii sum and experimental separations is given in Table V. Of the melts studied to date,[4] the only two that show large differences between the two figures are CuCl and $ZnCl_2$. This discrepancy is taken to indicate some degree of covalency in the melt. Overall, comparison studies have shown that cation size is the dominant parameter that scales the anion substructure, certainly in the case of the 2:1 salts.[27] The electronic structure of the cation obviously has a strong influence, as is the case with $ZnCl_2$, for example.

TABLE IV. Ionic Radii (Å)[a]

Ag^+	1.26	Li^+	0.68
Al^{3+}	0.51	Mg^{2+}	0.66
Ba^{2+}	1.34	Mn^{2+}	0.80
Ca^{2+}	0.99	Na^+	0.97
Cl^-	1.81	Nd^{3+}	1.00
Cs^+	1.67	Ni^{2+}	0.69
Cu^+	0.96	Rb^+	1.47
Fe^{2+}	0.74	Sr^{2+}	1.12
K^+	1.33	Zn^{2+}	0.74

[a] Reference 32.

TABLE V. Cation-Anion Separations for Some Melts

Melt	$r_+ + r_-$ (Å) (Table IV)	r_{+-} (Å) (neutron experiment)	Reference	r_{+-} (Å) (x-ray experiment)	Reference
$BaCl_2$	3.15	3.10	26	—	—
$CaCl_2$	2.80	2.78	18	—	—
CuCl	2.77	2.3	22	—	—
LiCl	2.49	—		2.40–2.47	33
$MgCl_2$	2.47	2.42	23	—	—
$MnCl_2$	2.61	2.50	23	2.51	34
NaCl	2.78	2.78	7	2.73	9
$SrCl_2$	2.93	2.90	27	—	—
$ZnCl_2$	2.55	2.29	25	2.29	35

As regards coordination number change on melting, this value can usually be predicted from volume change on melting, as discussed by Ohno and Furukawa.[9]

11. Summary

Neutron or x-ray scattering enables total ion–ion correlation functions to be determined for most molten salts. The use of isotopes in addition in neutron scattering enables extraction of partial ion–ion correlation functions. This specific information is necessary in order to determine the ion–ion correlation functions of the melt unambiguously. These functions are fundamental to the melt structure, giving ion–ion separation and coordination numbers, and any indication of covalency.

The technique, certainly at steady-state sources, is well established as regards theory and data processing. This is not yet the case for pulsed sources. The technique, too, is straightforward experimentally, once a reliable furnace has been designed. It is the planning and meeting of schedules of just a few days' beam time, the allocation of which depends competitively on submission of proposals, however, that present limitations to the day-to-day use of this powerful technique.

Acknowledgments

I would like to thank John Enderby, with whom I work, for general guidance and discussion, and Roger Sinclair at AERE Harwell for reading this manuscript and making useful suggestions; and to acknowledge the Science and Engineering Research Council (U.K.) for financial support.

References

1. G. Bacon, *Neutron Diffraction*, Oxford University Press, London (1975).
2. G. L. Squires, *Thermal Neutron Scattering*, Cambridge University Press, Cambridge, England (1978).
3. B. E. Warren, *X-ray Diffraction*, Addison-Wesley, Reading, Massachusetts (1969).
4. J. E. Enderby and S. Biggin, in: *Advances in Molten Salt Chemistry* (G. Mamantov, ed.), Vol. 5, pp. 1–25, Elsevier, Amsterdam (1983).
5. J. E. Enderby, *J. Phys. C* **15**, 4609 (1982).
6. R. L. McGreevy and E. W. J. Mitchell, *J. Phys. C* **17**, 775 (1984).
7. S. Biggin and J. E. Enderby, *J. Phys. C* **15**, L305 (1982).
8. A. H. Narten, private communication (1983).
9. H. Ohno and K. Furukawa, *J. Chem. Soc. Faraday Trans. I* **77**, 1981 (1981).
10. L. Koester and W. B. Yelon, *Compilation of Low Energy Neutron Scattering Lengths and Cross Sections*, ECN—Netherlands Energy Research Foundation, Petten, Netherlands (1982).
11. F. G. Edwards, J. E. Enderby, and D. I. Page, *J. Phys. C* **8**, 3483 (1975).
12. S. F. Mughabghab and D. I. Garber (eds.), *Neutron Cross Sections*, 3rd edn., Vol. I, BNL Report No. 325 (1973).
13. J. M. Bouchard, *J. Four Electr. Ind. Electrochim.* **10**, 16 (1979).
14. *Inorganic Syntheses*, McGraw-Hill, New York, Vol. 1, pp. 28–33 (1939).
15. G. Placzek, *Phys. Rev.* **86**, 377 (1952).
16. S. Biggin, J. E. Enderby, R. L. Hahn, and A. H. Narten, *J. Phys. Chem.* **88**, 3634 (1984).
17. Y. Waseda, *Structure of Non-Crystalline Materials: Liquids and Amorphous Solids*, McGraw-Hill, Maidenhead, U.K. (1980).
18. S. Biggin and J. E. Enderby, *J. Phys. C* **14**, 3577 (1981).
19. Y. Derrien and J, Dupuy, *J. Phys. Paris* **36**, 191 (1975).
20. E. W. J. Mitchell, P. F. J. Poncet, and R. J. Stewart, *Phil. Mag.* **34**, 721 (1976).
21. Y. Derrien and J. Dupuy, *Phys. Chem. Liquids* **5**, 71 (1976).
22. S. Eisenberg, J.-F. Jal, J. Dupuy, P. Chieux, and W. Knoll, *Phil. Mag.* **46A**, 195 (1982).
23. S. Biggin, M. Gay, and J. E. Enderby, *J. Phys. C* **17**, 977 (1984).
24. A. Howe, private communication (1983).
25. S. Biggin and J. E. Enderby, *J. Phys. C* **14**, 3129 (1981).
26. F. G. Edwards, R. A. Howe, J. E. Enderby, and D. I. Page, *J. Phys. C* **11**, 1053 (1978).
27. R. L. McGreevy and E. W. J. Mitchell, *J. Phys. C* **15**, 5537 (1982).
28. S. Biggin and M. Blander, IPNS Progress Report 1981–83, Argonne National Laboratory (1983), p. 109.
29. S. Biggin, J. E. Enderby, and M. Blander, Proc. 168th Meeting ECS, Las Vegas, October 13–18, 1985.
30. M.-L. Saboungi, A. Rahman, and M. Blander, *J. Chem. Phys.* **80**, 2141 (1984).
31. S. Takahashi, T. Muneta, N. Koura, and H. Ohno, *J. Chem. Soc. Faraday Trans. II* **81**, 319–326 (1985).
32. *CRC Handbook of Chemistry and Physics*, 63rd edn., 1982–1983, CRC Press, Boca Raton, Florida, p. F-179.
33. H. A. Levy and M. D. Danford, in: *Molten Salt Chemistry* (M. Blander, ed.), Wiley, New York (1964).
34. H. Ohno, F. Furukawa, K. Tanemoto, Y. Tagaki and T. Nakamura, *J. Chem. Soc. Faraday Trans. I* **74**, 804 (1978).
35. R. Triolo and A. H. Narten, *J. Chem. Phys.* **74**, 703 (1981).

6

Dry Boxes and Inert Atmosphere Techniques

Duane E. Bartak

1. Introduction

There is generally a need to conduct experiments with molten salts in an inert atmosphere owing to the chemical reactivity of these media. Most fused salt solvent systems show reactivity to water and/or oxygen, which results in significant changes in the chemical properties of these solvents. Although such solvent systems (i.e., binary or ternary salt mixtures) are usually operated at the lowest liquidus operating temperature possible, elevated temperatures result in increased reactivity. The lithium chloride-potassium chloride eutectic at 450°C, which is probably the most studied of fused salt systems, absorbs water very rapidly and therefore must be kept under a dry, inert atmosphere.[1] Although several studies on the nature of an oxygen/"oxide" electrode in this melt have been reported, the reactions have not been fully understood because of apparent water contamination in many cases.[2,3] Nitrate melt systems are also hygroscopic; for example, the $LiNO_3$-KNO_3 eutectic (177°C) absorbs water to at least 0.2% by weight.[4] The result is that the electrochemistry of heavier, electropositive metal ions can be significantly altered.[5] In addition, trace amounts of water have been shown to significantly affect the oxygen-oxide redox chemistry in $NaNO_3$-KNO_3 melts (250°C).[6] Electrochemical studies in these fused salts require specially designed, controlled-atmosphere cells.[7]

The haloaluminates, which include $AlCl_3$-NaCl (175°C) as well as $AlCl_3$-organic salt binaries (e.g., $AlCl_3$-*n*-butylpyridinium chloride, 40°C), are particularly sensitive to the presence of both oxygen and moisture. Hydrolysis with trace amounts of water results in the evolution of HCl and

Duane E. Bartak • Department of Chemistry, University of North Dakota, Grand Forks, North Dakota 58202.

probable formation of aluminum oxychlorides.[8,9] Subsequent formation of insoluble aluminum oxides will affect the melt composition with resultant changes in the acid-base chemistry of these melts.[10]

Both oxygen and water must be kept away from molten hydroxide systems (e.g., NaOH–KOH eutectic, 227°C). Oxygen is a strong oxidant in the fused hydroxide systems with formation of superoxide ion from either oxide or water. In addition to water being a major contaminant in these melts, it is the conjugate acid of hydroxide ion and thus is important in acid-base chemistry (i.e., $2OH^- \rightleftarrows H_2O + O^{-2}$).[11] Although many metal-corrosion studies have been carried out in fused hydroxides, little quantitative information can be deduced because of probable water contamination.[12] Thus, it is imperative that *well-controlled* and *inert atmospheres* are utilized with most molten salt systems.

There are several excellent monographs that deal in a general manner with inert atmosphere techniques. In particular, Shriver's text on techniques for handling air-sensitive materials remains a standard.[13] Additionally, there are published chapters in serial texts, which deal with general glove box techniques, particularly with regard to nuclear chemistry.[14] Furthermore, there are descriptions for the use of glove boxes and glove bags, which are available from the manufacturers of these devices.[15] The reader is referred to the above references for specific information on these subjects. The objective of this chapter is to describe general, inert atmosphere techniques, which can be used by the molten salt experimentalist. Because of the obvious limitations of volatility, vacuum manipulations will not be considered. Rather, the use of glove boxes, glove bags, and inert bench-top techniques will be discussed. Because of the background of the author much of the material described and the references utilized are from the electrochemical molten salt literature. However, the information should be of use to anyone who plans to carry out reactions in molten salt media. The areas to be covered are (1) glove box and bag equipment, (2) operation and maintenance of glove boxes and glove bags, and (3) common operations conducted inside glove boxes. This chapter complements that by Reavis (Volume 3, Chapter 2), which specifically deals with actinide handling within glove boxes.

2. *Equipment*

The first part of this section will describe briefly the basic components of an inert atmosphere system, starting with the simplest, least expensive components and continuing with increasingly complex systems. The second part will discuss some of the accessory components available that enhance operations of more complex systems. There are three basic components of any inert atmosphere system. The first component is an enclosure that is utilized to perform an experiment or operation that contains the desired inert atmosphere. The second component is an airlock or transfer port used for moving material into and out of the enclosure without contaminating the atmosphere within. The third component is a source of gas, which will constitute the atmosphere that will be maintained. Additional components for more complex systems include provisions for circulating and purifying the atmosphere being maintained and to regenerate the purification system when needed. Much accessory equipment is available for complex systems for such ancillary tasks as automatic pressure control, automatic cycling, atmosphere monitoring, and atmosphere analysis.

2.1. Glove Bags

The simplest enclosure for maintaining an inert atmosphere is a plastic bag that usually has attached gloves and hence is called a glove bag (Fig. 1). Usually there is a small opening in the bag to connect a gas source. In addition, a larger opening, called an equipment sleeve, that is used to pass material into and out of the bag can be obtained on larger glove bags (Fig. 1). This larger opening is sealed when not in use. The bag is kept inflated by the pressure of the filling gas or it can be supported by an internal rigid frame. A pamphlet called *85 Tips and New Uses for Glove Bag,* prepared by Instruments for Research and Industry (P.O. Box 159N, Cheltenham, Pennsylvania, 19012), provides an excellent discussion of glove bags and their uses.[15]

The major advantages of glove bags are their ease of use and low cost. Glove bags usually are individually packaged. The glove bag is simply

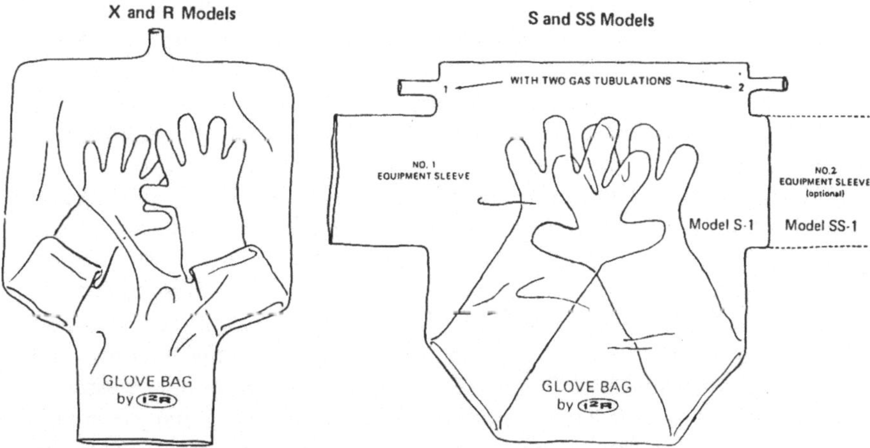

FIGURE 1. **Polyethylene glove bag designs which are commercially available. (Courtesy of Instruments for Research and Industry, 108 Franklin Ave., Cheltenham, Pennsylvania.)**

removed from the manufacturer's package and a gas cylinder is connected to the bag. The bag is purged with inert gas from the cylinder while introducing the necessary equipment into the bag through the equipment sleeve. The equipment sleeve is then sealed with a special sleeve seal, or by simply rolling the sleeve closed and clamping the fold when the enclosure is ready for use. In the case of glove bags, the transfer port is simply an opening large enough to pass equipment through. When the sleeve is open, gas flowing through the bag prevents contaminating atmosphere from entering the bag. The port is closed simply by clamping the sleeve closed. When not being used, the glove bag can easily be stored, or disposed of in the event of contamination. They are useful, in particular, when heavy contamination might result from a reaction or chemical manipulations. The major drawback of glove bags is their limited size. While they may be useful for transferring chemicals between containers or for small temporary experiments, they are not feasible for larger, long-term experiments.

2.2. Glove Boxes

The enclosure device for larger, continuous experiments is the glove box, which is a rigid box with openings for gloves and usually with an airlock to pass materials into and out of the box. In addition, valves necessary for filling and purging the box atmosphere are usually present (Fig. 2). Glove boxes are larger than glove bags and can be used on a more permanent basis. Glove boxes are constructed from many different kinds of materials including rigid plastic (e.g., Lucite), fiberglass-reinforced plastic, mild steel, stainless steel, and aluminum. Aluminum is probably the most common material used in the commercial glove boxes[(16)] that are utilized by molten salt experimentalists. In addition, stainless steel has been used in both commercial[(17)] and custom-designed glove boxes. For example, a unique custom-designed stainless steel box, which is evacuable, has been described in the literature for molten salt studies.[(18)] The commercial boxes (both stainless steel and aluminum) shown in Figs. 3 and 4 are typical examples of those used in several laboratories engaged in molten salt work.

Since viewing of the experiment in the box is necessary, an optically clear window is required. Typical materials for the windows are Lucite (polyacrylic), Lexan (a polycarbonate), and/or safety glass. Most commercial manufacturers use a combination of Lexan and safety glass to provide an inert, but flexible impact resistant window. An important consideration in glove box construction is the type of seals between the window, end panels, and other modules that make up the box. A high-quality elastomer that is relatively inert to many common chemicals is used in many commercial systems. Neoprene elastomers are probably utilized in most commercial systems. Gas-tight gaskets are also necessary to seal the glove ports to the

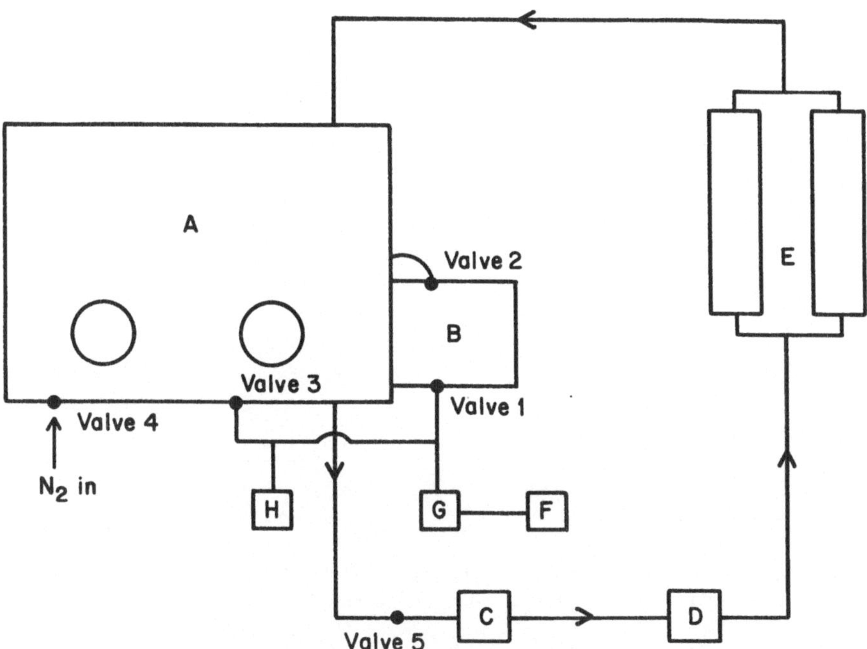

FIGURE 2. Schematic of a glove box system including a recirculating system. A, Glove box; B, entry port; C, dry ice–acetone traps; D, recirculating pump; E, oxygen scrubber; F, vacuum pump; G, Liquid nitrogen traps; H, McCleod gauge. [From Ref. 23, reprinted with permission from E. C. Ashy and R. D. Schwartz, *J. Chem. Ed.* **51,** 65 (1974).]

box. A light smear of a proprietary silcone vacuum grease or silicone rubber compound can be used when seals are replaced, for example, when major cleaning is done. A glove port, which is usually constructed of the same material as the box itself (e.g., aluminum), consists of a circular opening in the front of the box to which the gloves are attached.

The gloves are probably the source of most atmospheric contamination in glove box systems. Oxygen and/or moisture can diffuse through the glove walls as well as passing through small holes that have been created by usage. There have been several reported studies on the relative rates of diffusion of water vapor through various glove materials.[19-21] Butyl rubber has been shown to be the least permeable of the materials that have been tested. Other materials used for gloves, particularly by many European experimentalists, include neoprene rubber and natural rubber. Most chemists in the U.S. who work in the molten salt area use heavy-duty (e.g., 0.030-in. thickness) butyl rubber gloves. Clearly, a compromise between cost, availability, flexibility, and permeability, is called for. The types of boxes illustrated in Figs. 3 and 4 are the basic units around which more complex systems are built. In order to maintain an inert atmosphere for

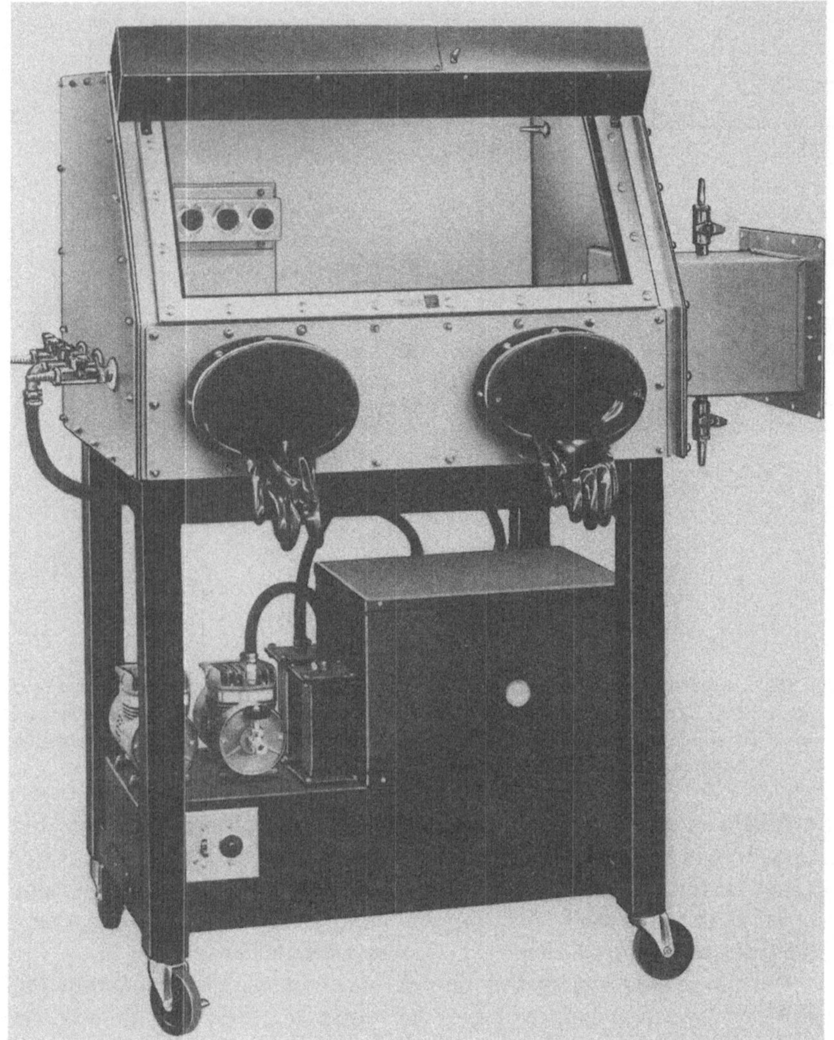

FIGURE 3. Stainless steel controlled-atmosphere dry box. (Courtesy of Kewaunee Equipment Corp., Laboratory Division, Statesville, North Carolina.)

long periods, additional equipment is required to maintain the inert atmosphere, thus adding cost and complexity to the simple glove box.

2.3. *Transfer Ports for Enclosures*

Glove boxes usually have attached to a side of the box a separate small compartment (often called an antechamber) with inner and outer doors.

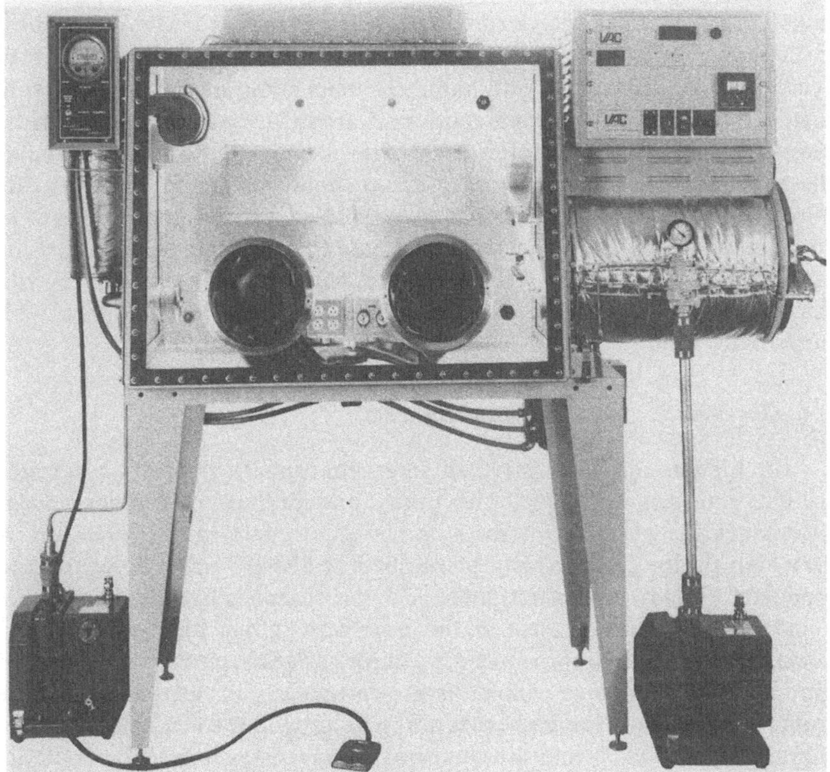

FIGURE 4. Aluminum controlled-atmosphere dry box. (Courtesy of Vacuum Atmospheres, 4652 W. Rosecrans Ave., Hawthorne, California.)

When passing material into the box, the inner door is kept closed while the material is placed in the transfer port. The outer door is closed and the antechamber, or port, is alternatively evacuated and filled with inert gas to establish an inert atmosphere in the port before opening the inner door to finally pass the material into the box. An important aspect of the antechamber is the degree of tightness of the doors, which is important in maintaining the integrity of the box. Most commercial port doors utilize O-ring seals, which can be treated with a high-vacuum grease (e.g., Apeizion N) to produce a high-vacuum (e.g., 10^{-4} mm) enclosure. A well-designed, heavy-duty clamping system on the port doors, which often consists of a cross-bar or several individual latches, is critical. Most ports should be constructed of heavy-duty aluminum or steel to withstand relatively good vacuums (e.g., 10^{-4} mm).

Commercial systems (Vacuum Atmospheres) are available to automatically control the evacuation and backfilling of the antechamber with the inert filling gas, which is taken directly from a gas tank or via the atmosphere in the box. The number and times of vacuum pumping cycles (up to four cycles) can be controlled with automatic backfilling of the chamber with inert atmosphere after each pumpdown. A secondary parameter to consider on such a system is the pumping speed of the vacuum pump used to evacuate the antechamber. Typical pumping speed should be greater than 160 liter min^{-1}, which is approximately 6 cfm. With this type of a pump, periods of 15–30 min usually are sufficient to evacuate the antechamber. Thus, using multiple pumping-refilling techniques (e.g., two to four), new materials with the exception of paper products can be introduced into the box with a 1–2-h turnaround time.

2.4. Gas Sources and Purification Systems

In molten salt chemistry, an inert atmosphere typically comprises helium, nitrogen, or argon. The purity and dryness of the source gas determines the ultimate inertness of the atmosphere in the enclosure. A very high purity gas (typically prepurified or 99.99% pure) is required to maintain a highly inert atmosphere. The gas source is usually drawn from a cylinder and passes through the necessary tubing and valves to the enclosure and transfer ports. In more complex systems, purifiers are included that continually remove contaminants, particularly oxygen, and moisture from the enclosure. The gases that are to be used in the box can be further purified prior to entry into the glove box system. Oxygen scrubber material including finely divided copper on an inert support (e.g., BTS catalyst) can be activated and placed in a column on the inlet gas line. Molecular sieves are often used to remove water. There are several commercial oxygen and moisture-removal systems which can be purchased in column-form or as gas-traps from gas chromatographic supplies (e.g., Alltech Associates, Deerfield, Illinois 60015). Most commercial gas purifier systems, which are enclosed in metal or glass tubing, have been designed to decrease the oxygen and moisture level to less than 1 ppm, a requirement for capillary gas chromatographic column operation.

Once the inert and purified gas has been introduced into the glove box system, it should be continually passed through a purification system. The need for a continuous-recirculating purification system arises owing to impurity introduction when new materials are brought into the antechamber. However, even if new material is not brought into the glove box, a recirculation–purification system is necessary because of diffusion of oxygen and moisture through the gloves and any leakages around gaskets and seals (e.g., O-ring seals on the antechamber). Figure 5 shows two schematic representations of inert atmosphere systems with purification–recirculation systems, which are commercially available. Purification–recirculation systems can be assembled using a large stainless steel (or glass) column or container (typically 15 cm × 40 cm) containing an oxygen-scrubber (e.g., BTS catalyst) and molecular sieves (Linde type). Gas blowers (Vacuum Atmospheres), which can be mounted in the glove box, are available with minimum flow rates of 5 cfm to recirculate the inert gas. These systems are claimed to be capable of reducing oxygen and moisture to 1 ppm or less by volume when recirculating argon, nitrogen, or helium. The authors have found that a 5 cfm flow rate train, as shown in Fig. 6, is sufficient for most aluminum chloride melt work. These systems are capable of recovery, after the introduction of several liters of air, to less than 1 ppm oxygen in a 25 ft^3 box volume after a few minutes (Fig. 6).

It is necessary to frequently regenerate or activate the oxygen-scrubber catalyst and molecular sieves. Thus a heater, which is mounted inside the

(a)

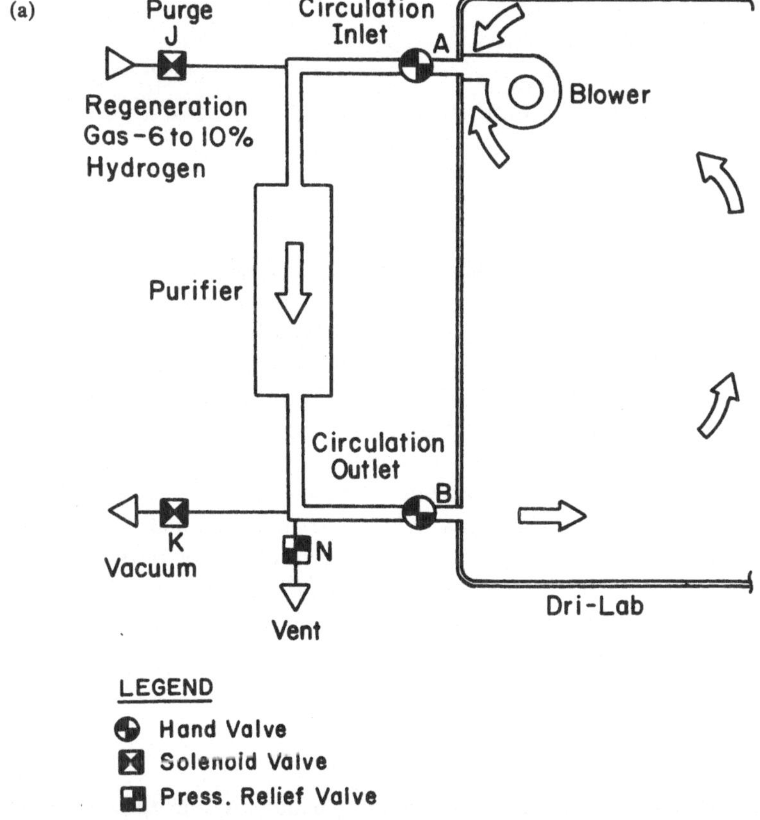

FIGURE 5. Schematic representations of two commercial purification–recirculation systems for dry boxes. (a) Vacuum Atmospheres' HE-493 recirculation system. (b) Kewaunee's dual-column recirculation system.

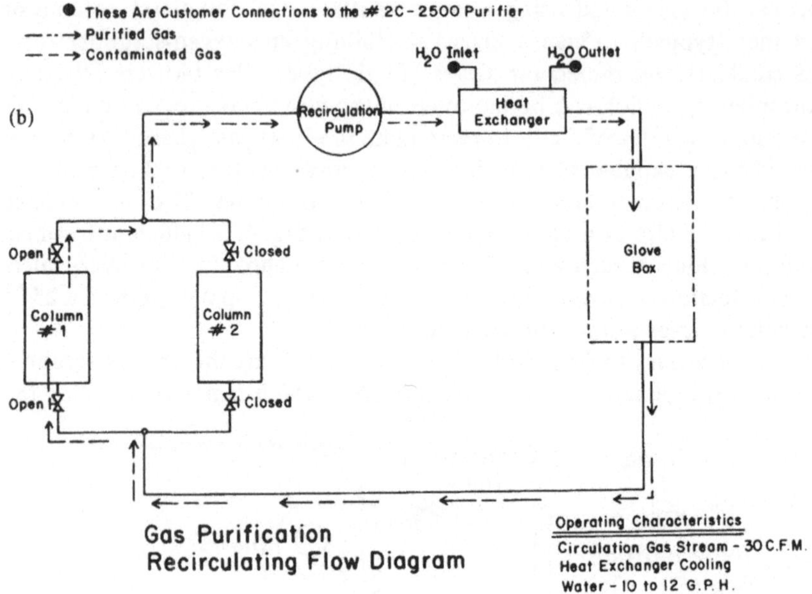

FIGURE 5. Continued.

column containing these materials, or which is wrapped around the outside of the column and well insulated, is necessary for activation. Most of the commercial purification systems have a manually initiated, but automatic programmer, which controls the heating time, purge time, and evacuation time on the purification train. A 10% hydrogen in nitrogen gas mixture is typically utilized in the purge cycle to reduce the copper oxide on the BTS catalyst to metallic copper. The final stage is an evacuation stage to remove water and other volatile material from the molecular sieves at 200°C.

Most molten salt experiments, which are to be carried out in glove boxes, should have a box equipped with a good recirculation purifier system. Without such a system, the quality of the atmosphere in an enclosure will degenerate with diffusion of oxygen and moisture through the gloves, gaskets, and any small holes in the overall box structure.

2.5. Accessories

Accessories are additional components, usually associated with more complex systems, that are not directly related to maintaining an inert atmosphere, but are necessary for automatic control or operation of the entire inert atmosphere system.

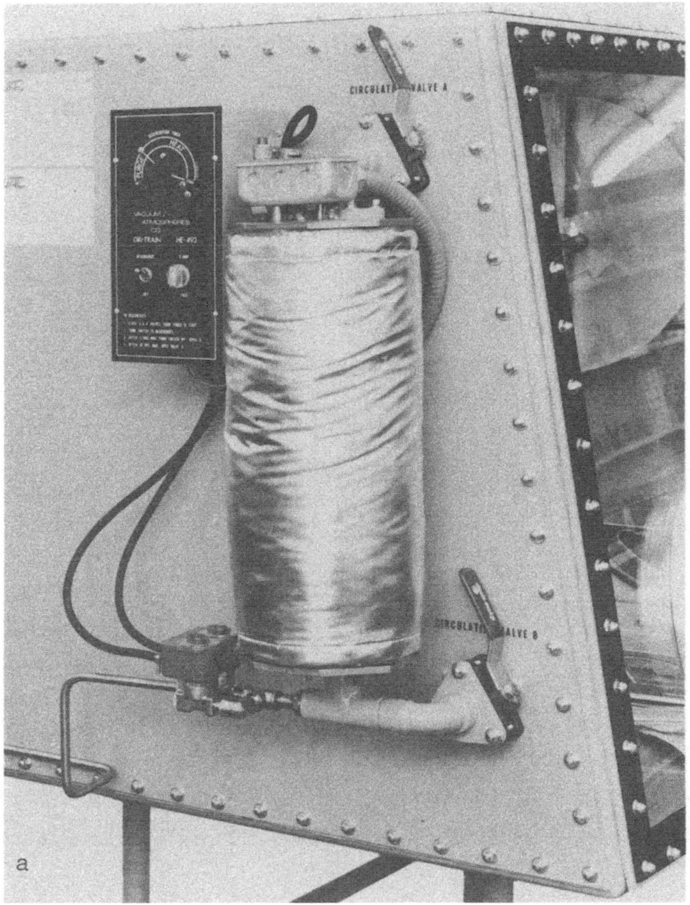

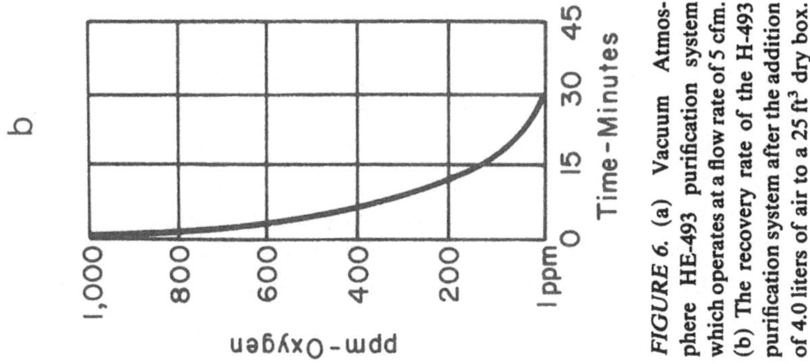

FIGURE 6. (a) Vacuum Atmosphere HE-493 purification system which operates at a flow rate of 5 cfm. (b) The recovery rate of the H-493 purification system after the addition of 4.0 liters of air to a 25 ft^3 dry box.

Common accessories include (1) pressure control systems, (2) temperature control systems, (3) gas impurity analysis systems, (4) electrical devices, and (5) shelves and other utilities. Pressure control systems have been an important factor in the development of commercial glove box systems for nuclear chemistry work. The most useful type of pressure controller is one that can be operated both in a manual and automatic mode. Figure 7 shows a relatively inexpensive pressure controller developed by Vacuum Atmospheres. The device in the automatic mode should be able to maintain a particular pressure range by either opening a valve to a vacuum pump to decrease pressure, or opening a second valve to the pressurized, purified-gas line to increase the pressure. Manual control using the same valving system can allow for pressure control during operations in the box. Positive pressure control of 3–5-in. water column pressure is a good practice when the box is not in use so as to minimize diffusion of impurities through the gloves and gaskets. Other pressure control devices are safety devices that will turn off all power to the glove box system in cases of excessive negative or positive pressures. An alarm will also sound when the pressure of the box is not within an adjustable pressure range (e.g., ±5 in. H_2O).

FIGURE 7. Pressure control device (top) and automatic evacuating device (bottom) for the Vacuum Atmospheres dry box. (Courtesy of J. Wilkes, Frank J. Seiler Laboratory, USAF Academy, Colorado.)

Temperature control systems include both ovens and refrigerator units. They can be self-contained units within the standard glove box, or the entire box volume can be controlled. Commercial cold storage box units are available with an ability to control temperatures down to −40°C. There are also temperature control systems, which are designed to cool the entire gas atmosphere in a glove box. These control devices are useful in preventing glove boxes from overheating, which can cause increased reactivity of materials in some cases and discomfort over long periods of time for an operator. These refrigerator units are commercially available from glove box manufacturers (Vacuum Atmospheres) with a capacity of 1,500–10,000 BTU/h. These devices can be useful for high-temperature melt work, particularly with regard to controlling perspiration of the operator's hands while in the gloves.

Convenient oven systems can be introduced into glove box systems to maintain the elevated temperatures necessary for temperature control. Furnace and oven systems with many configurations are available; however, a limiting factor is the size of the antechamber versus the size of the oven or furnace. Most ovens require maintenance and it is often necessary to bring these devices in and out of the box without affecting the integrity of the box's atmosphere. There are commercially available vacuum ovens which are mounted as an additional antechamber or port on the opposite end panel from the standard antechamber of a metal glove box. These devices can be useful in outgassing and drying material prior to box entry. Although these devices could be potentially useful in highly reactive melts, the author has obtained satisfactory drying by initially heating the glassware and material in a conventional laboratory oven, and then directly introducing the hot items in the antechamber or port for immediate evacuation.

Analysis of impurities in the atmosphere of the box is important. Several relatively facile approaches to monitoring the box's atmospheres are discussed in a later section. However, it should be noted that continuous monitoring analyzers are commercially available for moisture, oxygen, and nitrogen analysis (Vacuum Atmospheres). Moisture analyzers typically utilize an aluminum oxide sensor, which undergoes a decrease in resistance or increase in capacitance upon contact with water, either of which can be easily measured. Other devices in common usage include phosphoric acid-coated hygrometers, e.g., from Salford Electrical Instruments (Eccles, Manchester, U.K.). Oxygen analyzers are capable of ±1 ppm accuracy in oxygen analysis and usually are based on the electrolytic oxygen cell. An example of this type of an oxygen analyzer is the Chemtrix type 30 Oxygen Meter (Hillsboro, Oregon), which has been employed by electrochemists. In this case, the oxygen amperometric electrode is placed in the box and connected to the meter, which is outside the box, by electrical feed-throughs. The electrode should be brought into the box antechamber using a vacuum-tight vessel so as not to rupture the electrode membrane. In addition, this electrode requires a filling solution which needs to be periodically replenished. Nitrogen analysis can be accomplished by a gas chromatographic technique, which employs a thermal conductivity detector. As noted above, all of these commercial devices are mounted outside the box, except for the probe (e.g., aluminum oxide sensor) or a diaphragm pump to furnish the gas to the analyzer system. Additional gas and moisture sensors are currently under development.

Electrical feed-throughs are important to maintain control and to make physical and chemical measurements on melt systems in the box. Special high-current, standard 8-pin octal (Fig. 8) and general service port feed-throughs can be constructed, or are commercially available. Other accessories that are commercially available include rack and shelf assemblies, liquid nitrogen feed-throughs, water-line feed-throughs, floor liners, and both external and internal glove-port covers. The internal glove-port covers are useful in the event a glove is torn, cut, or punctured. The integrity of the box's atmosphere can be maintained by placing the glove-port cover in the port while a new glove is installed. External glove-port covers can

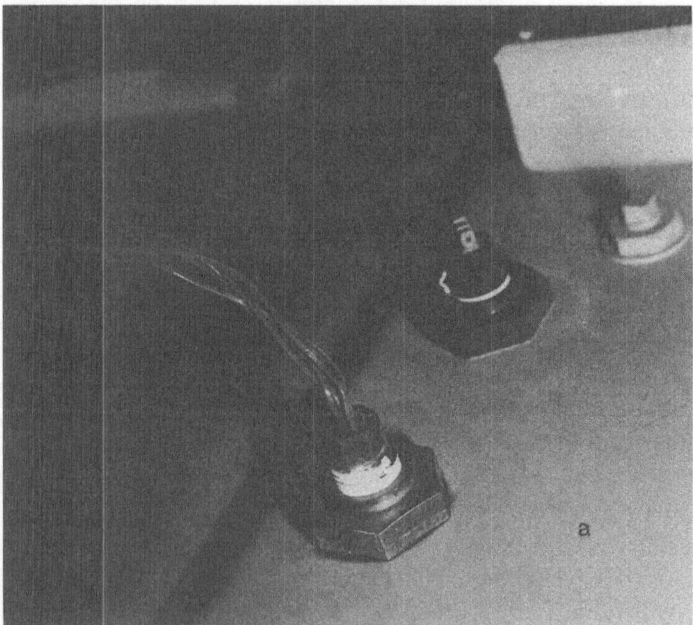

FIGURE 8. Electrical feed-through devices for dry boxes: (a) Homemade device constructed with modified plastic plug and epoxy. (Courtesy of J. Wilkes.) (b) Commercial high-current (500-A) device. (Courtesy of Vacuum Atmospheres, Hawthorne, California.)

FIGURE 8. Continued.

be utilized by installation over each glove, to maintain the box atmosphere over extended times when the box is not in use.

The brief discussion above shows that inert atmosphere systems range from simple glove bags that can be used for a few minutes and put away in a drawer, to complex room-sized systems, as in Figs. 9 and 10, which operate almost automatically for extended periods. Table I lists some of the manufacturers of the dry boxes, glove boxes, and glove bags that are currently available. The listing includes addresses, phone numbers, and the type of material used in the enclosure's fabrication.

3. Operation

The discussion on operation of inert atmosphere systems will begin with the procedures for attaining the initial inert atmosphere in an enclosure,

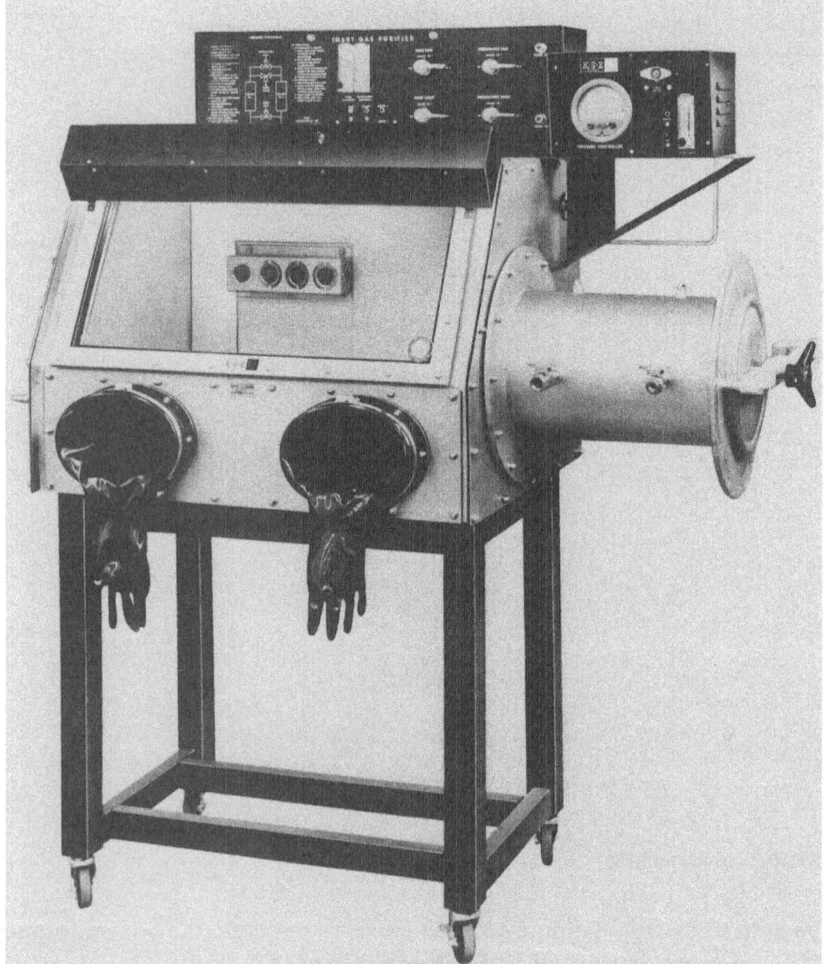

FIGURE 9. A stainless steel dry box with recirculating system and other accessories. The recirculating system is on top of the dry box. (Courtesy of Kewaunee Equipment Corp., Laboratory Division, Statesville, North Carolina.)

that is, purging the enclosure of the ambient air and replacing it with an oxygen and moisture-free atmosphere. The discussion will continue with the most important operation, especially with permanent installations, of maintaining the atmosphere at the desired level of inertness on a day-to-day basis. Related to the latter operation is a third area in this discussion, which is the monitoring of the enclosure atmosphere.

3.1. Startup Procedures

The following procedures assume that the enclosure has been properly assembled and tested for leaks. In order to maintain an inert atmosphere

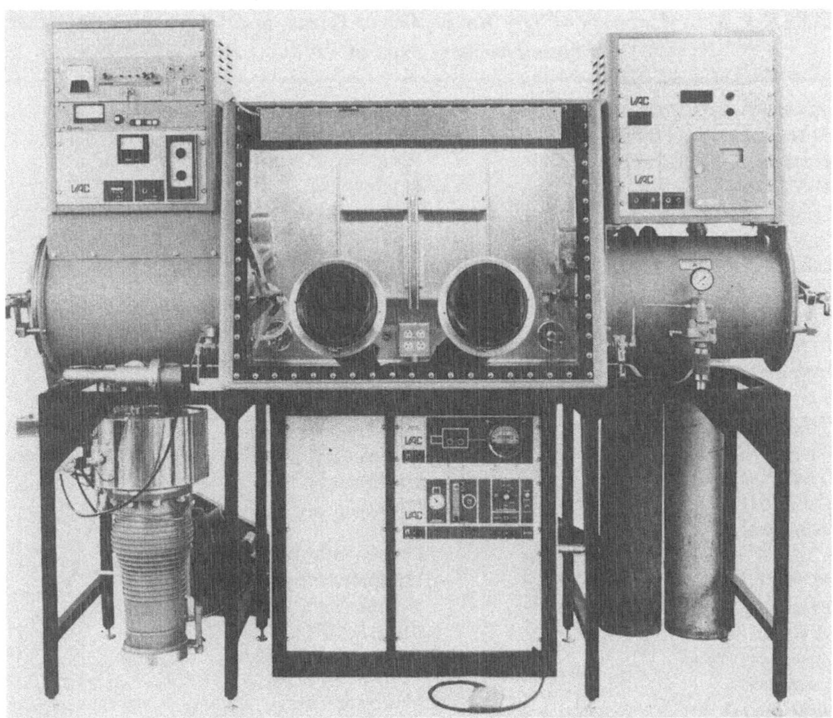

FIGURE 10. An aluminum dry box system complete with recirculation train and two antechambers. The recirculating system is underneath the dry box. The device above the antechambers on the left is an oven-controller for oven in left-hand antechamber. The enclosure above the right-hand antechamber contains the oxygen and moisture analyzers. (Courtesy of Vacuum Atmospheres, Hawthorne, California.)

for an extended period, the enclosure must be free from very small leaks. Seals and joints require special attention. Once the enclosure has been made completely leak-free, one can start the process of purging the enclosure and establishing an inert atmosphere. There are two common methods that can be used to purge the enclosure. One method is alternately (partially or completely) to evacuate the enclosure and refill it with the inert gas that will constitute the inert atmosphere. A second method is a lengthy continuous purge using the inert gas. The method used depends primarily on the size of the enclosure. Glove bags and small boxes can be purged by simply allowing the inert gas to flow in on one side and out the other side until the enclosure is fully flushed with the inert gas. For larger boxes, the first method is preferable. The box can be partially evacuated and refilled with the inert gas. The level of pressure in the enclosure can be monitored by the inflation of the attached gloves. Some authors suggest a purge time of 8 h.[23] The purging process should be continued until the amount of inert gas that has been added is at least equal to three times the volume of

TABLE I. Manufacturers of Dry Boxes, Glove Boxes, and Glove Bags (Address, Phone Number, Type of Product)

Benchmark Industries, Inc.
540 N. Commercial Street
Manchester, NH 03015
603-627-8484
(stainless-steel dry boxes)

Blickman Fabricating Co.
530 Gregory Avenue
Weehawken, NJ 07087
201-863-0800
(steel dry boxes and evacuable boxes)

BTU Engineering Corp.
Esquire Road
N. Billevica, MA 01862
617-677-5111
(stainless-steel dry boxes)

CIS (UK) Ltd.
Rex House
354 Ballards Lane
London N12 0EG
01-446-4405[a]
(glove boxes)

Coy Laboratory Products
22 Metty Drive
Ann Arbor, MI 48103
313-633-1320
(glove bag—anaerobic chambers)

Forma Scientific
Box 649
Marietta, OH 45750
800-848-3080
(steel anaerobic chambers)

A. G. Gallenkamp Ltd.
Belton Road West
Loughborough
LE11 0TR
UK
(small glove box and gloves)

Instruments for Research and Industry
108 Franklin Avenue
Cheltenham, PA 19012
215-379-3333
(glove bags)

High Vacuum Equipment Corp.
110 Industrial Park Drive
Hingham, MA 02043
617-749-9000
(custom-made steel or aluminum boxes)

Kewaunee Scientific Equipment Corp.
Laboratory Division, 2700 W. Front St.
Statesville, NC 28677
704-873-7202
(stainless-steel dry boxes)

Labconco Corp.
8811 Prospect Avenue
Kansas City, MO 64132
816-333-8811
(fibreglass dry boxes)

Liberty Industries, Inc.
133 Commerce Street
East Berlin, CT 06023
203-828-6361
(plastic dry boxes)

Manostat Corp.
519 8th Avenue
New York, NY 10018
212-594-6262
(plexiglass glove boxes)

Marine and Industrial Plastics Ltd.
Fort Fareham Industrial Estate
Newgate Lane
Fareham PO14 1BJ
UK
0329-280184[a]
(plastic dry boxes, ports, gloves, etc)

Miller–Howe Ltd.
Watlington Industrial Park
Watlington
Oxford OX9 5LU
UK
049161-3171/2[a]
(custom-made recirculating dry boxes)

Nuclear Associates
100 Voice Road
Carle Place, NY 11514-1593
516-741-2166
(plastic enclosures)

(*cont.*)

TABLE I. (cont.)

Nuclear Equipment Chemical Corp. 280 South Street Farmingdale, NY 11735 516-420-8000 (custom-made, stainless-steel dry boxes)	Vacuum Atmosphere Co. 4652 Rosecrans Avenue Hawthrone, CA 90255 213-644-0251 (aluminum dry boxes)
United States Testing Co. Inc. 1415 Park Avenue Hoboken, NJ 07030 201-792-2400 (fiberglass glove boxes)	Vertex Industries, Inc. 23 Carol Street Clifton, NY 07014 201-473-6900 (plastic glove boxes)

[a] To telephone the United Kingdom from the United States dial 01144 + number, but omitting the zero.

the enclosure. There are also automatic purge devices that are commercially available (Vacuum Atmospheres), which utilize a timer and valves to control the purging process. They will usually have a pressure safety shut-off switch on the gas supply. Commercial literature indicates that approximately 8 volumes of inert gas must be purged through the box in order to establish an atmosphere containing less than 50 ppm air.(24)

An alternative method utilizes an evacuation balloon, which is available from some glove box manufacturers (e.g., Labconco, Kansas City, MO). The evacuation balloon is inflated in the box with the gas which is to be used in the box. In the process of inflating the balloon, the volume to be purged in the box is physically displaced. After the volume within the box is completed displaced, the balloon is deflated and the volume is concurrently replaced by the new, controlled atmosphere.

The final traces of oxygen and moisture can be removed by circulating the purged atmosphere through a purifying system containing an oxygen getter and desiccant. For glove bags, or small boxes that do not have a circulating system, the inert gas can be passed through a purifier and drying agent prior to entering the enclosure. Table II contains a list of desiccants used for moisture removal and Table III contains a list of dry oxygen getters used in moisture free atmospheres.(22)

3.2. *Maintaining the Inert Atmosphere*

The most difficult operation for inert atmosphere work is the daily maintenance of the atmosphere at the desired level of inertness. The components that must be continually removed usually are oxygen, moisture, and sometimes nitrogen. In simple systems, such as glove bags or small boxes, the enclosure can merely be refilled with new inert gas that has been passed through the necessary purification apparatus. In larger, more com-

TABLE II. Desiccants

Agent	Equilibrium water vapour pressure (Torr)	Remarks
CaH_2	$<10^{-5}$	Evolves H_2; no regeneration; basic
P_2O_5	2×10^{-5}	Capacity limited by formation of surface film; acidic
$Mg(ClO_4)_2$	5×10^{-4}	Good capacity; regeneration at 250°C *in vacuo*; dangerous with reducing agents
BaO	4×10^{-4}	Small capacity; regeneration is unhandy; basic
Linde molecular sieves, 4 Å or 5 Å	ca. 1×10^{-3}	Good capacity; regeneration at 400°C *in vacuo* or in "dry" gas stream
Alumina (active)	ca. 1×10^{-3}	Fair capacity; regeneration at 500°C *in vacuo* or in "dry" gas stream, or 700°C in air
Silica gel	ca. 2×10^{-3}	Fair capacity; regeneration at 300°C
KOH	ca. 2×10^{-3}	Small capacity owing to coating of solid with solution; basic
CaO	3×10^{-3}	Limited capacity; especially in presence of CO_2; basic
H_2SO_4	ca. 3×10^{-3}	Oxidizing agent; acidic
H_3PO_4 (syrupy)	ca. 3×10^{-3}	Acidic
$CaSO_4$ (Drierite)	5×10^{-3}	Regenerated at 250°C
$CaCl_2$	0.2	Good capacity; slightly acidic

[a] References 13 and 22. See also Lantelme, Inman, and Lovering in Volume 2 of this series, as well as White in both Volume 1 of this series and in *Ionic Liquids* (D. Inman and D. G. Lovering, eds.), Plenum Publishing, New York (1981).

plex systems, the atmosphere is maintained by continuously circulating the atmosphere through a purification system. It is the opinion of this author that a recirculating system (vide supra) on a "good" metal box is essential to maintain the atmosphere necessary for good molten salt research. In particular, recirculation systems that consist of a column packed with an oxygen scavenger (Table III), such as BTS catalyst (copper form) or MnO_2, and a desiccant (Table II), such as molecular sieves (type 4A), are attractive because of their ability to operate at room temperature. However, it should be noted that MnO_2 can be poisoned by sulfur-containing compounds introduced in the box. The BTS catalyst, now known as BASF R3-11 Catalyst (BASF, P.O. Box 181, Parsippany, New Jersey 08054) is relatively easy to reduce to its oxygen-scavenger form with mixtures of H_2/N_2. However, the

TABLE III. Dry Oxygen Scavengers[a]

Agent	Description
Na-K (67-81 wt% K)	Liquid above 0°C and may be used in a U-tube gas bubbler. Removes O_2, H_2O, and Hg vapor.
Na or K supported on glass wool	Similar to above. Supported Na is prepared by embedding the metal chunks in glass wool, evacuating, and heating to 300°C
CoO	Removes O_2 at room temperature. Prepared by slowly heating $CoCO_3$ to 340°C *in vacuo.* Not easily regenerated.
Cr[II] in silica gel	Prepared by absorption of a Cr[III] Solution on silica gel followed by reduction at 500°C in H_2. Efficient low-capacity O_2 absorption at room temperature.
Palladized or platinized asbesti ("Deoxo" unit)	Removes traces of O_2 from H_2 at room temperature.
BASF catalyst R3-11 in reduced form (formly "BTS Catalyst")	Similar to above catalyst. 70°C required for removal of O_2 from an H_2 stream, 30-40°C required for CO removal.
Ba, Ca, Ca-10% Mg alloy, La, Mg, Th, and U	Removes O_2 from Ar stream at 300, 650, 475, 500, 600, 400, and 600°C, respectively. Also removes O_2 and N_2 at 400, 650, 500, 800, 640, 800, and 1000°C, respectively.
Brass, Cu, Ce, or U	Removes O_2 from Ar or N_2 stream at 500, 600, 300, 200°C, respectively.
Li	Similar to Ca. Removes N_2 and O_2 but reacts with quartz or glass.

[a] References 13 and 22. See also Lantelme, Inman, and Lovering in Volume 2 of this series.

reduced, active form of BTS catalyst can be inactivated by volatile sulfur and chloride-containing compounds.

For systems that are in daily continuous use, one must be concerned about the introduction of contaminants into the box. Glassware must be carefully dried before being introduced into the enclosure. Care must be exercised as to the type of items that go into the box. This is particularly true for absorbent types of materials such as paper towels, which can contribute to raising the moisture level when working in the range of a few parts per million water vapor. If any paper-type materials (e.g., towels or wipes) are introduced into a box, they should be pumped down in the antechamber for at least 24 h before they are actually brought into the box. In addition, storage of any hydrous or porous material (i.e., paper) in a vacuum desiccator over a good drying agent (Table II) should be done before they are placed in the antechamber.

Oxygen can diffuse into the box through the gloves and gaskets in the box. Very small holes in gloves that go unnoticed can contribute significantly to raising the oxygen level in the enclosure. It should be noted that all boxes become contaminated with oxygen and moisture even when not in use because of minor leakage and diffusion through the gloves. Since the diffusion through the gloves can be on the order of 10 ml/h of H_2O and O_2, continuous recirculation through a purification train is probably necessary to maintain less than 10 ppm levels of these contaminants over extended periods.

Therefore glove maintenance is extremely important in sustaining a quality atmosphere in an enclosure such as a dry box. In addition to the recirculation train, there are ancillary techniques to help remove oxygen and moisture. Placement of desiccants and/or dry oxygen scavengers, as described in Tables II and III, can further reduce the oxygen and moisture content in the box. For example, an open dish of barium oxide (finely divided) or phosphorus pentoxide is useful. A pan of Na–K alloy to remove oxygen and moisture has been described by Peak and Ritter.[25]

3.3. Monitoring the Atmosphere

The monitoring of enclosure atmosphere pertains primarily to larger systems that are to be used on a more permanent basis. When the desired atmosphere is measured in the few parts per million of oxygen and water vapor, the monitoring of the atmosphere becomes very important. It must be continuous in order to detect an increase in contaminants in time for corrective action. Elaborate (and expensive) instrumentation is available to continuously monitor the atmosphere within the enclosure and sound an alarm when the atmosphere becomes contaminated. Such equipment primarily includes oxygen analyzers, nitrogen analyzers, and moisture analyzers. However, simpler devices are available. One simple device to detect oxygen is a 25-W light bulb that has a hole filed or drilled through the globe.[26,27] The hole in the end of the glass envelope of the light bulb can be made by contacting it with the molten tip of a hot glass rod or by carefully filing a notch at the same point. Such a bulb with the filament exposed to the enclosure atmosphere will burn for various lengths of time depending on the oxygen and moisture levels in the enclosure. At a concentration of 1 ppm total oxygen and water vapor, the bulb should burn for 7 or more days. The bulb will burn for approximately 2 days at 5 ppm total ($O_2 + H_2O$) and for 2–8 h at 15 ppm. In addition, the color of the coating on the inside of the bulb can give some indication of the contaminant. In the presence of water or high oxygen partial pressure, white or yellowish-white WO_3 is formed, but at low oxygen partial pressure in the absence of water, blue W_2O_5 is formed.[28] Thus, a continuously burning light bulb allows an operator to constantly monitor the quality of the atmosphere and quickly alerts him to the need to regenerate the purifiers, or of leaks. However, it has been noted by this author that different sources of light bulbs can produce a variation in the "burn" time of the exposed filaments. This is probably due to different tungsten-filament sizes (diameters) in different light bulbs. Thus, this evaluation is a qualitative or semiquantitative one for oxygen contamination. Other methods of detecting oxygen include the use of tetrakis(dimethylamino)ethylene, which forms a luminescent compound with oxygen.[(29)] Although this compound is very sensitive to trace amounts of oxygen, it has a high affinity for silicone stopcock grease. Diethylzinc dissolved in heptane can also be used to detect oxygen since a cloud of vapour will form in atmosphere greater than 5 ppm oxygen. However, this material is pyrophoric and extreme care should be taken in handling diethylzinc. A simple and specific test for oxygen, which can be conducted in a dry box containing organic vapors (which would react with the hot tungsten filament) has been described by Sekutowski and Stucky.[(30)] A mixture of $(Cp)_2TiCl_2$ and Zn powder in freshly distilled tetrahydrofuran(THF) results in the formation of the compound $(Cp_2TiCl)_2ZnCl_2 \cdot THF$, which is specifically sensitive to oxygen at levels down to 5 ppm.

A simple test for water vapor in the atmosphere of a box is the placement of a few crystals of $TiCl_4$ in an open dish. If the water vapor content is greater than 10 ppm, a vapor cloud will appear above the crystals. An alternative method is to have a small dish of P_2O_5 which will form a "skin" when it reacts with moisture. In addition to detecting moisture, this technique can be used to remove moisture in the box atmosphere continuously by periodically mixing the P_2O_5 powder. Finally, both oxygen and water can be qualitatively monitored by observing the time required to "cloud" a fresh piece of sodium.

4. *Common Operations Conducted inside Dry Boxes*

Many different types of techniques or operations that are pertinent to molten salt chemistry can be carried out in a dry box. The operations that will be discussed herein include those that deal with (1) weighing and volume measurements, (2) electrochemical measurements, and (3) spectroscopic measurements. It should be noted that all glassware that is to be introduced into the box should be clean and dry to prevent atmospheric contamination in the box. All glassware should be taken out of the drying oven and placed directly in the antechamber. Glassware with frits or many compartments (e.g., electrochemical cells) should be pumped down in the antechamber for several hours.

4.1. Weighing and Volumetric Operations

Accurate weighings inside glove boxes can be easily accomplished with modern top-loading electronic balances. These types of balances can be obtained with an accuracy of ±0.1 mg, taring capability, and a digital readout. However, it should be noted that a balance with ±1 mg accuracy is probably more practical since vibrations and atmosphere movement (due to the recirculatory fan) limit the accuracy of most weighings in boxes. In order to obtain accurate weighings inside the box, it is sometimes necessary to temporarily turn off the recirculating fan and any pumps that are operating with the box system to minimize these vibrations. In addition, electronic balances are compact and easy to operate and thus they are ideal for a limited enclosed space. If funds are limited and high accuracy is not necessary, a simple metal triple-beam balance can suffice.

Static electricity in dry boxes can be a problem which may lead to poor stability of some balances and difficulty in transfer of dry materials from one container to another. The use of an alpha-particle type static eliminator (commonly sold in audio stores) is very helpful in this regard. In addition, good electrical grounding of the balance and the box itself is a means of minimizing this problem. If a balance is not available for work inside the box, a weighing-bottle difference approach can be utilized. In this method, the material is preweighed outside the box in a weighing bottle or some other small container. The material is brought into the box in the usual fashion through the antechamber. In the case of volatile liquids or sublimable solids, the material can be placed in a small air-tight container, which has been flushed or purged with the same type of inert gas used in the box. In some cases, a small vacuum desiccator can be used to introduce these types of materials into the box. After the material has been dispensed in the glove box, the weighing bottle is removed from the box and reweighed in the normal manner.

Volume measurements in a dry box can be carried out in a variety of ways. Accurate measurements can be accomplished with conventional glass pipettes. However, for smaller volume measurements, the adjustable pipettes with disposable tips are useful. These devices are capable of dispensing volumes from 5 ml down to 1 μl and are available from most chemical supply houses. The fixed microdispensers (e.g., Eppendorf-type) are easily capable of dispensing from 1000 to 1 μl with an accuracy of better than ±1%. These devices are particularly practical in boxes because of their relatively small size and their use of plastic tips, which are readily disposable.

4.2. Electrochemical Measurements

Many different types of electrochemical measurements on molten salts have been carried out in dry boxes. Different kinds of electrochemical cell designs have been used in this regard and have been reported in this series, as well as other molten salt literature. In addition, a general discussion of electrochemical experiments in a dry box can be found by Frank and Park in the text *Laboratory Techniques in Electroanalytical Chemistry.*[31]

There are three approaches which can be taken with regard to molten salt electrochemistry in a dry box. The first approach involves "loading" an air-tight cell (i.e., melt preparation and addition of solute) in the dry box and then transfer of the cell outside the box for conventional electrochemical experiments. The limitation in this case is the design of good air-tight cells using standard ground-glass fittings. Additionally, this approach requires two furnaces for elevated-temperature molten salts, one inside the box and one outside the box. The advantage in this approach is the ease of manipulation of the cell for various electrochemical measurements when it is outside the box.

The second approach is to "load" the cell and keep the cell in the box for all measurements. However, the electrochemical instrumentation is outside the box with electrical feedthroughs into the box. This is probably the most common arrangement because it permits many different types of cell/furnace configurations that can remain in the box. However, electrical leads that are too long can be a problem because of introduction of electrical noise to the analog signal that is to be measured. In order to minimize noise problems, shielded cable should be used in which the shield is either grounded or "driven" by conventional methods. In addition, IR-voltage drop due to long leads should be considered.

The third approach is to place the potentiostat-galvanostat in the box along with the cell. Because of size limitations, custom-made potentiostats are often used in this case. The primary advantage is that the electrode leads can be relatively short with a concomitant decrease in noise during the measurement process. In this case, the amplified analog signals from the potentiostat can be passed out of the box by the electrical feedthroughs to various data acquisition devices (i.e., recorders, microcomputers). A further improvement is to digitize the signal before transfer via the electrical leads to devices outside the box. However, because of limited space in the box and the possible presence of corrosive vapors from the experiments in the melt, the latter improvement in signal transfer is not always feasible.

4.3. Spectroscopic Measurements

The approaches that can be used for molten salt spectroscopic experimentation using dry boxes are similar to those described for electrochemical techniques.† The most common approach is preparation

† See also chapter by Griffiths in Vol. 2 of this series.

of the solute in the melt in the dry box. Transfer of this "solution" from the initial melt container to a spectroscopic cell is subsequently carried out in the box. The spectroscopic cell containing the melt solution is carefully sealed and then removed from the box for conventional spectroscopic measurements. An example of this approach is a study of the spectroelectrochemistry of niobium redox chemistry in molten chloroaluminates at 175°C.[32] The disadvantage to this approach is the need for two furnace systems for elevated-temperature melts, one in the dry box and one in the spectrometer sample compartment.

Because of the relatively large size of most spectrometers, it is not feasible in most cases to have the spectrometer inside the box. However, there are alternative approaches that basically "bring the spectrometer" into the dry box. The advantage to these approaches is that the melt remains in the box and therefore manipulations of the melt for spectroscopic measurements are easier. The first approach is effectively to attach the spectrometer or optical device to the box. This is illustrated in Fig. 11, in which a glove box–microscope system has been constructed.[33] In this case, a shallow perspex viewing container is sealed to the front face of the glove

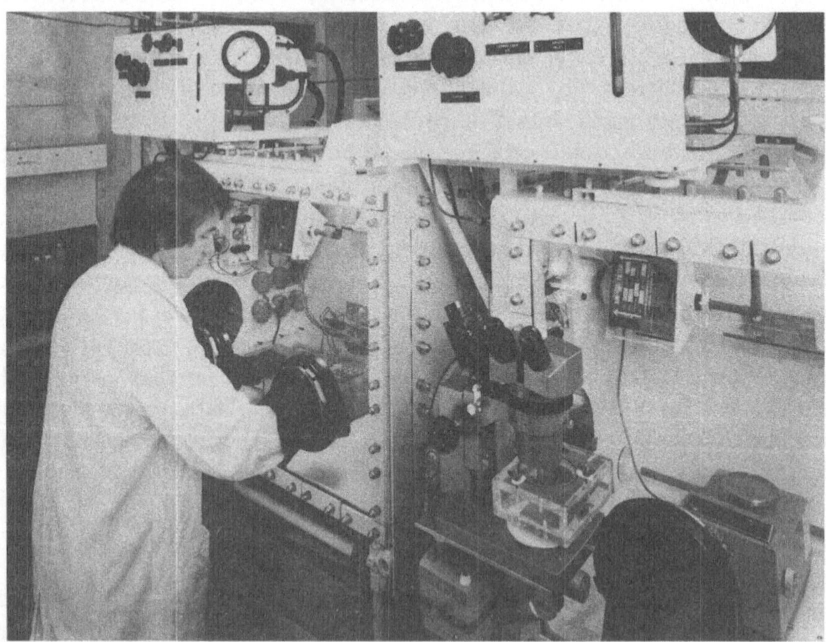

FIGURE 11. Dual dry box system. The right-hand dry box has a binocular microscopy which has been attached to the front face of the dry box. (Courtesy of J. M. North, Materials Development Division, A.E.R.E., Harwell, Oxfordshire, England.)

box. The binocular microscope has a sufficient optical working distance to accommodate the perspex viewing container. A related approach that "brings" the glove box to the spectrophotometer has been described by Oertel and Fehl.[34] In their case, a combination glove box (plexiglass) sample compartment cover was used for FT–IR measurements in oxygen-exchange kinetic studies.

A second approach that should be considered is more flexible and does not require the considerable modification of the glove box (*vide ante*) in order to carry out these measurements. In this approach spectroscopic measurements (in particular, uv, visible, ir) can be made on the optical cell, which remains in the glove box, using fiber optics.[35] This type of remote spectroscopy has been used in industrial on-line process monitoring and plantwide chemical monitoring. Fiber optics probes are available to withstand temperatures of 350°C, and if cooling is provided, temperatures of 700°C are possible (Guided Wave, Inc., 5200 Golden Foothills Parkway, Unit A, El Dorado, California 95630). The only modifications required for the dry box would be feedthroughs of the optical fibers, in the same fashion as electrical feedthroughs or lines. Thus, these devices would appear to have considerable potential in spectroscopic studies of reactive melts in dry boxes.

References

1. J. A. Plambeck in: *Encyclopedia of Electrochemistry of the Elements* (A. J. Bard, ed.), Vol. X, Chap. 2, Marcel Dekker, New York (1976).
2. D. M. Wrench and D. Inman, *J. Electroanal. Chem.* **17**, 319 (1968).
3. Y. Kanzaki and M. Takahashi, *J. Electroanal. Chem.* **58**, 349 (1975).
4. G. G. Bombi, M. Fiorani, and G. A. Mazzochin, *J. Electroanal. Chem.* **9**, 457 (1965).
5. J. A. Plambeck in: *Encyclopedia of Electrochemistry of the Elements* (A. J. Bard, ed.), Vol. X, p. 166, Marcel Dekker, New York (1976).
6. J. Jordan, *J. Electroanal. Chem.* **29**, 127 (1971); P. G. Zambonin, *J. Electroanal. Chem.* **45**, 451 (1973).
7. P. G. Zambonin, *Anal. Chem.* **41**, 868 (1969).
8. H. Linga, Z. Stojek, and R. A. Osteryoung, *J. Am. Chem. Soc.* **103**, 3754 (1981); also see Gale in Volume 1.
9. G. Letisse and B. Tremillon, *J. Electroanal. Chem.* **17**, 387 (1968).
10. B. Tremillon, A. Bermond, and R. Molina, *J. Electroanal. Chem.* **74**, 53 (1976).
11. J. Goret and B. Tremillon, *Bull. Soc. Chim. Fr.*, 67 (1966); also see Claes and Glibert in Volume 1.
12. J. A. Plambeck in: *Encyclopedia of Electrochemistry of the Elements* (A.J. Bard, ed.), Vol. X, Chap. 10, Marcel Dekker, New York (1976).
13. D. F. Shriver, *The Manipulations of Air-sensitive Compounds*, McGraw-Hill, New York (1969).
14. C. J. Barton in: *Technique of Inorganic Chemistry* (H. B. Jonassen and A. Weissberger, eds.), Vol. III, Chap. 4, Interscience, New York (1963).

15. *85 Tips and New Uses for Glove Bag*, Instruments for Research and Industry, P.O. Box 159N, Cheltenham, Pennsylvania 19012.
16. Vacuum Atmospheres Company, 4652 Rosecrans Ave., North Hollywood, California.
17. Kewaunee Manufacturing Company, Box 618, Adrian, Michigan.
18. Clampitt, B. H., Oak Ridge National Laboratory, unpublished work; C. J. Barton in: *Techniques of Inorganic Chemistry* (H. B. Janassen and A. Weissberger, eds.), Vol. 3, p. 316, Interscience, New York (1963).
19. J. H. Rowan, Anal. Chem. **28**, 402 (1956).
20. G. J. Amerongen, *J. Appl. Phys.* **17**, 972 (1946).
21. W. O. Greenhalgh, R. C. Smith, D. I. Powell, Report HEDL-TME-79-16; *Chem. Abstr.* **93**, 33519z (1980).
22. L. F. Druding, *J. Chem. Ed.* **47**, A815 (1970).
23. E. C. Ashby and R. D. Schwartz, *J. Chem. Ed.* **51**, 65 (1974).
24. Vacuum Atmospheres Technical Information on Quick Purge QP-30 Series.
25. R. Peak and D. M. Ritter, *Rev. Sci. Instrum.* **27**, 109 (1956).
26. I. D. Eubanks and F. J. Abbott, *Anal. Chem.* **41**, 1708 (1969).
27. D. E. Bartak and R. A. Osteryoung, *J. Electrochem. Soc.* **122**, 600 (1975).
28. R. A. Foust, Jr., memo, General Motors Corporation, Warren, MI, October 3, 1968.
29. R. L. Purett, J. T. Burr, K. E. Rapp, C. T. Bahner, J. D. Gibson, and R. H. Lafferty, Jr., *J. Am. Chem. Soc.* **72**, 3646 (1950).
30. D. G. Sekutowski and G. D. Stucky, *J. Chem. Ed.* **53**, 110 (1976).
31. S. N. Frank and S. M. Park, in: *Laboratory Techniques in Electroanalytical Chemistry* (P. T. Kissinger and W. R. Heineman, eds.), Chap. 15, Marcel Dekker, New York (1984).
32. G. Mamantov, V. E. Norvell, and L. Klatt, *J. Electrochem. Soc.* **127**, 1768 (1980); also see Chapter 7 in Vol 1 of this series by first two named authors.
33. J. M. North, Harwell Materials Development Division, AERE Harwell, Oxfordshire, England OX11 ORA, personal communication.
34. R. P. Oertel and A. J. Fehl, *Rev. Sci. Instrum.* **46**, 855 (1975).
35. P. Fitch and A. G. Gargus, *Am. Lab.* **17**(12), 64 (1985).

Index